A Concise Introduction to Measure Theory

Satish Shirali

A Concise Introduction
to Measure Theory

 Springer

Satish Shirali
Formerly of Panjab University
Chandigarh, India

and

The University of Bahrain
Zallaq, Bahrain

ISBN 978-3-030-03240-1   ISBN 978-3-030-03241-8   (eBook)
https://doi.org/10.1007/978-3-030-03241-8

Library of Congress Control Number: 2018960220

Mathematics Subject Classification (2010): 28-01, 28A05, 28A10, 28A12, 28A20, 28A25, 28A35

This Springer imprint is published by the registered company Springer Nature Switzerland AG
The registered company address is: Gewerbestrasse 11, 6330 Cham, Switzerland

Preface

Satish Shirali

The concept of measure in an abstract space, and of the integral with respect to it, has played a fundamental role in analysis and probability in the decades after the pathbreaking work of H. Lebesgue in 1902. It has therefore been an essential component of a mathematics curriculum, accessible to any student who has acquired a facility with basic analysis, including the Heine–Borel theorem, the theory of Riemann integration, infinite series, and also with the use of sets and quantifiers ("for all" and "there exists") in definitions and proofs. This book assumes such a facility on the part of the reader and also an understanding of the notions of countability and Cartesian product. Equivalence classes and the axiom of choice are needed only once.

It is well known that the abstract integral can be interpreted as an improper Riemann-type integral of a related function on the real line, thereby rendering it relatively more concrete. The related function is one that closely resembles what is called the cumulative distribution in probability and statistics and may reasonably be called simply the *distribution*. This book introduces the concept of the integral of a nonnegative measurable function on a measure space via the improper integral of the related distribution function, without the intermediacy of simple functions. Doing so brings out more vividly how the Lebesgue approach begins by partitioning the range rather than the domain of the integrand. Also, the concept is then amenable to a quick extension to "fuzzy" measures. It also helps to appreciate how the monotone convergence property is unrelated to the additivity of measure.

It is natural to begin by recasting the Riemann integral of a step function as the integral of its distribution function, which is essentially the Abel summation formula. This is what Chap. 1 starts with. For other kinds of functions, it is not always clear that a distribution function exists at all. A countably additive measure as the "size" of a set is then presented as motivated by the need for a limit of functions having distributions again to have a distribution.

The concept of a countably additive measure on a σ-algebra and measurability of functions are introduced in Chap. 2 against the backdrop of the motivation described earlier. The integral of a nonnegative measurable function is defined as the improper integral of its distribution function, and the monotone convergence theorem is proved. The integral of a function that may take negative values is introduced, but its properties are dealt with in the next chapter.

Simple functions are defined in Chap. 3 and used to establish the additivity of the integral. The dominated convergence theorem, for which the additivity of the integral is used, is also treated in this chapter. There is a discussion of the extension to subadditive fuzzy measures, but this material is optional.

Chapter 4 is about constructing Lebesgue measure. After defining Lebesgue outer measure, Carathéodory's ideas are applied to an abstract outer measure in the usual manner. It is shown that the classical Riemann integral agrees with the Lebesgue integral for Riemann integrable functions, and the existence of a nonmeasurable subset of the reals is discussed. The chapter ends with induced measures, which find application in connection with product measures and in identifying certain improper integrals as being Lebesgue integrals.

The counting measure and interchanging the order of summation in a repeated sum as amounting to interchanging the order of integration are treated at length in Chap. 5. In this connection, the unconditional sum is identified with the integral respect to the counting measure.

Product measures and the theorems of Tonelli and Fubini are taken up in Chap. 6.

In Chap. 7, the relation to differentiation is discussed in some detail. There are many approaches to the topic, the one adopted in this book being via the Vitali covering theorem and Tonelli's theorem. While the concept of total variation is presented ad hoc in most texts, here it is presented as a natural outcome of attempts to decompose a function as the sum of an increasing and a decreasing function.

The relation between differentiation and Lebesgue integration is not as straightforward as in the case of Riemann integration, and no discussion of the matter can be complete without the Cantor set and function. Their essentials are discussed in Chap. 8.

The symbols $\mathbb{R}$, $\mathbb{Q}$, $\mathbb{Z}$ and $\mathbb{N}$ are commonly understood these days, and no explanation of them has been given.

For matters that are treated differently or not treated at all in most other works covering measure and integration, a prospective reader having prior experience of the subject may wish to check out the following:

1.1.3, 1.1.6 (motivation), 1.2.4, 3.1.9, 3.1.10, 3.1.P1, 3.1.P7, 3.1.P8, 3.1.P9, Figures in Sect. 3.2, Sect. 3.3, 4.1.2(b), 4.2.P6, 4.2.P8, 4.2.P9, 4.4.1, 5.1.2, 5.1.4, 5.1.10, 5.2.5, 5.2.7, 5.2.8, 6.2.P4, 6.6.4, 7.1.3(b), 7.1.P3, beginning of Sect. 7.6 (motivation), 7.8.2, Sect. 8.2 (manner of constructing the Cantor function), 8.2.P3, 8.2.P4, 8.2.P5.

The author is grateful to Dr. H. L. Vasudeva for his valuable comments on the presentation, which contributed significantly to the improvement in readability.

Gurugram, India   Satish Shirali
September 2018

Contents

Chapter 1
Preliminaries

1.1 The Riemann Integral Revisited

It is elementary from the basic theory of the Riemann integral, for example as in Rudin [7] or Shirali and Vasudeva [9], that if a function on a closed bounded interval $[\alpha, \beta]$ is constant on (α, β), its integral is the product of that constant with the length of the interval. The values of the function at the endpoints make no difference. Each term in a lower or upper sum of a bounded function $f : [a, b] \to \mathbb{R}$ is a product of this form. To put it more precisely, denote the lower sum of a bounded function $f : [a, b] \to \mathbb{R}$ over a partition $P : a = x_0 < x_1 < \cdots < x_n = b$ of $[a, b]$ as usual by

$$L(f, P) = \sum_{j=1}^{n} m_j(x_j - x_{j-1})$$

where m_j means the infimum of f over the subinterval $[x_{j-1}, x_j]$ of P. The term $m_j(x_j - x_{j-1})$ is the Riemann integral over $[x_{j-1}, x_j]$ of the function that equals m_j on the subinterval (x_{j-1}, x_j), regardless of what value the function takes at the endpoints. Taking into account that the sum of the integrals of a function on contiguous intervals $[\alpha, \beta]$ and $[\beta, \gamma]$ is the integral of that function over $[\alpha, \gamma]$, it is seen that the lower sum is equal to the integral $\int_a^b m$ of any function $m : [a, b] \to \mathbb{R}$ which, for each j, equals the constant m_j on the interval (x_{j-1}, x_j). The values of m at the points x_j make no difference; but in order to be specific, we shall choose m to be m_j on the left closed subinterval $[x_{j-1}, x_j)$ of P and $m(b)$ to be 0. The argument why refining the partition increases the lower sum can then be interpreted in terms of such a function m as proving that refining increases the function m.

Similar observations apply to the upper sum, which will be denoted as usual by

$$U(f, P) = \sum_{j=1}^{n} M_j(x_j - x_{j-1})$$

© Springer Nature Switzerland AG 2018
S. Shirali, *A Concise Introduction to Measure Theory*,
https://doi.org/10.1007/978-3-030-03241-8_1

where M_j means the supremum of f over the interval $[x_{j-1}, x_j]$.

The notation $L(f, P)$ for the lower sum and $U(f, P)$ for the upper sum will be used in the sequel without explanation.

Functions that remain constant on the open intervals of a partition are lurking within the definition of upper and lower sums, although one may not make explicit mention of them. As we intend to consider them at some length now, we formally introduce the standard name by which such functions are known.

1.1.1. Definition. *A function $f : [a, b] \to \mathbb{R}$ is called a* **step function** *if there is some partition $P : a = x_0 < x_1 < \cdots < x_n = b$ of $[a, b]$ such that f is constant on each of the open intervals (x_{j-1}, x_j). It is said to be* **constant on the subintervals of P**.

It should be borne in mind that such a function is constant only on the *open* subintervals of P, although we shall usually suppress the word "open" for convenience. Evidently, the function is also constant on the subintervals of any refinement of P; thus the partition P mentioned in the definition is far from being unique.

Although we shall generally denote the Riemann integral over $[a, b]$ of a function ϕ by $\int_a^b \phi$, it will be convenient to use the more customary notation $\int_a^b \phi(x)dx$ when the function is known by its expression $\phi(x)$ without any symbol such as ϕ having been introduced for it.

The Riemann integral $\int_a^b f$ of a step function f is easy to compute: It is simply $\sum_{j=1}^n c_j(x_j - x_{j-1})$, where c_j denotes the value on (x_{j-1}, x_j). In the language of elementary calculus, the "area under the graph" of a nonnegative step function is the sum of the areas of rectangles stretching downward from the graph all the way to the horizontal axis. The terms in the sum represent the areas of the rectangles. If the graph is drawn with the axes interchanged, then the rectangles stretch leftward from the graph all the way to the vertical axis.

1.1.2. Exercise. For each of the functions below, determine whether it is a step function, and describe in terms of intervals the set S mentioned along with it:

(a) $f : [0, 4] \to \mathbb{R}$, where $f(x) = 6$ if $0 \le x < 1$, $f(x) = 2$ if $1 \le x \le 3$, $f(x) = 6$ if $3 < x \le 4$; $S = \{x \in [0, 4] : f(x) > 2\}$.
(b) $f : [0, 2] \to \mathbb{R}$, where $f(x) = x^2$; $S = \{x \in [0, 2] : f(x) > \frac{1}{4}\}$.
(c) $f : [0, \pi] \to \mathbb{R}$, where $f(x) = \sin x$; $S = \{x \in [0, \pi] : f(x) > t\}$, where $0 \le t$.
(d) $f : [0, \pi] \to \mathbb{R}$, where $f(x) = 1 + \sin x$; $S = \{x \in [0, \pi] : f(x) > t\}$, where $0 \le t$.

Solutions: (a) Consider the partition $P : 0 < 1 < 3 < 4$. On the (open) subintervals $(0, 1)$, $(1, 3)$ and $(3, 4)$, the function has the constant values 6, 2 and 6 respectively. Therefore f is a step function. $S = [0, 1) \cup (3, 4]$.

(b) The function takes infinitely many different values and therefore cannot be a step function. $S = (\frac{1}{2}, 2]$.

(c) The function takes infinitely many different values and therefore cannot be a step function. If $t < 1$, we have $S = (\sin^{-1} t, \pi - \sin^{-1} t)$, as a graph would reveal; we dispense with the formal argument. When $t \ge 1$, the set S is empty because of the inequality $\sin x \le 1$.

(d) The function takes infinitely many different values and therefore cannot be a step function. $S = [0, \pi]$ if $0 \leq t < 1$ and $S = (\sin^{-1}(t-1), \pi - \sin^{-1}(t-1))$ if $1 \leq t < 2$. For $t \geq 2$, the set S is empty.

In view of what has been noted in parts (c) and (d) of the foregoing Exercise, we shall henceforth consider the empty set as an interval having length 0. With this convention, we can say that the set S has turned out to be a finite union of disjoint intervals in each case. In part (b), if the number $\frac{1}{4}$ is replaced by an arbitrary $t \geq 0$, then S is the interval $(\sqrt{t}, 2]$ if $0 \leq t < 4$ and $\emptyset$ if $4 \leq t$.

Since the length of the interval $\emptyset$ is finite, we shall count it among bounded intervals; moreover, just as in the case of $\mathbb{R}$, it will be considered as closed and open at the same time. This changes the meaning of "interval" slightly, but all that is needed is to exercise some care in using results about intervals.

In part (a) of the exercise, one can ask for $S = \{x \in [0, 4]: f(x) > t\}$ for an arbitrary $t \geq 0$. The reader will find it helpful to draw a graph of the step function f to see that S varies with the value of t in the following manner:

$$
S = \begin{cases}
[0, 4] & \text{if } 0 \leq t < 2 \\
[0, 1) \cup (3, 4] & \text{if } 2 \leq t < 6 \\
\emptyset & \text{if } 6 \leq t
\end{cases}
$$

Thus the subset of the domain on which the function exceeds any given number $t \geq 0$ turns out to be a finite union of intervals in each of the instances in the exercise. There are many other functions for which this happens, for example, those that are monotone. When it does happen, the function has a *distribution*, to be denoted herein by $\tilde{f}$, which is defined to be *the function whose value at any $t \geq 0$ is the sum of the lengths of all the finitely many disjoint intervals whose union is the subset S of the domain on which the function exceeds t*. If the subset is empty, we take $\tilde{f}(t)$ to be 0. In particular, when $t \geq \sup f$, the set S is empty and $\tilde{f}(t) = 0$. A significant point to note is that the distribution, if it exists, is a decreasing function (which means nonincreasing in this book), and therefore Riemann integrable (see Theorem 9.3.3. of Shirali and Vasudeva [9] or 6.9 of Rudin [7]). It is decreasing because $t_1 > t_2$ implies $\{x: f(x) > t_1\} \subseteq \{x: f(x) > t_2\}$.

The reader may note however that disjoint intervals comprising the subset need not be unique—Example: $[0, 3]$ is the union of the disjoint intervals $[0, 1), [1, 3]$ as well as $[0, 1], (1, 2], (2, 3]$—and therefore one could legitimately ask whether their total length is unique, and consequently whether the concept of the distribution $\tilde{f}$ is unambiguous. Some authors may not share these doubts (see Rudin [7]) and in any case, the matter can be settled in the affirmative. However, we prefer to leave the technical details to the problems and proceed by taking the affirmative answer for granted. Readers who do not feel any need for the technical details of the matter would do well to avoid engaging with Problems 1.1.P6, 1.1.P7, 1.1.P8 and 1.1.P9.

For the function f in part (a) of the exercise above, the set $\{x \in [0, 4]: f(x) > t\}$ has been described at the beginning of the preceding paragraph and one can conclude therefrom that the distribution $\tilde{f}: [0, \infty) \to \mathbb{R}$ is given by

$$\tilde{f}(t) = \begin{cases} 4 \text{ if } 0 \leq t < 2 \\ 2 \text{ if } 2 \leq t < 6 \\ 0 \text{ if } 6 \leq t. \end{cases}$$

Since $\sup f = 6$, the last line in this description of $\tilde{f}$ reflects the earlier observation that, in general, $\tilde{f}(t) = 0$ when $t \geq \sup f$. We draw attention to the fact that

$$\int_0^4 f = 6 \cdot 1 + 2 \cdot 2 + 6 \cdot 1 = 16 \text{ and } \int_0^{\sup f} \tilde{f} = \int_0^6 \tilde{f} = 4 \cdot 2 + 2 \cdot 4 = 16,$$

so that $\int_0^4 f$ and $\int_0^{\sup f} \tilde{f}$ are equal to each other. The figures below are intended to illustrate how the two sums lead to the area of the same region broken up into rectangles in two different ways. The figure on the left shows the "area under the

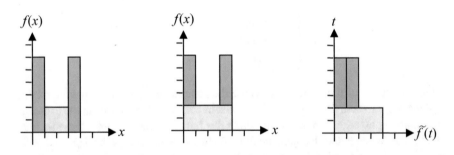

graph" of f in the usual manner as a union of rectangles. Adding up the areas of the rectangles is the same computation as in the computation of $\int_0^4 f$ above. The middle figure shows the same area broken up into rectangles differently. The figure on the right shows the graph of $\tilde{f}$ with the axes interchanged and the area that would have been under the graph if the axes had not been interchanged. The darker shaded rectangle has been broken up into two by a vertical line so as to show the relation to the rectangles in the middle figure. Again, adding up the areas of the rectangles is the same computation as in the computation of $\int_0^{\sup f} \tilde{f}$ above. One can visually confirm that the total shaded areas are the same in all three figures, which is a reflection of the equality of integrals that was obtained above by computation. The forthcoming proposition shows that the equality of the integrals of the nonnegative step function and its distribution is not a coincidence.

1.1.3. Proposition. *Any nonnegative-valued step function* $f : [a, b] \to \mathbb{R}$ *has a distribution* $\tilde{f}$ *and*

$$\int_a^b f = \int_0^{\sup f} \tilde{f}.$$

Proof: The case when f is 0 everywhere is trivial because $\sup f = 0$ in this situation. Therefore we consider only the case when f has at least one positive value.

The set $\{x \in [a, b]: f(x) > t\}$, even if empty, is the union of a finite family of disjoint bounded intervals. By Problem 1.1.P9, the total length of disjoint intervals comprising the set is then uniquely determined, so that the distribution indeed exists.

Let $P: a = x_0 < x_1 < \cdots < x_m = b$ be a partition of $[a, b]$ such that f has constant value c_j on the open interval (x_{j-1}, x_j). Then

$$\int_a^b f = \sum_{j=1}^m c_j(x_j - x_{j-1}).$$

By grouping together the terms for which the values c_j agree but are not 0, we rewrite the above sum as

$$\int_a^b f = \sum_{k=1}^n T_k \lambda_k, \tag{1.1}$$

where the $T_k (1 \le k \le n)$ are the distinct positive values among $c_1, \ldots, c_m$ and λ_k is the total length of the intervals on which f takes the value T_k. Any values of f that are not among the T_k are taken only at one of the points x_j of the partition P and therefore the total length of the intervals where such values are taken is 0. Thus we can include them in the sum with 0 as the corresponding λ_k. The numbers T_k then represent all the distinct positive values of f. By renumbering if necessary, we may assume that they are in increasing order. For reasons that will become clear in the next paragraph, it will be convenient to set $T_0 = 0$, although it does not appear in the sum.

Since $T_1, \ldots, T_n$ are the distinct positive values of f arranged in increasing order, we have

$$T_n = \sup f \quad \text{and} \quad \tilde{f}(T_n) = 0. \tag{1.2}$$

Consider any integer k such that $1 \le k \le n$. Surely, $f(x) > T_{k\_1} \Leftrightarrow f(x) \ge T_k$. In other words,

$$\{x \in [a, b]: f(x) > T_{k-1}\} = \{x \in [a, b]: f(x) > T_k\} \cup \{x \in [a, b]: f(x) = T_k\}. \tag{1.3}$$

Denote by $\mathcal{F}_1$ and $\mathcal{F}_2$ the families of disjoint intervals comprising the two sets on the right side. Since the sets are disjoint, we deduce that the intervals in $\mathcal{F}_1$ are disjoint from those in $\mathcal{F}_2$, so that $\mathcal{F}_1 \cup \mathcal{F}_2$ is a family of disjoint intervals comprising the set on the left side of (1.3). In particular, $\mathcal{F}_1$ and $\mathcal{F}_2$ can have no nonempty interval in common, so that the total length of the intervals in $\mathcal{F}_1 \cup \mathcal{F}_2$ is the sum of the separate total lengths. This means $\tilde{f}(T_{k-1}) = \tilde{f}(T_k) + \lambda_k$, and hence by (1.1),

$$\int_a^b f = \sum_{k=1}^n T_k(\tilde{f}(T_{k-1}) - \tilde{f}(T_k)). \tag{1.4}$$

If $n = 1$, then by (1.2), $\tilde{f}(T_1) = 0$; therefore, keeping in mind that we have set $T_0 = 0$, we find that the sum in (1.4) becomes

$$T_1(\tilde{f}(T_0) - \tilde{f}(T_1)) = (T_1 - T_0)\tilde{f}(T_0) = \sum_{k=1}^n \tilde{f}(T_{k-1})(T_k - T_{k-1}).$$

Hence,

$$\int_a^b f = \sum_{k=1}^n \tilde{f}(T_{k-1})(T_k - T_{k-1}) \tag{1.5}$$

when $n = 1$. Suppose $n \geq 2$. By (1.2), $\tilde{f}(T_n) = 0$, and therefore the equality (1.4) can, by separating the last term in the sum, be recast as

$$\int_a^b f = \sum_{k=1}^{n-1} T_k(\tilde{f}(T_{k-1}) - \tilde{f}(T_k)) + T_n \tilde{f}(T_{n-1}).$$

Also, $T_k = (T_k - T_{k-1}) + \cdots + (T_1 - T_0)$, remembering that $T_0 = 0$. Using the Abel summation formula of Problem 1.1.P5 (with $a_k = \tilde{f}(T_{k-1})$ and $b_k = T_k - T_{k-1}$), we find that (1.5) holds in the present case too, namely, $n \geq 2$.

It remains to show only that $\int_0^{\sup f} \tilde{f} = \sum_{k=1}^n \tilde{f}(T_{k-1})(T_k - T_{k-1})$.

In view of the first equality in (1.2), we have $\int_0^{\sup f} \tilde{f} = \int_0^{T_n} \tilde{f}$. Since f does not take any value between T_{k-1} and T_k, it is true of any $t \in [T_{k-1}, T_k)$ that $f(x) > t \Leftrightarrow f(x) > T_{k-1}$. It follows that $\tilde{f}(t) = \tilde{f}(T_{k-1})$ for every $t \in [T_{k-1}, T_k)$, which is to say, the restriction of $\tilde{f}$ to $[0, \sup f]$ is a step function having constant value $\tilde{f}(T_{k-1})$ on each interval $[T_{k-1}, T_k)$. The equality that remained to be shown is now an immediate consequence. □

For a monotone function f, the set $\{x: f(x) > t\}$ is an interval and therefore the function has a distribution $\tilde{f}$. The function of Exercise 1.1.2(d) is not even monotone and yet the aforementioned set is an interval. It is natural to ask whether the equality proved above for step functions is valid for other (nonnegative) functions that have a distribution. We shall not prove this yet, because more technical hassles are involved than proving that the total length of a disjoint union of intervals is determined uniquely, i.e. independently of the particular representation as such a union. One cannot imitate the proof for the case of step functions because the integral is no longer a finite sum. Although the proof is deferred until a proper formulation and argument are possible (appearing as Theorem 4.3.4 later on), we shall illustrate some cases now.

1.1.4. Example. (a) As noted in the paragraph following Exercise 1.1.2, for the function $f:[0, 2] \to \mathbb{R}$ such that $f(x) = x^2$, the set $S = \{x \in [0, 2]: f(x) > t\}$ is the interval $(\sqrt{t}, 2]$ if $0 \le t < 4$ and $\emptyset$ if $4 \le t$. Also sup $f = 4$. Therefore $\tilde{f}:[0, 4] \to \mathbb{R}$ is given by $\tilde{f}(t) = (2 - \sqrt{t})$ for $0 \le t \le 4$, and thus

$$\int_a^b f = \int_0^2 x^2 dx \quad \text{and} \quad \int_0^{\text{sup } f} \tilde{f} = \int_0^4 (2 - \sqrt{t}) dt.$$

Elementary computations show that both are $\frac{8}{3}$.

(b) As seen earlier in Exercise 1.1.2(d), for the function $f:[0, \pi] \to \mathbb{R}$, where $f(x) = 1 + \sin x$, the set $S = \{x \in [0, \pi]: f(x) > t\}$ is the

interval $[0, \pi]$ when $0 \le t < 1$,

interval $(\sin^{-1}(t - 1), \pi - \sin^{-1}(t - 1))$ when $1 \le t < 2$ and

$\emptyset$ when $2 \le t$.

Of course, sup $f = 2$. Therefore $\tilde{f}:[0, \text{sup } f] \to \mathbb{R}$ is given on $[0, \text{sup } f)$ by

$$\tilde{f}(t) = \begin{cases} \pi & \text{if } 0 \le t < 2 \\ \pi - 2\sin^{-1}(t - 1) & \text{if } 1 \le t < 2. \end{cases}$$

Thus

$$\int_a^b f = \int_0^\pi (1 + \sin x) dx \quad \text{and} \quad \int_0^{\text{sup } f} \tilde{f} = \pi(1 - 0) + \int_1^2 (\pi - 2\sin^{-1}(t - 1)) dt.$$

Elementary computations show that both are $\pi + 2$.

When f is bounded, the distribution $\tilde{f}$ is 0 on $[\text{sup } f, \infty)$ and the integral $\int_0^M \tilde{f}$ is the same for all $M \ge \text{sup } f$. Thus the restriction of $\tilde{f}$ to $[0, \text{sup } f]$ can "do everything for us" that the unrestricted function can and therefore will also be called the distribution $\tilde{f}$.

The Exercise about to be presented will be used to obtain a glimpse into the technical hassles mentioned above. *The concept of countability will play an essential role in the exercise as well as further on.* In the terminology of this book, "countable" includes finite. The elements of a countably infinite set A can by definition be arranged in a sequence with no repetitions and range equal to the entire set A. Any such sequence will be called an *enumeration* of A.

1.1.5. Exercise. Let $\{r_k\}_{k\geq 1}$ be an enumeration of the set of all rational numbers in $(0, 1)$. For each $n \in \mathbb{N}$, let $f_n: [0, 1] \to \mathbb{R}$ be the function which vanishes at the first n numbers of the sequence $\{r_k\}_{k\geq 1}$ and also at 0 and 1, while having the value 1 at every other point of its domain. Also, let $f: [0, 1] \to \mathbb{R}$ be the function for which $f(x) = \inf\{f_n(x): n \in \mathbb{N}\}$ for each $x \in [0, 1]$. Obviously, $0 \leq f(x) \leq 1$. Show that

(i)   for each $x \in [0, 1]$ and $n \in \mathbb{N}$, the inequality $f_n(x) \geq f_{n+1}(x)$ holds;
(ii) for each $x \in [0, 1]$, $f(x) = \lim_{n \to \infty} f_n(x)$;
(iii) each f_n is a step function with $\tilde{f}_n(t) = 1$ when $0 \leq t < 1$;
(iv) $S = \{x \in [0, 1]: f(x) > t\}$ is not a countable union of intervals when $0 \leq t < \sup f$.

Solution: The functions f_n and f_{n+1} differ only at r_{n+1} and $f_n(r_{n+1}) = 1 > 0 = f_{n+1}(r_{n+1})$. This proves (i). It now follows from elementary facts about monotone sequences that (ii) holds (see Theorem 3.3.3 of Shirali and Vasudeva [9] or 3.14 of Rudin [7]). The first n numbers of the sequence $\{r_k\}_{k\geq 1}$ provide a partition of $[0, 1]$ when arranged in increasing order. By definition, f_n is 1 on each open subinterval of the partition, so that it is a step function. Besides, for any t such that $0 \leq t < 1$, the set $S_n = \{x \in [0, 1]: f_n(x) > t\}$ is the union of the open subintervals of the partition, which means its total length is 1. This proves (iii).

    To prove (iv), we first note that $f(x) = 0$ if x is rational and 1 otherwise. This is because, on the one hand, since the range of $\{r_k\}_{k\geq 1}$ is the set of *all* rational numbers in $(0, 1)$, any rational x equals r_k for some k and hence we have $f_n(x) = f_n(r_k) = 0$ for every $n \geq k$; on the other hand, since any irrational x differs from r_k for every k, we have $f_n(x) = 1$ for every $n \in \mathbb{N}$. From the foregoing description of f, we observe that $\sup f = 1$. Moreover, when $0 \leq t < 1$, the set $S = \{x \in [0, 1]: f(x) > t\}$ consists of all the irrational numbers in $[0, 1]$. Thus $S = \cap_{n=1}^{\infty} S_n$. If S were to be a countable union of intervals, then each interval would have to consist of at most one point (because any interval consisting of more than one point contains a rational number, which cannot be in S), thereby making their union a countable set, which S is not.

    The above exercise illustrates that it can happen for a sequence $\{f_n\}_{n\geq 1}$ of step functions converging at each point to a limit function f that all functions of the sequence have perfectly simple distributions and yet the limit function has none. This jarring situation has arisen because although (in the notation of the solution presented above) $S = \cap_{n=1}^{\infty} S_n$, where each S_n is a finite union of intervals, the set S is not even a countable union of intervals. Of course, one has the option of switching over to the complement of S in $[0, 1]$, which consists of all the rational numbers in $[0, 1]$ and is thus a countable union of the single point intervals $[0, 0]$, $[1, 1]$ and $[r_n, r_n]$, whose total length can be taken as $0 + 0 + \ldots = 0$. This makes it reasonable to think of the "total length" of S as being 1, although it has not been arrived at as a sum of lengths.

    Whatever we understand by total length of a more general kind of a set has to go *beyond a mere totaling process.*

    The obstacle will eventually be surmounted in Chap. 4, but only after we have looked at some other benefits that will accrue from considering distributions, something we begin doing in the next section.

1.1.6. Remark. In summary: In order to arrange for a more general class of functions to have distributions that are sure to be decreasing functions, what we would like is that the total length idea be extended—if necessary by going beyond a totaling process—so as to ensure that:

(1) it is applicable to a class of sets that is broad enough to include (i) unions of countable families of sets in the class and (ii) complements of sets in the class; (intersections of countable families of sets in the class will then necessarily be included);

(2) the total length (in the extended sense) of a set is never less than that of a subset, so that a distribution is sure to be a decreasing function.

The complement of the set S occurring in Exercise 1.1.5 was seen to be the union of a countable number of disjoint single point intervals, and the totaling process to obtain its length led to a series with its nth partial sum representing the sum of the lengths of the first n intervals. This would have been so even if the lengths of the intervals were not 0. This suggests that the total length in the extended sense should also have the property that

(3) the (extended) total length of a set $S = \cup_{n=1}^{\infty} S_n$, where any two among the sets S_n are disjoint, agrees with the sum of the series formed by (extended) total lengths of S_n.

The other concepts that we shall develop by using distributions have far-reaching consequences. For one of them, the reader may go through the statement of Exercise 3.1.10 right away.

Problem Set 1.1
1.1.P1. For each $t \geq 0$, find $S = \{x \in [0, 4]: f(x) > t\}$, where $f: [0, 4] \to \mathbb{R}$ is the function given by:

$$f(x) = 6 \text{ if } 0 \leq x < 1, \ f(1) = 5, \ f(x) = 2 \text{ if } 1 < x < 3, \ f(3) = 1,$$
$$f(x) = 6 \text{ if } 3 < x \leq 4.$$

1.1.P2. For $f: [1, 2] \to \mathbb{R}$ given by $f(x) = e^x$, find the distribution $\tilde{f}: [0, \sup f] \to \mathbb{R}$ and show that $\int_0^{\sup f} \tilde{f}$ agrees with $\int_1^2 f$.

1.1.P3. Let $A > 0$. For the function $f: [0, A] \to \mathbb{R}$ given by $f(x) = e^{-x^2}$, find the distribution $\tilde{f}: [0, \sup f] \to \mathbb{R}$ and show that $\int_0^{\sup f} \tilde{f}$ agrees with $\int_1^A f$.
Hint: Use an appropriate substitution and integrate by parts.

1.1.P4. Abel summation formula. If $a_1, \ldots, a_n$ and $b_1, \ldots, b_n$ are two finite sequences of n terms each ($n \geq 2$) and $S_k = b_1 + \cdots + b_k$, then $b_k = S_k - S_{k-1}$ for $k = 2, \ldots, n$ and therefore

$$a_1 b_1 + \cdots + a_n b_n = a_1 S_1 + a_2(S_2 - S_1) + a_3(S_3 - S_2) + \cdots + a_n(S_n - S_{n-1})$$
$$= (a_1 - a_2)S_1 + (a_2 - a_3)S_2 + (a_3 - a_4)S_3$$

$$+ \cdots + (a_{n-1} - a_n)S_{n-1} + a_n S_n.$$

In Σ notation:

Let $\{a_k\}_{k=1}^n$ and $\{b_k\}_{k=1}^n$ be finite sequences of numbers, where $n \geq 2$, and let $S_k = \sum_{j=1}^k b_j$ for $1 \leq k \leq n$. Then

$$\sum_{k=1}^n a_k b_k = \sum_{k=1}^{n-1} (a_k - a_{k+1})S_k + a_n S_n.$$

Prove this in Σ notation without resorting to algebraic manipulations concealed behind an ellipsis $\cdots$ How should T_k be defined in order to obtain the alternative version $\sum_{k=1}^n a_k b_k = a_1 T_1 + \sum_{k=2}^n (a_k - a_{k-1})T_k$?

1.1.P5. (Relevant for Problem 1.1.P7 and to be used in Theorem 4.3.1) If one visualizes two disjoint intervals on the real line, one of them lies entirely on the left of the other. Prove the following precise formulation of this idea for bounded intervals: If two bounded nonempty intervals are disjoint, then the right endpoint of one of them is less than or equal to the left endpoint of the other.

Deduce that, if the union of two disjoint intervals is an interval, then the sum of their lengths is equal to the length of the union.

1.1.P6. Suppose $\mathcal{F}$ is a finite family of disjoint bounded intervals. Show that there exists a nonempty finite family $\mathcal{G}$ of bounded intervals such that

(a) a union of two distinct intervals in $\mathcal{G}$ is never an interval;
(b) $\cup\mathcal{G} = \cup\mathcal{F}$ (Here "$\cup\mathcal{H}$" means union of all the intervals in $\mathcal{H}$);
(c) the total length of the intervals in $\mathcal{G}$ is the same as for $\mathcal{F}$.

[Note that the intervals in $\mathcal{G}$ are necessarily disjoint in view of (a).]

1.1.P7. If an interval I has an endpoint that belongs to an interval J, show that the union $I \cup J$ is an interval. (Only a left endpoint need be considered.)

1.1.P8. Let $\mathcal{G}$ be a finite nonempty family of bounded intervals such that a union of two distinct intervals of $\mathcal{G}$ is never an interval. Suppose $\mathcal{F}$ is a finite family of disjoint bounded intervals such that $\cup\mathcal{F} = \cup\mathcal{G}$. Show that every interval of $\mathcal{F}$ is contained in some interval of $\mathcal{G}$. What more can be said if $\mathcal{F}$ too is nonempty and has the property that a union of two distinct intervals of it is never an interval?

1.1.P9. Let $\mathcal{F}_1$ and $\mathcal{F}_2$ be finite families of bounded intervals, each having the property that its intervals are disjoint. If $\cup\mathcal{F}_1 = \cup\mathcal{F}_2$, show that the total length of the intervals of each family is the same.

1.2 Improper Integrals

Towards the end of the preceding section, we encountered some difficulty in assigning a size, or total length, to certain subsets of $\mathbb{R}$. What we did have was an instance of a family $\mathcal{F}$ of subsets of an interval, with each set $A \in \mathcal{F}$ having an associated nonnegative real number; in effect, a function $\mu: \mathcal{F} \to \mathbb{R}$ such that $\mu(A) \geq 0$ for every $A \in \mathcal{F}$. It is easy to conceive of other instances of a family $\mathcal{F}$ of subsets of some set X with a nonnegative-valued function $\mu: \mathcal{F} \to \mathbb{R}$. In fact, here is a trivial example:

Let X consist of two distinct elements a and b, and let $\mathcal{F}$ consist of all subsets. Define μ by setting $\mu(\emptyset) = 0$, $\mu(\{a\}) = 2$, $\mu(\{b\}) = 3$ and $\mu(X) = 5$.

The fact that so trivial an example, lacking any ostensible purpose, is possible testifies to the wide sweep of the idea. It turns out however that examples that do have an ostensible purpose and are equally far removed from the original one in which X is an interval are also possible. Readers who have worked with probability will recognize that, in what is called a sample space, the family $\mathcal{F}$ of subsets called "events" has a probability $\mu: \mathcal{F} \to \mathbb{R}$ defined on it. In recent years, there has been great interest in what are called belief measure, plausibility measure and possibility measure, all of which are cases of a number $\mu(A)$ being assigned to every subset A of some set. For details, the reader may consult the book "Fuzzy Measure Theory" by Wang and Klir [11].

A function whose domain consists of some or all subsets of a set is sometimes called a *set function*. All the functions in the above three paragraphs are set functions.

What is of immediate interest against the background of the preceding section is that the concept of *distribution* makes sense in such a general framework. If a nonnegative-valued function $f: X \to \mathbb{R}$ has the property that

$$\{x \in X : f(x) > t\} \text{ belongs to the family } \mathcal{F} \text{ for every } t \geq 0,$$

then $\mu(\{x \in X : f(x) > t\})$ is meaningful and we can take it as $\tilde{f}(t)$ for $t \geq 0$.

If f takes only finitely many values, then as observed before for step functions, the set $\{x \in X : f(x) > t\}$ remains unchanged as t varies in the open interval between two consecutive values; thus $\tilde{f}: [0, \infty) \to \mathbb{R}$ is a step function and therefore has a Riemann integral on any bounded subinterval of its domain. Since it is 0 on $[\sup f, \infty)$, the integral $\int_0^M \tilde{f}$ is the same for all $M \geq \sup f$. Therefore

$$\lim_{M \to \infty} \int_0^M \tilde{f} = \int_0^{\sup f} \tilde{f}.$$

The limit here is of course the improper integral $\int_0^M \tilde{f}$.

To accommodate functions that may take infinitely many values, in which case the distribution may fail to be a step function, we would like μ to be *monotone* in the following sense: $A \subseteq B \Rightarrow \mu(A) \leq \mu(B)$; this ensures that $\tilde{f}$ is decreasing, because

$$t_1 > t_2 \Rightarrow \{x \in X: f(x) > t_1\} \subseteq \{x \in X: f(x) > t_2\}.$$

Since $\tilde{f}$ is decreasing, we know that the integral $\int_0^M \tilde{f}$ exists for all M (see Theorem 9.3.3 of Shirali and Vasudeva [9] or 6.9 of Rudin [7]) and increases with M in view of the nonnegativity of $\tilde{f}$. Consequently, the limit $\int_0^\infty \tilde{f} = \lim\limits_{M\to\infty} \int_0^M \tilde{f}$ exists either as a real number or as ∞.

1.2.1 Remark. We have not ruled out the possibility that $\int \tilde{f} = \infty$. We also do not wish to rule out the possibility that the size of a subset is ∞, so that we can take X to be $\mathbb{R}$ or $[0, \infty)$; this means allowing $\tilde{f}$ to take ∞ as a value. We shall therefore need some conventions about ∞ when it appears in a computation, and we state them below. If one or more of the terms in a statement or computation turn out to be ∞, then the validity of the assertion will generally have to be checked separately by applying the conventions we are about to articulate. Doing so is usually straightforward and details will be left to the reader. In adopting these conventions, we are introducing ∞ as a mathematical object and no longer regarding it as just convenient shorthand.

(a) $x < \infty$ for every $x \in \mathbb{R}$; $\infty \not< \infty$;
(b) $x + \infty = \infty + x = \infty$ for every $x \in \mathbb{R}$;
(c) $\infty + \infty = \infty$;
(d) $\infty \cdot \infty = \infty$; $\infty^p = \infty$ for every $p > 0$;
(e) $x \cdot \infty = \infty \cdot x = \infty$ for every positive $x \in \mathbb{R}$;
(f) $\infty \cdot 0 = 0 \cdot \infty = 0$.

Because of (a), the statement $\int \tilde{f} = \infty$ [or $\tilde{f}(t) < \infty$] is equivalent to $\int \tilde{f} \neq \infty$ [or $\tilde{f}(t) \neq \infty$], which is also expressed by saying that $\int \tilde{f}$ [or $\tilde{f}(t)$] is finite. Furthermore, $0 < \infty$ and therefore ∞ is regarded as positive but only "extended real". Thus, a function taking values in $\{x \in \mathbb{R} : x \geq 0\} \cup \{\infty\}$, a set to be denoted henceforth by $\mathbb{R}^{*+}$, is said to be *nonnegative extended real-valued*. Note that such a function need not actually take the value ∞. The symbol $\mathbb{R}^+$ will denote the set $\{x \in \mathbb{R} : x \geq 0\}$ of nonnegative real numbers; a function taking values in $\mathbb{R}^+$ will be described as *nonnegative real-valued*.

We also adopt the convention that the supremum of a nonempty set that is not bounded above is ∞. Moreover, for a sequence $\{s_n\}_{n \geq 1}$ in $\mathbb{R}^{*+}$, we take $\lim_{n\to\infty} s_n = \infty$ to mean that

for every $A > 0$ there exists a natural number n_0 such that $n > n_0 \Rightarrow s_n > A$.

Then it is easy to see that, as far as an increasing sequence $\{s_n\}_{n \geq 1}$ is concerned, the statements

$$\lim_{n\to\infty} s_n = \infty \quad \text{and} \quad \sup\{s_n : n \geq 1\} = \infty$$

both mean that the sequence is unbounded above. Therefore, for such a sequence, we have

$$\lim_{n\to\infty} s_n = \sup\{s_n : \geq 1\}.$$

It is elementary that this equality holds good for a bounded increasing sequence (see Theorem 3.3.3 of Shirali and Vasudeva [9] or 3.14 of Rudin [7]). Thus it holds for any increasing sequence, whether bounded or not. In particular, the sum of a series of nonnegative terms is the supremum of its partial sums, and the sum is ∞ if and only if the partial sums are unbounded. We shall later need the consequence of the "rearrangement theorem" (see Theorem 4.2.4 of Shirali and Vasudeva [9] or 3.55 of Rudin [7]) that the sum of a series of nonnegative terms is independent of the order of terms even if the sum is ∞.

For increasing sequences, we have $\lim_{n \to \infty}(s_n + t_n) = \lim_{n \to \infty} s_n + \lim_{n \to \infty} t_n$ whether the limits are finite or not; similarly, for nonempty sets, we have $\sup \{s + t : s \in S, t \in T\} = \sup S + \sup T$, whether the suprema are finite or not (obvious when one supremum is ∞; for finite suprema, see Problem 1.6.P6 of Shirali and Vasudeva [9]). The first part can be summarized as "the limit of a finite sum of increasing sequences is the sum of their separate limits".

If one term in a series of nonnegative terms is ∞, then every partial sum up to that term and beyond is ∞ and hence so is the sum of the series.

Suppose that for each $n \in \mathbb{N}$, we have a series $\sum_{k=1}^{\infty} a_{n,k}$ of nonnegative terms in $\mathbb{R}^{*+}$. Then their sums form a sequence of nonnegative terms, using which one could set up a series $\sum_{n=1}^{\infty} \sum_{k=1}^{\infty} a_{n,k}$. What this means is that we have a function $a : \mathbb{N} \times \mathbb{N} \to \mathbb{R}^{*+}$, called a *double sequence*, and an associated *repeated sum* $\sum_{n=1}^{\infty} \sum_{k=1}^{\infty} a_{n,k}$. By reversing the roles of n and k, we get another repeated sum $\sum_{k=1}^{\infty} \sum_{n=1}^{\infty} a_{n,k}$. It will be useful to note that values of the two sums are the same. This is of course trivial when one of the terms $a_{n,k}$ is ∞. To see why it is true when every term is finite, consider any $N, K \in \mathbb{N}$. It is elementary that $\sum_{n=1}^{N} \sum_{k=1}^{K} a_{n,k} = \sum_{k=1}^{K} \sum_{n=1}^{N} a_{n,k}$. Therefore

$$\sum_{k=1}^{\infty} \sum_{n=1}^{N} a_{n,k} = \lim_{K \to \infty} \sum_{k=1}^{K} \sum_{n=1}^{N} a_{n,k} = \lim_{K \to \infty} \sum_{n=1}^{N} \sum_{k=1}^{K} a_{n,k} = \sum_{n=1}^{N} \sum_{k=1}^{\infty} a_{n,k},$$

where the final equality is based on the fact that the limit as $K \to \infty$ of the sum of the N sequences $\sum_{k=1}^{K} a_{n,k}$ is the sum of their separate limits. Since each $a_{n,k}$ is nonnegative, we can now argue that $\sum_{k=1}^{\infty} \sum_{n=1}^{\infty} a_{n,k} \geq \sum_{k=1}^{\infty} \sum_{n=1}^{N} a_{n,k} = \sum_{n=1}^{N} \sum_{k=1}^{\infty} a_{n,k}$. Upon taking the limit as $N \to \infty$, we obtain $\sum_{k=1}^{\infty} \sum_{n=1}^{\infty} a_{n,k} \geq \sum_{n=1}^{\infty} \sum_{k=1}^{\infty} a_{n,k}$. The reverse inequality can be established by an analogous argument and therefore $\sum_{n=1}^{\infty} \sum_{k=1}^{\infty} a_{n,k} = \sum_{k=1}^{\infty} \sum_{n=1}^{\infty} a_{n,k}$ ("interchanging the order of summation"). For a detailed discussion of this topic, particularly when $a_{n,k}$ is real or complex and not restricted to be nonnegative when real, the reader may consult the article "Double sequences and double series" by Habil [4].

The concepts of liminf and limsup carry over quite smoothly and we shall encounter the liminf in Theorem 3.2.3.

We have not introduced $-\infty$, although we could have done so by extending appropriately the list (a)–(e) above. Therefore, care must be taken to avoid such expressions as $-\int \tilde{f}$ or $-\tilde{f}$ without ensuring first that the integral or function concerned is finite-valued.

We have encountered $\int_0^{\infty} \tilde{f}$ before and now we need to lay down what such an "improper" integral means when the integrand can have an infinite value and need not vanish beyond some $M > 0$. However, we need consider only functions that are decreasing and take nonnegative values, ∞ included; in other words, decreasing nonnegative extended real-valued functions. They need not be distributions of any functions, and accordingly, we shall dispense with the tilde in the rest of this section.

1.2.2. Definition. *For a decreasing nonnegative extended real-valued function f on* $[0, \infty)$, *we define the* **integral from 0 to** ∞ *as*

$$\int_0^\infty f = \sup\{\int_a^b g : 0 < a < b < \infty; g : [a, b] \to \mathbb{R} \text{ Riemann integrable}$$

and $0 \le g \le f$ on $[a, b]\}$. \hfill (1.6)

It is understood here that g is real-valued, because otherwise it would not be Riemann integrable.

This is similar to the "extended integral" defined by Munkres [5, p. 121], except that the domain of integration here is always $[0, \infty)$ and the integrand must be decreasing, but the integrand as well as the integral may take ∞ as a value.

Consider the case when f is real-valued except possibly at 0. Since it is decreasing, its restriction to any $[a, b]$, where $0 < a < b < \infty$, is Riemann integrable. Moreover, $\int_a^b f \ge \int_a^b g$ for any g of the kind mentioned in (1.6). It follows that

$$f \text{ real-valued on } (0, \infty) \Rightarrow \int_0^\infty f = \sup\{\int_a^b f : 0 < a < b < \infty\}. \quad (1.7)$$

If f is not real-valued on $(0, \infty)$, that is to say, $f(\alpha) = \infty$ for some real $\alpha > 0$, then by considering a sequence of nonnegative functions $g_n \le f$ such that $g_n(x) = n$ on $[0, \alpha]$, we find that $\int_0^\infty f = \infty$. Therefore

$$\int_0^\infty f < \infty \Rightarrow f \text{ real-valued on } (0, \infty). \quad (1.8)$$

Now suppose f is real-valued on $(0, \infty)$. If $f(0)$ is also real, then f is Riemann integrable over $[0, A]$ for every $A > 0$, because it is decreasing. The usual meaning of the improper integral $\int_0^\infty f$ is then $\lim_{A \to \infty} \int_0^A f$ which agrees with the supremum in (1.6) because f is nonnegative-valued. If $f(0) = \infty$, then f is Riemann integrable over $[x, B]$ whenever $0 < x < B < \infty$. Two possibilities arise. One is that f vanishes beyond some $B > 0$; the usual meaning of the improper integral $\int_0^\infty f$ is then $\lim_{x \to 0} \int_x^B f$, which again agrees with the supremum in (1.6). The other possibility is that there is no such B. In this situation, the usual meaning of the improper integral $\int_0^\infty f$ is $\lim_{x \to 0} \int_x^B f + \lim_{A \to \infty} \int_B^A f$, which is independent of the chosen $B > 0$. With a little effort, one can see that this too agrees with the supremum in (1.6). Here, the result that the sum of the integrals of a function on contiguous intervals $[\alpha, \beta]$ and $[\beta, \gamma]$ equals the integral of that function over $[\alpha, \gamma]$ is relevant.

We conclude from the above paragraph that when f is real-valued on $(0, \infty)$, the integral $\int_0^\infty f$ in the sense of Definition 1.2.2 can be computed as usual.

1.2.3. Lemma. *If $\{g_n\}_{n\geq 1}$ is a sequence of real-valued functions on an interval $[a, b]$ such that*

(a) g_n *is a decreasing function on $[a, b]$ for each $n \in \mathbb{N}$,*
(b) $g_n(x) \leq g_{n+1}(x)$ *for each $x \in [a, b]$ and each $n \in \mathbb{N}$,*
(c) $\lim_{n\to\infty} g_n(x) = g(x)$ *for each $x \in [a, b]$ and $g(a) < \infty$,*

  then g is Riemann integrable and

$$\int_a^b g = \lim_{n\to\infty} \int_a^b g_n.$$

Proof: Since (a) and (c) together imply that g is a real-valued decreasing function on $[a, b]$, it follows that it is Riemann integrable. We need prove only the limit. In view of (b) and (c), we already know that $\lim_{n\to\infty} \int_a^b g_n$ exists and

$$\int_a^b g \geq \lim_{n\to\infty} \int_a^b g_n. \tag{1.9}$$

The proof will be complete as soon as we prove the reverse inequality. For this purpose, consider any $\varepsilon > 0$. There exists a partition $P: a = x_0 < x_1 < \ldots < x_m = b$ of $[a, b]$ such that the lower sum $L(g, P)$ of g over P satisfies

$$\int_a^b g - L(g, P) < \frac{\varepsilon}{2}. \tag{1.10}$$

Since g is decreasing, its infimum on any interval $[x_{j-1}, x_j]$ is $g(x_j)$ and therefore

$$L(g, P) = \sum_{j=1}^m g(x_j)(x_j - x_{j-1}),$$

and correspondingly for each g_n. Hence for each $n \in \mathbb{N}$, the following equality holds:

$$L(g, P) - L(g_n, P) = \sum_{j=1}^m (g(x_j) - g_n(x_j))(x_j - x_{j-1}). \tag{1.11}$$

In view of (c), there exists an $N \in \mathbb{N}$ such that $N \leq n \in \mathbb{N}$ implies

$$g(x_j) - g_n(x_j) < \frac{1}{2} \cdot \frac{\varepsilon}{b - a} \quad \text{for } 1 \leq j \leq m.$$

Together with (1.11), this implies

$$L(g, P) - L(g_n, P) < \frac{\varepsilon}{2},$$

which, together with (1.10), yields $\int_a^b g - L(g_n, P) < \varepsilon$ for $n \geq N$. Consequently,

$$\int_a^b g - \varepsilon < L(g_n, P) \leq \int_a^b g_n \quad \text{for } n \geq N.$$

Since $\lim_{n\to\infty} \int_a^b g_n$ exists, it follows that

$$\int_a^b g - \varepsilon \leq \lim_{n\to\infty} \int_a^b g_n.$$

This has been shown to be true for every $\varepsilon > 0$. Therefore the reverse of the inequality (1.9) must hold. As noted earlier, this is all that needed to be proved. □

1.2.4. Proposition. *If $\{f_n\}_{n\geq 1}$ is a sequence of nonnegative extended real-valued functions on the interval $[0, \infty)$ such that*

(a) *f_n is a decreasing function on $[0, \infty)$ for each $n \in \mathbb{N}$,*
(b) *$0 \leq f_n(t) \leq f_{n+1}(t)$ for each $t \in [0, \infty)$ and each $n \in \mathbb{N}$,*
(c) *$\lim_{n\to\infty} f_n(t) = f(t)$ for each $t \in [0, \infty)$,*

  then $\int_a^\infty f = \lim_{n\to\infty} \int_a^\infty f_n$.

Proof: First suppose $\int_a^\infty f$ is not ∞. Then by the consequence (1.8) of Definition 1.2.2, f is real-valued on $(0, \infty)$ and hence by (1.7), for any $\varepsilon > 0$, there exist a and b such that $0 < a \leq b < \infty$ and

$$0 \leq \int_0^\infty f - \int_a^b f < \frac{\varepsilon}{2}.$$

Since f is real-valued on $(0, \infty)$, it follows by (b) and (c) of the hypothesis that each f_n is real-valued on $(0, \infty)$. Therefore by Lemma 1.2.3 and by (a) of the hypothesis, there exists an $N \in \mathbb{N}$ such that

$$n \geq N \Rightarrow 0 \leq \int_a^b f - \int_a^b f_n < \frac{\varepsilon}{2} \Rightarrow 0 \leq \int_0^\infty f - \int_a^b f_n < \varepsilon.$$

But $0 \leq \int_0^\infty f - \int_0^\infty f_n \leq \int_0^\infty f - \int_a^b f_n$. Therefore

$$n \geq N \Rightarrow 0 \leq \int_0^\infty f - \int_0^\infty f_n < \varepsilon.$$

This proves the contention when $\int_a^\infty f$ is not ∞.

Now, suppose $\int_a^\infty f = \infty$ and that f is real-valued on $(0, \infty)$. Then the same is true of each f_n in view of (b) and (c) of the hypothesis. By (1.7), for any $K > 0$, there exist a and b such that $0 < a \leq b < \infty$ and $\infty > \int_a^\infty f > K$. The foregoing lemma yields an integer $N \in \mathbb{N}$ such that

$$n \geq N \Rightarrow 0 \leq \int_a^b f - \int_a^b f_n < \int_a^b f - K \Rightarrow \int_a^b f_n > K.$$

Therefore $\int_a^\infty f = \lim_{n\to\infty} \int_a^\infty f_n$ also when $\int_a^\infty f = \infty$ and f is real-valued on $(0, \infty)$.

Finally, suppose $\int_a^\infty f = \infty$ but f is not real-valued on $(0, \infty)$. Then $f(\alpha) = \infty$ for some real $\alpha > 0$. From (b) and (c) of the hypothesis, we deduce that for any $M > 0$, there exists an $N \in \mathbb{N}$ such that $n \geq N \Rightarrow f_n(\alpha) \geq M$. It follows from (a) that $n \geq N \Rightarrow f_n \geq M$ on $[\frac{\alpha}{2}, \alpha]$, so that $\int_0^\infty f_n \geq \int_{\alpha/2}^\infty M = \frac{\alpha}{2} M$. Since such an N exists for any $M > 0$, we have $\lim_{n\to\infty} \int_a^\infty f_n = \infty = \int_a^\infty f$. □

Problem Set 1.2
1.2.P1. For the double sequence $a: N \times N \to \mathbb{R}$ given by

$$a_{n,k} = 1 \text{ if } n = k, \quad a_{n,k} = -1 \text{ if } k = n+1 \quad \text{and} \quad a_{n,k} = 0 \text{ in all other cases,}$$

show that the repeated sums $\sum_{n=1}^\infty \sum_{k=1}^\infty a_{n,k}$ and $\sum_{k=1}^\infty \sum_{n=1}^\infty a_{n,k}$ are unequal, although $\sum_{n=1}^\infty |\sum_{k=1}^\infty a_{n,k}|$ and $\sum_{k=1}^\infty |\sum_{n=1}^\infty a_{n,k}|$ are convergent.
Hint: One can think of $a_{n,k}$ as representing an infinite matrix with first three rows:

$$\begin{array}{cccccc} 1 & -1 & 0 & 0 & \cdots \\ 0 & 1 & -1 & 0 & 0 & \cdots \\ 0 & 0 & 1 & -1 & 0 & 0 & \cdots \end{array}$$

1.2.P2. Let $b > 0$ and g be a decreasing nonnegative function on $[0, b]$ that is real-valued except perhaps at 0. Denote by f the decreasing nonnegative function on $[0, \infty)$ obtained by extending g to be equal to 0 to the right of b. Show that

$$\int_0^\infty f = \lim_{x\to 0} \int_x^b g.$$

Note: If $g(0) < \infty$, the function g is Riemann integrable on $[0, b]$ and the limit here is equal to $\int_a^b g$ (see Problem 9.3.P2 of Shirali and Vasudeva [9]).

1.2.P3. In Lemma 1.2.3, show that the condition $g(a) < \infty$ does not follow from the remaining conditions in the hypothesis even if we impose the additional condition that g is real-valued on $(a, b]$. If $g(a) = \infty$ but g (which must be decreasing) is real-valued on $(a, b]$, then the improper integral $\int_a^b g$ in the usual sense is $\lim_{x \to 0} \int_x^b g$, though it may be ∞; show that $\int_0^\infty g = \lim_{n \to \infty} \int_a^b g_n$. If g is not real-valued on $(a, b]$, which to say, $g(\alpha) = \infty$ for some $\alpha > a$, show that $\lim_{n \to \infty} \int_a^b g_n = \infty$.

Chapter 2
Measure Space and Integral

2.1 Measure and Measurability

The intent behind the definitions that are about to follow may be easier to appreciate if Remark 1.1.6 is kept in view. Symbol-free descriptions of the properties in the definitions are given alongside.

The complement of a set will be indicated by a superscript "c"; thus, the symbol A^c will mean the complement of the set A.

2.1.1. Definition. *A family $\mathcal{F}$ of subsets of a nonempty set X is called a* **σ-algebra** (or a **σ-field**) *if*

(a) $\emptyset \in \mathcal{F}$ (contains the empty set),
(b) *for every sequence $\{A_k\}_{k\geq 1}$ of sets, each belonging to $\mathcal{F}$,*
   $\bigcup_{k=1}^{\infty} A_k \in \mathcal{F}$ (closed under countable unions),
(c) $A^c \in \mathcal{F}$ *whenever* $A \in \mathcal{F}$ (closed under complements).

Since $X = \emptyset^c$, conditions (a) and (c) in the definition together imply that $X \in \mathcal{F}$. Also, conditions (b) and (c) together imply closure under countable intersections, considering that

$$\bigcap_{k=1}^{\infty} A_k = \left(\bigcup_{k=1}^{\infty} A_k^c\right)^c.$$

From conditions (a) and (b), we infer that a σ-algebra is closed under finite unions as well:

$$\text{if } A_k \in \mathcal{F} \text{ for } 1 \leq k \leq n, \text{ then } \bigcup_{k=1}^{n} A_k \in \mathcal{F}.$$

© Springer Nature Switzerland AG 2018
S. Shirali, *A Concise Introduction to Measure Theory*,
https://doi.org/10.1007/978-3-030-03241-8_2

This is true because we may take $A_k = \emptyset$ for $k > n$. The same goes for finite intersections as well. In particular, if A and B belong to $\mathcal{F}$, then so does the set-theoretic difference $A \backslash B$, which is defined to be $A \cap B^c$. These facts and their simple consequences will be used without further ado.

We shall speak of three or more sets as being *disjoint* when no two of them have an element in common. This is the same as what many authors call "mutually disjoint" or "pairwise disjoint".

2.1.2. Definition. *A nonnegative extended real-valued set function* $\mu\colon \mathcal{F} \to \mathbb{R}^{*+}$, *where* $\mathcal{F}$ *is a σ-algebra of subsets of a set X, is called a* **measure** *if*

(a) $\mu(A) \geq 0$ *for every* $A \in \mathcal{F}$ (nonnegative),
(b) $\mu(\emptyset) = 0$ (vanishes at the empty set),
(c) *for every sequence* $\{A_n\}_{n \geq 1}$ *of disjoint sets, each belonging to* $\mathcal{F}$,

$$\mu\left(\bigcup_{n=1}^{\infty} A_n\right) = \sum_{n=1}^{\infty} \mu(A_n)\text{(countably additive)}.$$

The triplet $(X, \mathcal{F}, \mu)$ *is called a* **measure space**.

The backdrop for countable additivity is requirement (3) of Remark 1.1.6.

Given any finite sequence $A_1, \ldots, A_n$ of disjoint sets belonging to $\mathcal{F}$, it is possible to extend it to an infinite sequence of disjoint sets belonging to $\mathcal{F}$ by appending $\emptyset$, $\emptyset, \ldots$, and hence properties (b) and (c) of Definition 2.1.2 justify the assertion that

$$\mu\left(\bigcup_{k=1}^{n} A_k\right) = \sum_{k=1}^{n} \mu(A_n)\text{(finitely additive)}.$$

This has the further consequence that $A \subseteq B \Rightarrow B = A \cup (B\backslash A) \Rightarrow \mu(B) = \mu(A) + \mu(B\backslash A)$ by finite additivity, because $A \cap (B\backslash A) = \emptyset$; monotonicity (which means $A \subseteq B \Rightarrow \mu(A) \leq \mu(B)$) now follows from property (a) of Definition 2.1.2.

In the terminology of these definitions, Remark 1.1.6 says that we would like the total length idea to be extended so as to become a measure μ on a σ-algebra $\mathcal{F}$ that includes all those subsets of $[a, b]$ that are finite unions of disjoint intervals. By "extend" we mean that for such finite unions, μ must agree with total length.

Elements of a σ-algebra $\mathcal{F}$ are called $\mathcal{F}$-**measurable**, or just **measurable** especially when the σ-algebra is understood and has not been denoted by any symbol. When this is so, it is easier to speak of a measure on a set X, with the tacit understanding that some σ-algebra is intended but may or may not have been named. Similarly, it is often convenient to speak of X as a measure space, with the tacit understanding that some σ-algebra and measure are intended but may or may not have been named.

It will be convenient to abbreviate $\mu(A)$ as μA when there is no risk of confusion.

2.1.3. Example. (a) Consider the example mentioned at the beginning of Sect.
1.2, in which X consists of two distinct elements a and b, and $\mathcal{F}$ consists of all
subsets. The set function μ is defined by setting $\mu(\emptyset)=0$, $\mu(\{a\})=2$, $\mu(\{b\})=$
3 and $\mu(X)=5$. It is easy to check that μ is a measure. Let the function f on X
be defined as $f(a)=1, f(b)=3$. Then the set $S=\{x \in X: f(x)>t\}$ is seen to be
as described in the next three lines:

$$X \quad \text{if } 0 \leq t < 1$$
$$\{b\} \text{ if } 1 \leq t < 3 = \sup f$$
$$\emptyset \quad \text{if } 3 \leq t.$$

Since $\mathcal{F}$ consists of all subsets of X, there is such a thing as μS whatever nonnegative number t may be. Therefore there exists a distribution $\tilde{f}$, as explained
in Sect. 1.2.

(b) Let X consist of three distinct elements a, b, c and let $\mathcal{F}$ consist of all subsets.
Define μ by setting $\mu\emptyset = 0$, $\mu\{a\}=2$, $\mu\{b\}=3$, $\mu\{c\}=4$, $\mu\{b, c\}=6$, $\mu\{c,$
$a\}=5, \mu\{a, b\}=4$ and $\mu X = 8$. Obviously, μ is nonnegative and vanishes at the
empty set; one can easily check case by case that it is also monotone. However,
it is not a measure, as it fails to be finitely additive: $\mu\{a\}=2, \mu\{b\}=3$ but $\mu\{a,$
$b\}=4 \neq 2+3$. Next, consider the function f on X defined as $f(a)=1, f(b)=3$,
$f(c)=4$. For this function, the set $S=\{x \in X: f(x)>t\}$ is seen to be as described
in the next four lines:

$$X \qquad \text{if } 0 \leq t < 1$$
$$\{b, c\} \text{ if } 1 \leq t < 3$$
$$\{c\} \qquad \text{if } 3 \leq t < 4 = \sup f$$
$$\emptyset \qquad \text{if } 4 \leq t.$$

Since $\mathcal{F}$ consists of all subsets of X, there is such a thing as μS whatever
nonnegative number t may be. Therefore there is a distribution $\tilde{f}$. If $\mathcal{F}$ were
to consist of selected subsets, then f would have to have the property that S is
always one of the selected subsets before we could speak of its distribution.

2.1.4. Definition. *Let $\mathcal{F}$ be a σ-algebra of subsets of a set X. A function $f: X \rightarrow$
$\mathbb{R} \cup \{\infty\}$ is said to be $\mathcal{F}$-**measurable** (or just **measurable**) if,*

for every $t \in \mathbb{R}$, the set $S = \{x \in X: f(x) > t\}$ is $\mathcal{F}$-measurable,

i.e. belongs to the σ-algebra $\mathcal{F}$.
   If there is also an extended real-valued function $\mu: \mathcal{F} \rightarrow \mathbb{R}^{*+}$, then the **distribution** of a measurable function f on X is the function $\tilde{f}: [0, \infty) \rightarrow \mathbb{R}^{*+}$ such
that

$$\tilde{f}(t) = \mu(\{x \in X: f(x) > t\}).$$

The distribution of a sum $f+g$ will be denoted by $\widetilde{(f+g)}$ or $\widetilde{(f+g)}$, depending on convenience. When we speak of a distribution, μ will usually be understood to be a measure but will occasionally be just nonnegative, monotone and vanishing at $\emptyset$, as was the case in Example 2.1.3(b). As was indicated just before Remark 1.2.1, this is enough to ensure that the distribution is nonnegative and decreasing.

If f takes only negative values, then $\{x \in X: f(x)>t\}$ is empty for any $t \geq 0$ and hence $\tilde{f}$ is 0 everywhere; however, we shall have no occasion to consider such a situation. In fact, we shall be dealing with distributions only of measurable nonnegative functions although measurability has been defined to be applicable to functions that may take negative values.

For reasons set forth in Sect. 1.1, when f is nonnegative and bounded, the restriction of $\tilde{f}$ to $[0, \sup f]$ will also be called the distribution.

The set $\{x \in X: f(x)>t\}$, sometimes called the "t-cut" of f, will be denoted by $X(f>t)$ when convenient. Without explicit mention of $\mathcal{F}$, one can state that an $\mathbb{R} \cup \{\infty\}$-valued function f on a set X is measurable if its t-cut is measurable for every real number t. Thus, when we speak of a measurable function on a set X, it is understood that some σ-algebra of subsets of X is intended.

2.1.5. Exercise. Let f, g be measurable functions on a space X with measure μ, and $f(x)=g(x)$ for all $x \in A$, where A is a measurable subset of X satisfying $\mu(A^c)=0$. Show that $\tilde{f} = \tilde{g}$.

Solution: Consider any real t and the t-cuts B and C of f and g respectively. Since a measure is monotone and $\mu(A^c)=0$, it follows that $\mu(B \cap A^c)=0$. Now $B =(B \cap A^c)\cup(B \cap A)$ and additivity of measure implies $\mu(B)=\mu(B \cap A^c)+\mu(B \cap A)= \mu(B \cap A)$. This means $\tilde{f}(t) = \mu(B \cap A)$. Similarly, $\tilde{g}(t) = \mu(C \cap A)$. By hypothesis, f and g agree on the set A, which implies that the t-cuts B and C satisfy $B \cap A = C \cap A$.

2.1.6. Definition. *For any subset $A \subseteq X$, the* **characteristic function** χ_A *is the function with domain X, for which*

$$\chi_A(x) = \begin{cases} 1 \text{ if } x \in A \\ 0 \text{ if } x \notin A. \end{cases}$$

For the characteristic function χ_A of any subset $A \subseteq X$, the supremum is 1, unless A is empty, in which case the supremum is 0. Thus the interval $[0, \sup \chi_A]$ is $[0, 1]$ if A is nonempty and is the single point interval $[0, 0]$ when A is empty.

It is easy to verify that (i) if $A \subseteq B$, then $\chi_A \leq \chi_B$ (ii) if $A \cap B =\emptyset$, then $\chi_A \chi_B = 0$ everywhere and $\chi_A+\chi_B = \chi_{A\cup B}$.

In Exercise 1.1.5, the relation between the function f_n and the complement S_n of the set formed by the first n terms of the sequence $\{r_k\}_{k\geq 1}$ is described. In the language of the above definition, the relation is that $f_n = \chi_{S_n}$. The relation described there between the function $f = \lim_{n\to\infty} f_n$ and the set $S = \bigcap_{n=1}^{\infty} S_n$ is that $f = \chi_S$. On the basis of the inclusion $S_n \supseteq S_{n+1}$, we obtain $\chi_{S_n} \geq \chi_{S_{n+1}}$, which is simply a restatement of the inequality $f_n \geq f_{n+1}$ of that exercise (cf. Problem 2.1.P1).

2.1.7. Proposition. *The characteristic function of a subset $A \subseteq X$ is measurable if and only if the subset A is measurable. When there is a measure μ on X, the characteristic function has a distribution $\widetilde{\chi_A}$, whose restriction to $[0, 1]$ is a step function having value A on $[0, 1)$ and 0 at 1. When $\mu A = 0$, this means $\widetilde{\chi_A}$ is 0 everywhere. Moreover,*

$$\int_0^\infty \widetilde{\chi_A} = \int_0^{\sup \chi_A} \widetilde{\chi_A} = \mu A.$$

Proof: For $t < 0$, the t-cut $X(\chi_A > t)$ is X, which is always measurable (because any σ-algebra has X as an element). For $0 \leq t < 1$, the set $X(\chi_A > t)$ is A and for $t \geq \sup \chi_A$, the set $X(\chi_A > t)$ is $\emptyset$, which is always measurable (because any σ-algebra has $\emptyset$ as an element). One consequence of this observation is that χ_A is measurable if and only if A is. Another consequence is that $\widetilde{\chi_A}(t)$ is μA on $[0, 1)$ and 0 at 1. Thus $\widetilde{\chi_A}$ is a step function on $[0, 1]$.

If $A \neq \emptyset$, then $\sup \chi_A = 1$ and hence $\int_0^{\sup \chi_A} \widetilde{\chi_A} = \int_0^1 \widetilde{\chi_A} = \mu A$; if $A = \emptyset$, then $\sup \chi_A = 0$ and hence $\int_0^{\sup \chi_A} \widetilde{\chi_A} = \int_0^0 \widetilde{\chi_A} = 0 = \mu A$. □

It is worth noting that the above proposition is valid for any set function μ: $\mathcal{F} \to \mathbb{R}^{*+}$ that vanishes at the empty set.

2.1.8. Example. (a) Let X consist of two elements a and b and $\mathcal{F}$ consist of only $\emptyset$ and X. It is trivial that $\mathcal{F}$ is a σ-algebra. The function f for which $f(a) = 1$, $f(b) = 3$, is not measurable because $X(f > 1) = \{b\} \notin \mathcal{F}$. Therefore it has no distribution.

(b) Let X, $\mathcal{F}$ and μ be as in Example 2.1.3(a). Define the nonnegative functions f and g as below:

$$f(a) = 1, \quad f(b) = 3 \quad \text{and} \quad g(a) = 5, \quad g(b) = 1.$$

Then $(f+g)(a) = 6$, $(f+g)(b) = 4$. Also,

$$\tilde{f}(t) = \begin{cases} 5 & 0 \leq t < 1 \\ 3 & 1 \leq t < 3 \\ 0 & 3 \leq t < \infty \end{cases}, \quad \tilde{g}(t) = \begin{cases} 5 & 0 \leq t < 1 \\ 2 & 1 \leq t < 5 \\ 0 & 5 \leq t < \infty \end{cases}$$

$$\text{and } \widetilde{(f+g)}(t) = \begin{cases} 5 & 0 \leq t < 4 \\ 2 & 4 \leq t < 6 \\ 0 & 6 \leq t < \infty. \end{cases}$$

Hence

$$\int_0^{\sup f} \tilde{f} = 5+6 = 11, \quad \int_0^{\sup g} \tilde{g} = 5+8 = 13,$$

and

$$\int_0^{\sup(f+g)} \widetilde{(f+g)} = 20+4 = 24.$$

Thus

$$\int_0^{\sup f} \tilde{f} + \int_0^{\sup g} \tilde{g} = \int_0^{\sup(f+g)} \widetilde{(f+g)}.$$

(c) Let X, $\mathcal{F}$ and μ be as in Example 2.1.3(b). As noted there, μ is not a measure but is monotone. It also provides an example of what we shall call an *outer measure* later on. Define the nonnegative functions f and g as below:

$$f(a) = 1, \ f(b) = 3, \ f(c) = 4 \ \text{and} \ g(a) = 2, \ g(b) = 1, \ g(c) = 4.$$

Then $(f+g)(a)=3, (f+g)(b)=4, (f+g)(c)=8$. Also,

$$\tilde{f}(t) = \begin{cases} 8 & 0 \le t < 1 \\ 6 & 1 \le t < 3 \\ 4 & 3 \le t < 4 \\ 0 & 4 \le t < \infty \end{cases}, \quad \tilde{g}(t) = \begin{cases} 8 & 0 \le t < 1 \\ 5 & 1 \le t < 2 \\ 4 & 2 \le t < 4 \\ 0 & 4 \le t < \infty \end{cases}$$

and

$$\widetilde{(f+g)}(t) = \begin{cases} 8 & 0 \le t < 3 \\ 6 & 3 \le t < 4 \\ 4 & 4 \le t < 8 \\ 0 & 8 \le t < \infty. \end{cases}$$

Hence

$$\overset{\sup f}{\underset{0}{\int}} \tilde{f} = 8 + 12 + 4 = 24, \quad \overset{\sup g}{\underset{0}{\int}} \tilde{g} = 8 + 5 + 8 = 21,$$

and

$$\overset{\sup(f+g)}{\underset{0}{\int}} \widetilde{(f+g)} = 24 + 6 + 16 = 46.$$

Thus

$$\overset{\sup f}{\underset{0}{\int}} \tilde{f} + \overset{\sup g}{\underset{0}{\int}} \tilde{g} < \overset{\sup(f+g)}{\underset{0}{\int}} \widetilde{(f+g)}.$$

The fact that we do not have equality here has to do with μ not being a measure. One might have liked the integral of the sum to appear on the smaller side of the inequality. However, the inequality

$$\sqrt{\overset{\sup(f+g)}{\underset{0}{\int}} \widetilde{(f+g)}} \leq \sqrt{\overset{\sup f}{\underset{0}{\int}} \tilde{f}} + \sqrt{\overset{\sup g}{\underset{0}{\int}} \tilde{g}},$$

in which the integral of the sum does appear on the smaller side, is found to hold in the present instance. Its validity for general nonnegative f and g can be traced to a property that μ has, called *subadditivity* (see Sect. 3.3).

Problem Set 2.1

2.1.P1. Let $\{S_n\}_{n\geq 1}$ be a sequence of subsets of a set X such that $S_n \subseteq S_{n+1}$ for every $n \in \mathbb{N}$ (no σ-algebra intended) and let $S = \cup_{n=1}^{\infty} S_n$. Denote the characteristic functions of S_n and of S respectively by f_n and f. For each $x \in X$, show that

$$\text{(i) } f_n(x) \leq f_{n+1}(x) \text{ and (ii) } f(x) = \lim_{n\to\infty} f_n(x).$$

2.1.P2. If f and g are functions such that $f \leq g$ everywhere, show for any real t that the t-cut of f is a subset of the t-cut of g. If f and g are also measurable and μ is a measure on their domain, show that $\tilde{f}(t) \leq \tilde{g}(t)$ for every t.

2.1.P3. Let $\{f_n\}_{n\geq 1}$ be a sequence of functions on X such that, for every $x \in X$ and $n \in \mathbb{N}$, the inequality $f_n(x) \leq f_{n+1}(x)$ holds. For any $t \in \mathbb{R}$, denote the t-cut of f_n by $S_n(t)$. Show that, $S_n(t) \subseteq S_{n+1}(t)$ for every $n \in \mathbb{N}$. If furthermore, f is a function on X such that $f(x) = \lim_{n\to\infty} f_n(x)$ for every $x \in X$, show that $X(f > t) = \cup_{n=1}^{\infty} S_n(t)$. Hence conclude that if each f_n is measurable, then f is measurable. (The latter part is true without the hypothesis that $f_n(x) \leq f_{n+1}(x)$, as is to be proved in Problem 2.2.P4.)

2.1.P4. (Note: This problem is of interest in Combinatorics rather than Measure Theory, because it relates to the *Inclusion-Exclusion Principle* there; see Problem 3.1.P1.) Let $A_1, \ldots, A_n$ be subsets of X. Using the symbol $\chi[B]$ to denote the characteristic function of a set B, show that

$$\chi[A_1 \cup \cdots \cup A_n] = \sum_{r=1}^{n} \left((-1)^{r-1} \sum_{1 \le j_1 < \cdots < j_r \le n} \chi[A_{j_1} \cap \cdots \cap A_{j_r}] \right).$$

2.1.P5. If $\mu \colon \mathcal{F} \to \mathbb{R}$ is monotone as well as finitely additive, then show that it is also *finitely subadditive* in the sense that $\mu(A \cup B) \le \mu(A) + \mu(B)$ for measurable sets A and B, disjoint or not.

2.1.P6. Let μ be as in Example 2.1.3(b) and the functions f, g be defined on $X = \{a, b, c\}$ as

$$f(a) = 1, \ f(b) = \sqrt{3}, \ f(c) = 2, \ g(a) = \sqrt{2}, \ g(b) = 1, \ g(c) = 2.$$

Show that

$$\sqrt[3]{\int_0^{\sup(f+g)^2} \widetilde{(f+g)}^2} < \sqrt[3]{\int_0^{\sup f^2} \widetilde{(f^2)}} + \sqrt[3]{\int_0^{\sup g^2} \widetilde{(g^2)}}$$

but

$$\sqrt{\int_0^{\sup(f+g)^2} \widetilde{(f+g)}^2} > \sqrt{\int_0^{\sup f^2} \widetilde{(f^2)}} + \sqrt{\int_0^{\sup g^2} \widetilde{(g^2)}}.$$

2.1.P7. Show that a measure is *countably subadditive* in the following sense: If $(X, \mathcal{F}, \mu)$ is a measure space and $\{A_j\}_{j \in \mathbb{N}}$ is a sequence of measurable sets, then $\mu(\cup_{j=1}^{\infty} A_j) \le \sum_{i=1}^{\infty} \mu(A_j)$. Is the corresponding statement true regarding a finite sequence of measurable sets?

2.1.P8. Let $\mathcal{F}$ be a σ-algebra of subsets of a set X. Suppose $\mu \colon \mathcal{F} \to \mathbb{R}^{*+}$ is finitely additive and vanishes at $\emptyset$. If it is also countably subadditive (as defined in Problem 2.1.P7), show that it is a measure.

2.1.P9. Suppose that, for each $j \in \mathbb{N}$, $\mathcal{F}_j$ is a σ-algebra of subsets of X_j. Define the family $\mathcal{F}$ of sets to consist of those subsets A of the union $X = \cup_{j=1}^{\infty} X_j$ for which $A \cap X_j \in \mathcal{F}_j$ for each $j \in \mathbb{N}$. Show that $\mathcal{F}$ is a σ-algebra.

2.1.P10. For each $j \in \mathbb{N}$, let $(X_j, \mathcal{F}_j, \mu_j)$ be a measure space and denote the union $\cup_{j=1}^{\infty} X_j$ by X. Set $\mathcal{F} = \{A \subseteq X : A \cap X_j \in \mathcal{F}_j \text{ for each } j \in \mathbb{N}\}$ and for each $A \in \mathcal{F}$, set $\mu(A) = \sum_{i=1}^{\infty} \mu_j(A \cap X_j)$. Show that μ is a measure.

2.1.P11. Let $\mathcal{F}$ be a σ-algebra of subsets of a nonempty set X and Y any set with a map $F\colon X \to Y$. Take $\mathcal{G}$ to be the family of subsets $\{B \subseteq Y\colon F^{-1}(B) \in \mathcal{F}\}$ of Y. Here, $F^{-1}(B)$ means $\{x \in X\colon F(x) \in B\}$, as usual. Show that $\mathcal{G}$ is a σ-algebra. If μ is a measure on $\mathcal{F}$, show that the map $\nu\colon \mathcal{G} \to \mathbb{R}^{*+}$ defined by $\nu(B) = \mu(F^{-1}(B))$ is a measure on $\mathcal{G}$.

2.1.P12. Let $\mathcal{F}$ be a σ-algebra of subsets of a set X, and $E \in \mathcal{F}$ have the property that every subset of it belongs to $\mathcal{F}$.

(a) If $B \subseteq A \cup E$, show that $A^c \cap B \in \mathcal{F}$.
(b) If $B \subseteq A \cup E$, $B \in \mathcal{F}$, show that $A \cap B \in \mathcal{F}$.
(c) If $B \subseteq A \cup E$, $A \subseteq B \cup E$, $B \in \mathcal{F}$, show that $A \in \mathcal{F}$.

2.1.P13. Prove that the result of Exercise 2.1.5 holds even if the set function $\mu\colon \mathcal{F} \to \mathbb{R}^{*+}$ is assumed to be only finitely additive provided it is monotone and vanishes at $\varnothing$.

2.2 Measurability and the Integral

As noted just after Definition 2.1.4, the monotonicity of measure guarantees that a distribution of a measurable nonnegative function f is a decreasing nonnegative function and therefore $\int_0^\infty \tilde{f}$ exists either as a real number (we then call it *finite*) or as ∞. Therefore the following definition makes sense.

2.2.1. Definition. *Suppose* μ *is a measure on a set X. For a measurable nonnegative function $f\colon X \to \mathbb{R}^{*+}$, the* **measure space integral** *is defined to be*

$$\int f = \int_0^\infty \tilde{f}.$$

Often it makes for easier reading if we write $\int_X f$, $\int_X f\,d\mu$ or $\int_X f(x)d\mu(x)$ in place of $\int f$, especially if two or more measure spaces are involved in the discussion.

For a discussion of the equivalence of this definition with the more traditional one, see 4.7.1 of Craven [1, p. 128]. The Abel summation formula is used.

We shall abbreviate the name to simply *integral*. When it is necessary to specify the measure μ being used, as may happen if there is another measure relevant to the context, we write the integral as $\int f d\mu$. Sometimes, the function is known by its expression $f(x)$ without any symbol such as f having been introduced for it; when this is so, the integral may be more conveniently denoted by $\int f(x)dx$ or $\int f(x)d\mu(x)$.

In terms of the integral, the conclusion of Example 2.1.8(b) can be rephrased as $\int f = 11$, $\int g = 13$ and $\int (f + g) = 24 = \int f + \int g$. Also, Proposition 2.1.7 can be phrased as: $\int \chi_A = \mu A$ for any measurable set $A \subseteq X$.

If the desired extension indicated in Remark 1.1.6 works out to be a measure on a σ-algebra, then Proposition 1.1.3 says that the Riemann integral of a step function agrees with its measure space integral. Moreover, Examples 1.1.4 would show that the same is true of the functions discussed there.

Henceforth a measurable nonnegative-valued function will be understood to be extended real-valued unless the context indicates that the function is not permitted to take the value ∞.

2.2.2. Proposition. (a) *Measurability of $f \colon X \to \mathbb{R} \cup \{\infty\}$ is equivalent to each of the following*:

>   (i) $X(f \geq t)$ *is measurable for every $t \in \mathbb{R}$*;
>   (ii) $X(f < t)$ *is measurable for every $t \in \mathbb{R}$*;
>   (iii) $X(f \leq t)$ *is measurable for every $t \in \mathbb{R}$*.

(b) *Measurability of $f \colon X \to \mathbb{R}$ implies that of $-f$; measurability of $f \colon X \to \mathbb{R} \cup \{\infty\}$ implies that of $|f|$.*

Proof: (a) Assume f measurable and let $t \in \mathbb{R}$. Then for each $n \in \mathbb{N}$, the set $X(f > t - \frac{1}{n})$ is measurable. Now, $f(x) \geq t \Leftrightarrow f(x) > t - \frac{1}{n}$ for every $n \in \mathbb{N}$. Therefore $X(f \geq t) = \bigcap_{n=1}^{\infty} X(f > t - \frac{1}{n})$. Since a σ-algebra is closed under countable intersections, the set $X(f \geq t)$ is measurable. Thus measurability of f implies the statement (i).

(i) $\Rightarrow$ (ii) because a σ-algebra is closed under complements.

By an argument analogous to the one above, we can prove (ii) $\Rightarrow$ (iii).

The reason why (iii) implies measurability of f is the same as why (i) $\Rightarrow$ (ii), namely, that a σ-algebra is closed under complements.

(b) Left as Problem 2.2.P1.   □

We shall next show that certain simple algebraic combinations of measurable function are measurable ("algebra of measurable functions"). Their limits are deferred to a later section.

2.2.3. Theorem. *If $f \colon X \to \mathbb{R} \cup \{\infty\}$ and $g \colon X \to \mathbb{R} \cup \{\infty\}$ are measurable functions and $c \in \mathbb{R}$, then the following functions are also measurable*:

(a) *cf provided that either f is real-valued or $c \geq 0$*;
(b) *$f+g$*;
(c) *fg provided that either both are real-valued or both are nonnegative*;
(d) *$\min \{f, g\}$ and $\max \{f, g\}$.*

Proof: (a) If $c = 0$, the statement is trivial. If $c > 0$, then

$$X(cf > t) = X(f > \frac{t}{c})$$

and the set on the right side is measurable. For $c < 0$, apply Proposition 2.2.2(b).

(b) Let $X_1 = X(g < \infty)$ and X_2 be the complement. Consider any $x \in X_1$. If there exists a rational number r such that

$$f(x) > r \quad \text{and} \quad g(x) > t - r, \tag{2.1}$$

then the inequality

$$(f + g)(x) > t \tag{2.2}$$

also holds. Conversely, if (2.2) holds, then $f(x) > t - g(x)$ and so, there exists a rational number r such that $f(x) > r > t - g(x)$, i.e. (2.1) holds. Thus the existence of a rational number r satisfying (2.1) is equivalent to (2.2). This means

$$X_1 \cap X(f + g > t) = X_1 \cap \left(\bigcup_{r \in \mathbb{Q}} (X(f > r) \cap X(g > t - r)) \right).$$

On the other hand, the same equality holds with X_1 replaced by its complement X_2 because both sides are then equal to X_2. Upon equating the unions of the left sides and the right sides of the two equalities, we obtain

$$X(f + g > t) = \bigcup_{r \in \mathbb{Q}} (X(f > r) \cap X(g > t - r)).$$

The sets $X(f > r) \cap X(g > t - r)$ on the right side of this equality are measurable and are countable in number (i.e. can be arranged in a sequence). Therefore their union, which has been shown to be $X(f + g > t)$, must be measurable.

(c) Here we shall prove only the case when both functions are real-valued. The case when both are nonnegative extended real-valued is left as Problem 2.2.P11. First we shall prove that measurability of f implies that of f^2. If $t < 0$, then $X(f^2 > t) = X$, which is measurable. If $t \geq 0$, then $f(x)^2 > t$ is equivalent to the assertion that either $f(x) > \sqrt{t}$ or $f(x) < -\sqrt{t}$ and hence we have $X(f^2 > t) = X(f > \sqrt{t}) \cup X(f > -\sqrt{t})$, which is measurable by Proposition 2.2.2(a). To prove measurability of the product fg, we note that

$$fg = \frac{1}{2}\left((f + g)^2 - f^2 - g^2 \right)$$

when f and g are both real-valued, and apply parts (a) and (b).

(d) For an arbitrary real number t, we have

$$\max\{f, g\}(x) > t \quad \text{if and only if} \quad f(x) > t \text{ or } g(x) > t.$$

Therefore

$$X(\max\{f, g\} > t) = X(f > t) \cup X(g > t).$$

Since both sets on the right are measurable, the one on the left is measurable. It follows that $\max\{f, g\}$ is measurable. The argument for $\min\{f, g\}$ is similar. $\square$

We need the following simple fact to proceed further.

2.2.4. Proposition. *For any $x \in \mathbb{R} \cup \{\infty\}$, let*

$$x^+ = \max\{x, 0\}, \quad x^- = \max\{-x, 0\} \quad if \ x \in \mathbb{R},$$

and

$$\infty^+ = \infty, \quad \infty^- = 0.$$

(a) *Then*

$$x^+ \geq 0, \quad x^- \geq 0,$$

$$x = x^+ - x^- \quad and \quad |x| = x^+ + x^-,$$

where $|\infty|$ means ∞. In particular,

$$x^+ \leq |x| \quad and \quad x^- \leq |x|.$$

(b) *If $x \in \mathbb{R}$,*

$$x^+ = \frac{1}{2}(|x| + x) \quad and \quad x^- = \frac{1}{2}(|x| - x).$$

(c) *Moreover, if $x = y - z$, where $y \geq 0$ and $z \geq 0$, then $y \geq x^+$ and $z \geq x^-$. (Obviously, z cannot be ∞ here.)*

Proof: (a) Since this is trivial when $x = \infty$, we need only consider $x \in \mathbb{R}$. If $x \geq 0$, then $x^+ = \max\{x, 0\} = x \geq 0$, $x^- = \max\{-x, 0\} = 0 \geq 0$ and $|x| = x$. Therefore $x^+ - x^- = x - 0 = x$ and $x^+ + x^- = x + 0 = x = |x|$. In the contrary case when $x < 0$, we have $x^+ = \max\{x, 0\} = 0 \geq 0$, $x^- = \max\{-x, 0\} = -x \geq 0$ and $|x| = -x$. Therefore $x^+ - x^- = 0 - (-x) = x$ and $x^+ + x^- = 0 + (-x) = -x = |x|$. Thus the two equalities hold in both cases.

(b) This follows from (a) by an elementary computation.

(c) Since z cannot be ∞ here (otherwise "$y - z$" would make no sense), x and y are both ∞ or both finite. If both are ∞, the truth of the assertion is trivial. In the contrary case when x, y, z are all finite, $y = x + z \geq x$ and $y \geq 0$, whence $y \geq \max\{x, 0\} = x^+$. Also, $z = y - x \geq -x$ and $z \geq 0$, whence $z \geq \max\{-x, 0\} = x^-$. $\square$

Using Proposition 2.2.4, for any real-valued function f on any set X, one can form the associated functions $f^+ = \max\{f,0\}$ and $f- = \max\{-f,0\}$, which then have the property that

$$f = f^+ - f^- \quad \text{and} \quad |f| = f^+ + f^-.$$

When f is nonnegative, we have $f^+ = f$ everywhere and $f^- = 0$ everywhere. It follows that if X is a measure space and $f \geq 0$ is measurable, then $\int f^+ = \int f$ and $\int f^- = 0$, so that $\int f = \int f^+ - \int f^-$. For a general measurable f that need not be nonnegative, it is still true in view of Theorem 2.2.3 that the nonnegative functions f^+ and f^- are measurable and $\int f^+ - \int f^-$ makes sense at least when both integrals are real numbers.

2.2.5. Definition. *Suppose* μ *is a measure on a set* X. *For any measurable real-valued function* $f : X \to \mathbb{R}$, *the* **measure space integral** *is defined to be*

$$\int f = \int f^+ - \int f^-,$$

provided that $\int f^-$ *is finite (i.e. is not* ∞). *The function f is said to be* **summable** *or* **integrable** *if both* $\int f^+$ *and* $\int f^-$ *are finite, or equivalently, if* $\int f$ *is finite.*

Often it makes for easier reading if we write $\int_X f$, $\int_X f\, d\mu$, or $\int_X f\, d\mu(x)$ in place of $\int f$, especially if two or more measure spaces are involved in the discussion.

Until we reach Sect. 3.2, we shall be dealing mainly with integrals of nonnegative (extended real-valued) measurable functions. Definition 2.2.5 has been stated here only for completeness.

An $\mathbb{R}^{*+}$-valued measurable function with a finite integral will not be called integrable unless it is known to take only finite values. The restriction that a function be finite-valued in order to be regarded as integrable can be removed but will be retained here so that we can consider $-f$ without having to introduce $-\infty$ and handle the attendant complications with "$\infty - \infty$".

At this juncture, it would be well to recall that for a bounded function $f: [a, b] \to \mathbb{R}$, the supremum (resp. infimum) of lower (resp. upper) sums, taken over all partitions of $[a, b]$, is called the *lower* (resp. *upper*) *integral* of the function over the interval and is denoted by $\underline{\int}_a^b f \left(\text{resp.} \overline{\int}_a^b f\right)$. In general, $\underline{\int}_a^b f \leq \overline{\int}_a^b f$. If equality holds, the function is said to be *Riemann integrable* and the common value of the lower and upper integrals is the Riemann integral $\int_a^b f$.

2.2.6. Proposition. *Let* $f: [a, b] \to \mathbb{R}$ *be Riemann integrable and* $c > 0$. *Then the function* $g: [ca, cb] \to \mathbb{R}$ *given by* $g(u) = f(\frac{u}{c})$ *is Riemann integrable and*

$$\int_{ca}^{cb} g = c \int_a^b f,$$

which is to say,

$$\int_{ca}^{cb} f\left(\tfrac{u}{c}\right)du = c\int_{a}^{b} f(u)du.$$

Proof: Let $P: a = x_0 < x_1 < \cdots < x_n = b$ be any partition of $[a, b]$. Since $c > 0$,

$$Q: ca = cx_0 < cx_1 < \cdots < cx_n = cb$$

is a partition of $[ca, cb]$. For any $x \in [x_{j-1}, x_j]$, we have $u = cx \in [cx_{j-1}, cx_j]$ and $f(x) = f\left(\tfrac{cx}{c}\right) = g(cx) = g(u)$. Conversely, for any $u \in [cx_{j-1}, cx_j]$, we have $x = \tfrac{u}{c} \in [x_{j-1}, x_j]$ and $g(u) = f\left(\tfrac{u}{c}\right) = f(x)$. Therefore

$$\{f(x): x \in [x_{j-1}, x_j]\} = \{g(u): u \in [cx_{j-1}, cx_j]\},$$

so that

$$\inf\{f(x): x \in [x_{j-1}, x_j]\} = \inf\{g(u): u \in [cx_{j-1}, cx_j]\}$$

and similarly for the suprema. Upon multiplying by $cx_j - cx_{j-1} = c(x_j - x_{j-1})$, and taking the summation over all j from 1 to n, we get

$$cL(f, P) = L(g, Q) \le \underline{\int_{ca}^{cb}} g \quad \text{and} \quad cU(f, P) = U(g, Q) \ge \overline{\int_{ca}^{cb}} g.$$

Upon taking the supremum and infimum respectively over all P, it follows from here that

$$c\underline{\int_a^b} f \le \underline{\int_{ca}^{cb}} g \quad \text{and} \quad c\overline{\int_a^b} f \ge \overline{\int_{ca}^{cb}} g.$$

Since $\underline{\int_{ca}^{cb}} g \le \overline{\int_{ca}^{cb}} g$ and $\underline{\int_a^b} f = \overline{\int_a^b} f = \int_a^b f$ by hypothesis, we conclude that

$$c\int_a^b f = \underline{\int_{ca}^{cb}} g = \overline{\int_{ca}^{cb}} g. \quad \square$$

2.2.7. Proposition. *Suppose* μ *is a measure on a set X. For a measurable $f: X \to \mathbb{R}^{*+}$ and for $0 \le c < \infty$, we have*

$$\int (cf) = c\int f.$$

Proof: For $c = 0$, the function cf is 0 everywhere and equality is seen to hold trivially, keeping in mind that $\infty \cdot 0 = 0 \cdot \infty = 0$ (Remark 1.2.1(f)). So, suppose $c > 0$.

We have $(cf)(x) > t \Leftrightarrow f(x) > \frac{t}{c}$, so that $X(cf > t) = X(f > \frac{t}{c})$, which implies that cf is measurable and that $\widetilde{(cf)}(t) = \tilde{f}(\frac{t}{c})$. In particular, $\tilde{f}$ is real-valued wherever $\widetilde{(cf)}$ is. If $\tilde{f}$ is not real-valued on $(0, \infty)$, that is to say, $\tilde{f}(\alpha) = \infty$ for some real $\alpha > 0$, then the same is true of $\widetilde{(cf)}$, in which case the required equality holds because both sides are equal to ∞. Consider the case when $\tilde{f}$ is real-valued except possibly at 0. The same must be true of $\widetilde{(cf)}$, and it follows by equality (1.7) of Sect. 1.2 and from Proposition 2.2.6 that the required equality holds. □

It is an immediate consequence of Proposition 2.1.7 and Definition 2.2.1 that for any measurable set, $\int \chi_A = \mu A$; it now follows from Proposition 2.2.7 that $\int (c\chi_A) = c(\mu A)$ when $0 \le c < \infty$.

2.2.8. Proposition. *Suppose μ is a measure on a set X and the nonnegative measurable functions f and g on X satisfy $f(x) \le g(x)$ for every $x \in X$. Then $\int f \le \int g$*

Proof: From the hypothesis that $f(x) \le g(x)$ for every $x \in X$, it follows that $\tilde{f}(t) \le \tilde{g}(t)$ for every t. The required inequality follows directly from this. □

2.2.9. Proposition. *Suppose μ is a measure on a set X and $f: X \to \mathbb{R}^{*+}$ is measurable. If $\int f < \infty$, then there exists a (measurable) set A such that $\mu A = 0$ and $f(x) < \infty$ for $x \notin A$.*

Proof: Let $A = \{x \in X : f(x) = \infty\}$, so that $f(x) < \infty$ for $x \notin A$. The set A is measurable because it equals $\cap_{n=1}^{\infty} X(f > n)$. We need only prove that $\mu A = 0$. Now, $n\chi_A(x) \le f(x)$ for every $x \in X$ and every positive integer n. It follows in view of Proposition 2.2.8 that $\int n\chi_A \le \int f$. But $\int n\chi_A = n(\mu A)$ and therefore $n(\mu A) \le \int f$. Since this must hold for every positive integer n and $\int f < \infty$, we get a contradiction unless $\mu A = 0$. □

2.2.10. Proposition. *Suppose μ is a measure on a set X and $f: X \to \mathbb{R}^{*+}$ is measurable. If $\int f < \infty$, then there exists a sequence $\{A_n\}_{n \ge 1}$ of measurable sets such that*

(a) $\mu A_n < \infty$ *for every n,*
(b) $f(x) = 0$ *for $x \notin \cup_{n=1}^{\infty} A_n$,*
(c) $f = f\chi_A$, *where $A = \cup_{n=1}^{\infty} A_n$.*

Proof: Put $A_n = X(f > \frac{1}{n})$, so that $\frac{1}{n}\chi_{A_n}(x) \le f(x)$ for every $x \in X$. By Proposition 2.2.8, we obtain $\int(\frac{1}{n}\chi_{A_n}) \le \int f$. But $\int(\frac{1}{n}\chi_{A_n}) = \frac{1}{n}(\mu A_n)$ and therefore $\frac{1}{n}(\mu A_n) \le \int f$. Since $\int f < \infty$, it follows that $\frac{1}{n}(\mu A_n) < \infty$, and hence $(\mu A_n) < \infty$. Thus (a) must hold.

If $x \notin \cup_{n=1}^{\infty} A_n$, then $x \notin A_n$ for every n, which means $f(x) \le \frac{1}{n}$ for every n. Since $f(x) \ge 0$, it follows that $f(x) = 0$. So, (b) must hold.

Consider any $x \in X$. If $x \in A = \cup_{n=1}^{\infty} A_n$, we have $\chi_A(x) = 1$ and hence $f(x) = f(x)\chi_A(x) = (f\chi_A)(x)$. In the contrary case, $\chi_A(x) = 0$ by definition of a characteristic function and $f(x) = 0$ by (b), which leads to $f(x) = 0 = f(x)\chi_A(x) = (f\chi_A)(x)$. Thus (c) must also hold. □

Problem Set 2.2

2.2.P1. Prove Proposition 2.2.2(b).

2.2.P2. If $f: X \to \mathbb{R}^{*+}$ has the property that $X(f \geq t)$ is measurable for every $t \in \mathbb{Q}$, show that f is measurable.

2.2.P3. Let $\{f_n\}_{n\geq 1}$ be a sequence of extended real-valued measurable functions on X. Then the function f defined on X as $f(x) = \inf f_n(x)$, i.e. $\inf\{f_n(x): n \in \mathbb{N}\}$, is measurable.

2.2.P4. In Problem 2.1.P3, show that the measurability of the limit function f holds without the hypothesis that $f_n(x) \leq f_{n+1}(x)$.

2.2.P5. If $f: X \to \mathbb{R}$ is measurable and $\phi: \mathbb{R} \to \mathbb{R}$ is increasing, show that the composition $\phi \circ f: X \to \mathbb{R}$ is measurable.

2.2.P6. Suppose μ is a measure on a set X. Prove that

(a) For a measurable function $f: X \to \mathbb{R}^{*+}$ and any real $p > 0$, the nonnegative function $f^p: X \to \mathbb{R}^{*+}$ is measurable; moreover,

$$\int f = 0 \quad \text{if and only if} \quad \tilde{f}(t) = 0 \text{ for every } t > 0,$$

which, in turn, is equivalent to $\int f^p = 0$ for any $p > 0$.

(b) If $f: X \to \mathbb{R}^{*+}$ and $g: X \to \mathbb{R}^{*+}$ are measurable functions satisfying $\int f = 0 = \int g$, then $\int (f + g) = 0$.

2.2.P7. (a) Let f, g be nonnegative measurable functions on a space X with measure μ, and $f(x) = g(x)$ for all $x \in A$, where A is a measurable subset of X satisfying $\mu(A^c) = 0$. Show that $\int f = \int g$.

(b) Let f, g be measurable functions on a space X with measure μ, and $f(x) = g(x)$ for all $x \in A$, where A is a measurable subset of X satisfying $\mu(A^c) = 0$. If g is integrable, show that f is also integrable and $\int f = \int g$.

2.2.P8. Let f, g be nonnegative measurable functions on a space X with measure μ, and $f(x) \leq g(x)$ for all $x \in A$, where A is a measurable subset of X satisfying $\mu(A^c) = 0$. Show that $\int f \leq \int g$.

2.2.P9. For a function $f: X \to \mathbb{R} \cup \infty$, let A be the set $X(f < \infty)$. Show that

(a) If $t < 0$, then $X(f\chi_A > t) = A^c \cup (A \cap X(f > t))$.

(b) If $t \geq 0$, then $X(f\chi_A > t) = A \cap X(f > t)$.

(c) The product $f\chi_A$ is real-valued.

(d) If f is measurable, then A is measurable and $f\chi_A$ is measurable.

2.2.P10. Let the function $f: X \to \mathbb{R} \cup \infty$ have the property that $X(t < f < \infty)$ is measurable for every $t \in \mathbb{R}$. Show that f is measurable. (The converse is trivial.)

2.2.P11. If $f\colon X \to \mathbb{R}^{+*}$ and $g\colon X \to \mathbb{R}^{+*}$ are both measurable, show that their product $fg\colon X \to \mathbb{R}^{+*}$ is measurable.

2.2.P12. For an integrable function $f\colon X \to \mathbb{R}^*$, show that

$$\alpha \cdot \mu(X(|f| > \alpha)) \le \int |f| \quad \text{for all } \alpha > 0.$$

2.2.P13. Suppose $f\colon X \to \mathbb{R}^{+*}$ is a measurable function such that, for some $\alpha > 0$, the function $\min\{f, \alpha\}$ has a finite integral. Show that f has a finite integral if and only if the series $\sum_{k=1}^{\infty} \tilde{f}(k)$ converges.

2.3 The Monotone Convergence Theorem

It follows from the result of Problem 2.1.P2 that, for any sequence $\{f_n\}_{n\ge 1}$ of measurable functions such that $f_n \le f_{n+1}$ everywhere, we have $\tilde{f}_n \le \tilde{f}_{n+1}$ everywhere. If $\lim_{n\to\infty} f_n(x)$ is a real-valued function f, then it follows from Problem 2.1.P3 that it is measurable, so that it has a distribution $\tilde{f}$. Since $f_n \le f$ everywhere, it further follows from the result of Problem 2.1.P2 that the distribution satisfies $\tilde{f} \ge \tilde{f}_n$. The fact that a measure has a property called *inner continuity* makes it possible to obtain something more satisfying; see Theorem 2.3.2 further below.

2.3.1. Proposition. *Let μ be a measure. Then for a sequence of measurable sets* $\{S_n\}_{n\ge 1}$, *we have*
$\mu(\bigcup_{n=1}^{\infty} S_n) = \lim_{n\to\infty} \mu(S_n)$ *provided that $S_n \subseteq S_{n+1}$ for every n.* (inner continuous)

Proof: As observed immediately after the definition of measure (Definition 2.1.2), μ is finitely additive. Now consider a sequence $\{S_n\}_{n\ge 1}$ of measurable sets such that $S_n \subseteq S_{n+1}$ for every $n \in \mathbb{N}$. Let $A_1 = S_1$ and $A_n = S_n \backslash S_{n-1}$ for $n > 1$. Then each A_n is measurable and the condition $S_n \subseteq S_{n+1}$ has the consequence that the sets A_n are disjoint and that

$$S_n = \bigcup_{j=1}^{n} A_j. \tag{2.3}$$

Therefore by finite additivity, we have

$$\mu(S_n) = \sum_{j=1}^{n} \mu(A_j). \tag{2.4}$$

Using countable additivity and then using (2.4), we get

$$\mu\left(\bigcup_{j=1}^{\infty} A_j\right) = \sum_{j=1}^{\infty} \mu(A_j)$$

$$= \lim_{n\to\infty} \mu(S_n). \tag{2.5}$$

But

$$\bigcup_{n=1}^{\infty} S_n = \bigcup_{j=1}^{\infty} A_j$$

in view of (2.3). It follows from this equality and from (2.5) that

$$\mu\left(\bigcup_{n=1}^{\infty} S_n\right) = \lim_{n\to\infty} \mu(S_n).$$

□

The functions μ in all our examples so far are inner continuous, including the one in Example 2.1.3(b), although it is not even a measure (it fails to be countably additive). The proof below does not use additivity directly but depends crucially on inner continuity.

2.3.2. Monotone Convergence Theorem. *Let μ be a measure on a set X. Suppose f is a nonnegative extended real-valued function on X and $\{f_n\}_{n\geq 1}$ a sequence of measurable nonnegative extended real-valued functions such that*
   (i) $f_n \leq f_{n+1}(x)$ *and* (ii) $fx = \lim_{n\to\infty} f_n(x)$ *for every $x \in X$.*
Then f is measurable, $\tilde{f}(t) = \lim_{n\to\infty} \tilde{f}_n(t)$ for every $t \geq 0$, and also $\int f(t) = \lim_{n\to\infty} \int f_n(t)$.

Proof: Measurability of f is a simple consequence of Problem 2.2.P4. In the sequence $\{\tilde{f}_n\}_{n\geq 1}$ of nonnegative extended real-valued functions, each term $\tilde{f}_n$ is a decreasing function on $[0, \infty)$. By Problem 2.1.P3, we know that $0 \leq \tilde{f}_n(t) \leq \tilde{f}_{n+1}(t)$ for each $t \in [0, \infty)$ and each $n \in \mathbb{N}$. By inner continuity (Proposition 2.3.1), in conjunction with Problem 2.1.P3, we also know that $\lim_{n\to\infty} \tilde{f}_n(t) = \tilde{f}(t)$ for each $t \in [0, \infty)$. Therefore Proposition 1.2.4 yields $\int_0^\infty \tilde{f} = \lim_{n\to\infty} \int_0^\infty \tilde{f}_n$. But by Definition 2.2.1, this means $\int f = \lim_{n\to\infty} \int f_n$. □

2.3.3. Exercise. Show that the "monotonicity" hypothesis (i) of the Monotone Convergence Theorem 2.3.2 cannot be dropped.

Solution: Let $X = \mathbb{N}$ and $\mu(A) = \sum_{j\in A}\left(\frac{1}{2^j}\right)$. Then μ is a measure. One way to see this is to note that it is obtained by applying Problem 2.1.P10 with $X_j = \{j\} \subset \mathbb{N}$ and μ_j defined by $\mu_j(X_j) = \frac{1}{2^j}$. Define f_n to be 2^n times the characteristic function of $\{n\}$. Then $\int f_n = 1$ for each n. Also, for each $j \in X = \mathbb{N}$, $n \geq j + 1 \Rightarrow f_n(j) = 0$; so $\lim_{n\to\infty} f_n(j) = 0$ for each $j \in X$. Thus $\int \lim_{n\to\infty} f_n = 0$ although $\lim_{n\to\infty} \int f_n = 0$.

Problem Set 2.3

2.3.P1. (Needed in Problem 3.2.P1 and Theorem 4.3.4) Let μ be a measure on X and f a nonnegative measurable function such that $\int f = 0$. Prove $\mu(X(f > 0)) = 0$.

2.3.P2. Let $\{f_n\}_{n \geq 1}$ be a sequence of real-valued functions on an arbitrary nonempty set X. Suppose f is a real-valued function on X such that, for every $x \in X$, some subsequence of the real sequence $\{f_n(x)\}_{n \geq 1}$ converges to $f(x)$. Show for any real t that $X(f > t) \subseteq \bigcup_{n=1}^{\infty} X(f_n > t)$.

2.3.P3. In a measure space, suppose that every set of infinite measure contains a subset of finite positive measure. Show that every set of infinite measure contains a subset of arbitrarily large finite measure.

Hint: For a given set of infinite measure, show that there exists a subset whose measure is the supremum of the measures of all subsets of finite measure. If its measure is finite, then the measure of its complement in the given set must be infinite.

2.3.P4. Let μ be a measure. For a sequence of measurable sets $\{S_n\}_{n \geq 1}$ with $\mu(S_1) < \infty$, show that

$\mu(\cap_{n=1}^{\infty} S_n) = \lim_{n \to \infty} \mu(S_n)$ provided that $S_n \supseteq S_{n+1}$ for every n. (outer continuous).

Chapter 3
Properties of the Integral

3.1 Simple Functions

So far, we have proved the Monotone Convergence Theorem but not the seemingly simpler property $\int (f + g) = \int f + \int g$. In order to prove this, we need to make full use of the finite additivity of measure, which we did not have to do for the Monotone Convergence Theorem, because inner continuity of measure, a consequence of its countable additivity, was sufficient.

In this section, we shall establish the property in question for a restricted class of functions and extend it to a broader class only later.

3.1.1. Definition *A measurable function* $f : X \to \mathbb{R}$ *is said to be* **simple** *if its range is finite.*

Simple functions are a generalization of step functions, which also take finitely many values, each value being taken on a union of finitely many intervals; thus, step functions are linear combinations of characteristic functions of intervals. For example, if $s : [1, 3] \to \mathbb{R}$ is the step function that equals 3 on $[1, 2)$, 4 on $[2, 3)$ and $s(3) = 3$, then $s = 3\chi_{[1,2)\cup[3,3]} + 4\chi_{[2,3)}$. The generalization consists in replacing unions of finitely many intervals by measurable sets.

3.1.2. Examples. (a) The characteristic function χ_A of a nonempty measurable set $A \subset X$ is a simple function with range $\{0, 1\}$. If the measurable set A is empty or equals X, the range of χ_A consists of only 0 or only 1.

(b) Suppose A and B are nonempty measurable sets, both distinct from X. As observed in (a), each of χ_A and χ_B has range $\{0, 1\}$; but the range of the sum $\chi_A + \chi_B$ may consist of 1, 2 or 3 numbers. Indeed, the range of $\chi_A + \chi_B$ is

$$\{1\} \text{ if } A \cap B = \emptyset \text{ and } A \cup B = X$$
$$\{0, 1\} \text{ if } A \cap B = \emptyset \text{ and } A \cup B \neq X$$
$$\{0, 2\} \text{ if } A = B \text{ (in this event, } A \cap B = A \neq X)$$

© Springer Nature Switzerland AG 2018
S. Shirali, *A Concise Introduction to Measure Theory*,
https://doi.org/10.1007/978-3-030-03241-8_3

$$\{1, 2\} \text{ if } A \cup B = X \text{ and } A \cap B \neq \emptyset$$

$$\{0, 1, 2\} \text{ if } A \cup B \neq X, A \neq B \text{ and } A \cap B \neq \emptyset.$$

(c) The sum of two functions with finite range has a finite range. Since the sum of two measurable functions is measurable (Theorem 2.2.3(b)), the sum of two simple functions is simple. Therefore a finite sum $\sum_{j=1}^{n} \alpha_j \chi_{A_j}$, where the sets $A_1, \ldots, A_n$ are measurable and $\alpha_1, \ldots, \alpha_n$ are real numbers, is always a simple function. It can have a value which is none of the numbers α_j, as illustrated in part (b), where $\chi_A + \chi_B = 1 \cdot \chi_A + 1 \cdot \chi_B = \alpha_1 \cdot \chi_A + \alpha_2 \cdot \chi_B$ with $\alpha_1 = 1 = \alpha_2 \neq 2$, but $\chi_A + \chi_B$ can have 2 as a value.

3.1.3. Remarks. (a) Let $\alpha_1, \ldots, \alpha_n$ be the *distinct* values assumed by a simple function s. Then the sets $A_j = X(s = \alpha_j)$ are nonempty and disjoint, and have union $\bigcup_{j=1}^{n} A_j = X$; moreover, each of them is measurable by virtue of Proposition 2.2.2(a), and

$$s(x) = \sum_{j=1}^{n} \alpha_j \chi_{A_j}(x).$$

Conversely, suppose $A_1, \ldots, A_n$ are any n measurable sets that are nonempty and disjoint, having union $\bigcup_{j=1}^{\infty} A_j = X$, and let $\alpha_1, \ldots, \alpha_n$ be any n distinct numbers. Then the simple function $s = \sum_{j=1}^{n} \alpha_j \chi_{Aj}$ has the numbers $\alpha_1, \ldots, \alpha_n$ as its distinct values and satisfies $X(s = \alpha_j) = A_j$. In particular, the range $s(X)$ of the function is $\{\alpha_j : 1 \leq j \leq n\}$. If the n distinct values $\alpha_1, \ldots, \alpha_n$ are arranged in increasing order, we shall refer to the sum as the **canonical representation** of the simple function s.

(b) The canonical representations of $\chi_A + \chi_B$ in the five illustrations in Example 3.1.2(b) are respectively

$$1 \cdot \chi_X, \ 0 \cdot \chi_{(A \cup B)^c} + 1 \cdot \chi_{A \cup B}, \ 0 \cdot \chi_{A^c} + 2 \cdot \chi_A, \ 1 \cdot \chi_{(A \cap B)^c} + 2 \cdot \chi_{A \cap B},$$

$$0 \cdot \chi_{(A \cup B)^c} + 1 \cdot \chi_{A \triangle B} + 2 \cdot \chi_{A \cap B}.$$

Here $A \triangle B$ means the "symmetric difference" $(A \cap B^c) \cup (A^c \cap B)$.

3.1.4. Exercise. When $X = \{a, b, c, d\}$ and $A = X$, $B = \{c, d\}$, $C = \{d\}$, find the canonical form of the simple function $s = \chi_A + \chi_B + \chi_C$.

Solution: Since a and b belong to precisely one among the three sets A, B, C (namely, A), we have $s(a) = s(b) = 1$. Since c belongs to precisely two of the sets (namely, A and B), $s(c) = 2$. Lastly, since d belongs to all three sets, $s(d) = 3$. To summarize,

$$s(a) = s(b) = 1, \quad s(c) = 2 \text{ and } s(d) = 3.$$

Thus s takes the three values 1, 2, 3. Now,

$$A_1 = X(s = 1) = \{a, b\}, \quad A_2 = X(s = 2) = \{c\} \text{ and } A_3 = X(s = 3) = \{d\}.$$

So, the canonical form is

$$s = 1 \cdot \chi_{A_1} + 2 \cdot \chi_{A_2} + 3 \cdot \chi_{A_3} = 1 \cdot \chi_{\{a,b\}} + 2 \cdot \chi_{\{c\}} + 3 \cdot \chi_{\{d\}}.$$

The next proposition involving a canonical representation will be used to prove the stronger result that the same conclusion is true with a representation that is not canonical. The latter will make it transparent why integrals of simple nonnegative functions add up as expected.

3.1.5. Proposition. *Suppose $(X, \mathcal{F}, \mu)$ is a measure space and $s : X \to \mathbb{R}$ a nonnegative simple function with canonical representation*

$$s(x) = \sum_{j=1}^{n} \alpha_j \chi_{A_j}(x).$$

Then

$$\int s = \sum_{j=1}^{n} \alpha_j \mu(A_j).$$

Proof: The case when s takes only one value is trivial. Therefore we consider only the case when s has at least two values, i.e. $n \geq 2$.

Since the numbers $\alpha_1, \ldots, \alpha_n$ are in increasing order (by definition of canonical), $\sup s = \alpha_n$. Therefore

$$\int s = \int_0^{\sup s} \tilde{s} = \int_0^{\alpha_n} \tilde{s}. \tag{3.1}$$

Another consequence of the numbers $\alpha_1, \alpha_2 \ldots, \alpha_n$ being in increasing order is that, for $j > 1$ and $\alpha_{j-1} \leq t < \alpha_j$, the inequality $s(x) > t$ holds if and only if $s(x) = \alpha_k$ for some $k \geq j$, i.e. $x \in A_j \cup \cdots \cup A_n$, which is to say, $\{x \in X : s(x) > t\} = A_j \cup \cdots \cup A_n$. Therefore $\tilde{s}(t) = \mu(\{x \in X : s(x) > t\}) = \mu(A_j \cup \cdots \cup A_n)$ when $\alpha_{j-1} \leq t < \alpha_j$. Since the sets A_j are disjoint, the additivity of μ yields

$$\mu(A_j \cup \cdots \cup A_n) = \mu(A_j) + \cdots + \mu(A_n).$$

Thus the distribution $\tilde{s}$ satisfies

$$\tilde{s}(t) = \mu(A_j) + \cdots + \mu(A_n) \text{ for } t \in [\alpha_{j-1}, \alpha_j) \text{ and } 1 < j \leq n, \tag{3.2}$$

and if $\alpha_1 > 0$, also satisfies

$$\widetilde{s}(t) = \mu(A_1) + \cdots + \mu(A_n) \text{ for } t \in [0, \alpha_1). \tag{3.3}$$

If each $\mu(A_j)$ is finite, $\widetilde{s}$ is a step function and it follows from (3.2) and (3.3) that, whether or not $\alpha_1 > 0$, we have

$$\int_0^{\alpha_n} \widetilde{s} = \alpha_1(\mu(A_1) + \cdots + \mu(A_n)) + \sum_{j=2}^n (\alpha_j - \alpha_{j-1})(\mu(A_j) + \cdots + \mu(A_n)). \tag{3.4}$$

One can transform the right side here by using the Abel summation formula of Problem 1.1.P5 with

$$b_1 = \alpha_1, \ b_j = \alpha_j - \alpha_{j-1}, (j \geq 2) \quad \text{and} \quad a_j = \mu(A_j) + \cdots + \mu(A_n),$$

thereby obtaining

$$\int_0^{\alpha_n} \widetilde{s} = \sum_{j=1}^n \alpha_j \mu(A_j). \tag{3.5}$$

Now (3.1) and (3.5) together imply the required equality for the case when each $\mu(A_j)$ is finite.

Next, suppose $\mu(A_j) = \infty$ for some $j > 1$. In this event, $\widetilde{s}(\alpha) = \infty$ for $\alpha \in [\alpha_{j-1}, \alpha_j)$ and hence both sides of the required equality are ∞.

Finally, suppose $\mu(A_j)$ is finite for all $j > 1$ but $\mu(A_1) = \infty$. If $\alpha_1 > 0$, then $\widetilde{s}(t) = \infty$ for $t \in [0, \alpha_1)$ and again both sides of the equality in question are ∞. If $\alpha_1 = 0$, then (3.2) shows that $\widetilde{s}$ is a step function and (3.4) holds with $\mu(A_1)$ replaced by any real number. As before, we arrive at (3.5) but with $\mu(A_1)$ replaced by a real number. However, (3.5) continues to hold if $\mu(A_1)$ is changed back to ∞ because now $\alpha_1 = 0$. Once again, (3.1) and (3.5) together imply the required equality.   □

When $\mu(A)$ is the number of elements in A, the function μ is a measure and is called the **counting measure**; it is understood that $\mu(A) = \infty$ when A is an infinite set. In Example 3.1.4, if we use the counting measure, $\int s$ can be computed on the strength of the preceding proposition as

$$\int s = 1 \cdot 2 + 2 \cdot 1 + 3 \cdot 1 = 2 + 2 + 3 = 7.$$

We note that $s = \chi_A + \chi_B + \chi_C$ and $\int \chi_A + \int \chi_B + \int \chi_C = \mu(A) + \mu(B) + \mu(C) = 4 + 2 + 1 = 7$, which agrees with $\int s$.

3.1.6. Exercise. Let $X = \{a, b, c, d\}$ and A, B, C be subsets of X such that $s = \int \chi_A + \int \chi_B + \int \chi_C$ works out to be the same as in Exercise 3.1.4. If $A = \{b, c, d\}$ and $B = \{a, d\}$, find C and compare $\int \chi_A + \int \chi_B + \int \chi_C$ with $\int s$ when μ is the counting measure.

Solution: Since $s(a)=s(b)=1$, the elements a and b must belong to precisely one among the three sets A, B, C. It is given that they belong to B and A respectively. Therefore they do not belong to C. Since $s(c)=2$, c must belong to precisely two among the sets A, B, C. It is given that it belongs to A but not B. It follows that it belongs to C. A similar argument shows that d too belongs to C. Thus $C = \{c, d\}$. Therefore

$$\int \chi_A + \int \chi_B + \int \chi_C = \mu(A) + \mu(B) + \mu(C) = 3 + 2 + 2 = 7,$$

which agrees with the value of $\int s$ computed above (just before this exercise).

Before proceeding, we remind the reader (see Remark 3.1.3(a)) that $\sum_{j=1}^{n} \alpha_j \chi_{A_j}$, where $\alpha_1, \alpha_2, \ldots, \alpha_n$ are in increasing order, is a canonical representation of a simple function on a measurable set X if and only if

(a) the measurable sets A_j are disjoint with union X, while
(b) the real numbers α_j are distinct and the sets A_j nonempty.

We shall say that the sets A_j **form a partition** of X when they satisfy (a), regardless of whether some of them are empty. It will be understood of course that the sets are measurable.

3.1.7. Proposition. *Let* $(X, \mathcal{F}, \mu)$ *be a measure space and let* $s\colon X \to \mathbb{R}$ *be a nonnegative simple function such that*

$$s = \sum_{i=1}^{m} \alpha_i \chi_{A_i} = \sum_{j=1}^{n} \beta_j \chi_{B_j},$$

where the sets $A_i (1 \le i \le m)$ *form a partition of* X *and so do the sets* $B_j (1 \le j \le n)$, *which is to say,*

$$A_i \cap A_{i_1} = \emptyset \text{ when } i \ne i_1, \quad B_j \cap B_{j_1} = \emptyset \text{ when } j \ne j_1 \qquad (3.6)$$

and

$$\bigcup_{i=1}^{m} A_i = \bigcup_{j=1}^{n} B_j = X. \qquad (3.7)$$

Then

$$\sum_{i=1}^{m} \alpha_i \mu(A_i) = \sum_{j=1}^{n} \beta_j \mu(B_j) = \int s. \qquad (3.8)$$

Proof: In view of (3.6), the sets $A_i \cap B_j$, $(1 \le j \le n)$, are disjoint for each fixed i, and (3.7) shows that their union is

$$\bigcup_{j=1}^{n}(A_i \cap B_j) = A_i \cap \bigcup_{j=1}^{n} B_j = A_i \cap X = A_i.$$

Therefore by finite additivity of μ,

$$\mu(A_i) = \sum_{j=1}^{n} \mu(A_i \cap B_j) \text{ for each } i. \tag{3.9}$$

Similarly, $\mu(B_j) = \sum_{i=1}^{m} \mu(A_i \cap B_j)$ for each j, and together with (3.9), this leads to

$$\sum_{i=1}^{m}\alpha_i\mu(A_i) = \sum_{i=1}^{m}\sum_{j=1}^{n}\alpha_i\mu(A_i \cap B_j) \text{ and } \sum_{j=1}^{n}\beta_j\mu(B_j) = \sum_{j=1}^{n}\sum_{i=1}^{m}\beta_j\mu(A_i \cap B_j). \tag{3.10}$$

Now consider any i $(1 \leq i \leq m)$ and any j $(1 \leq j \leq n)$. If $A_i \cap B_j \neq \emptyset$, the function s takes the value α_i as well as β_j on this nonempty set and therefore $\alpha_i = \beta_j$, so that

$$\alpha_i\mu(A_i \cap B_j) = \beta_j\mu(A_i \cap B_j).$$

In the contrary case, i.e. $A_i \cap B_j \neq \emptyset$, the equality still holds, because both sides are then zero. Thus it holds in both cases for all i and j. Using this in (3.10), we arrive at the first equality in (3.8).

Finally, since (3.6) and (3.7) are fulfilled when $\sum_{j=1}^{n} \alpha_j \chi_{A_j}$ is a canonical representation of s, the second equality in (3.8) follows from Proposition 3.1.5. $\square$

3.1.8. Proposition. *Let $(X, \mathcal{F}, \mu)$ be a measure space and suppose $s : X \to \mathbb{R}$ and $t : X \to \mathbb{R}$ are nonnegative simple functions. Then*

$$\int (s+t) = \int s + \int t.$$

Proof: Let $s = \sum_{i=1}^{m} \alpha_i \chi_{A_i}$ and $t = \sum_{j=1}^{n} \beta_j \chi_{B_j}$, where the sets $A_i (1 \leq i \leq m)$ as well as the sets $B_j (1 \leq j \leq n)$ form partitions of X. Then the sets $A_i \cap B_j$ also form a partition of X, and moreover, $s + t$ takes the value $\alpha_i = \beta_j$ on $A_i \cap B_j$, so that

$$s + t = \sum_{i=1}^{m}\left(\sum_{j=1}^{n}\left((\alpha_i + \beta_j)\chi_{A_i \cap B_j}\right)\right).$$

By Proposition 3.1.7,

$$\int s = \sum_{i=1}^{m}\alpha_i\mu(A_i), \quad \int t = \sum_{j=1}^{n}\beta_j\mu(B_j) \tag{3.11}$$

and

$$\int (s+t) = \sum_{i=1}^{m} \left(\sum_{j=1}^{n} \left((\alpha_i + \beta_j)\mu(A_i \cap B_j) \right) \right)$$

$$= \sum_{i=1}^{m} \left(\sum_{j=1}^{n} \left(\alpha_i \mu(A_i \cap B_j) \right) \right) + \sum_{j=1}^{n} \left(\sum_{i=1}^{m} \left(\beta_j \mu(A_i \cap B_j) \right) \right)$$

$$= \sum_{i=1}^{m} \alpha_i \left(\sum_{j=1}^{n} \mu(A_i \cap B_j) \right) + \sum_{j=1}^{n} \beta_j \left(\sum_{i=1}^{m} \mu(A_i \cap B_j) \right). \qquad (3.12)$$

For each i, the set A_i is the disjoint union $\bigcup_{j=1}^{n} (A_i \cap B_j)$ and therefore by additivity of μ,

$$\sum_{j=1}^{n} \mu(A_i \cap B_j) = \mu(A_i).$$

Similarly, for each j,

$$\sum_{i=1}^{m} \mu(A_i \cap B_j) = \mu(B_j).$$

These equalities together with (3.12) and (3.11) yield

$$\int (s+t) = \sum_{i=1}^{m} \alpha_i \mu(A_i) + \sum_{j=1}^{n} \beta_j \mu(B_j) = \int s + \int t. \qquad \square$$

3.1.9. Exercise. Three junior managers in a company submit lists of three cities each. When the lists are compiled, it turns out that they together mention only five different cities, a, b, c, d, e. The cities a and b appear in 2 lists each and c, d in 1 list each. In how many lists does the city e appears?

Solution: Let the sets of cities submitted by the junior managers be A, B, C. With the counting measure as μ, we have $\mu(A) = \mu(B) = \mu(C) = 3$. Therefore by Proposition 3.1.8, the simple function $s = \chi_A + \chi_B + \chi_C$ satisfies $\int s = \mu(A) + \mu(B) + \mu(C) = 3 + 3 + 3 = 9$. Note that $s(x)$ tells us how many among the sets A, B, C contain the element x, i.e. how many lists the city x appears in.

Now, it is given that $a, b \in X(s = 2)$ and $c, d \in X(s = 1)$. If e appears in only one list, then $X(s = 1) = \{c, d, e\}$ and hence $\mu(X(s = 1)) = 3$; also, $X(s = 2) = \{a, b\}$ and hence $\mu(X(s = 2)) = 2$. Therefore $\int s = 1 \cdot 3 + 2 \cdot 2 = 7 \neq 9$. Consequently, e cannot appear in only one list. By similar considerations, we deduce that if e were to appear in precisely two lists, then we would get $\int s = 1 \cdot 2 + 2 \cdot 3 = 8 \neq 9$.

Consequently, e cannot appear in precisely two lists. This shows that e must appear in all three lists. (Indeed, when e appears in all three lists, we have $\mu(X(s=1)) = 2$, $\mu(X(s=2)) = 2$ and $\mu(X(s=3)) = 1$, leading to $\int s = 1 \cdot 2 + 2 \cdot 2 + 3 \cdot 1 = 9$, as required. Problem 3.1.P3 is relevant here.)

3.1.10. Exercise. Each of four junior managers in a company submits a list of three cities to the manager, because the senior manager wants them to. The manager goes through the lists and informs the senior manager that there are seven cities in all the four lists taken together, and that three cities appear in two lists each and the remaining four in one list each. When the senior manager (who had studied a certain measure theory book very carefully) hears this, she is furious—Why?

Solution: $2 \cdot 3 + 1 \cdot 4 \neq 3 + 3 + 3 + 3$—that's why!

Problem Set 3.1

3.1.P1. Let μ be a measure on X and $A_1,\dots,A_n$ be measurable subsets of X such that $\mu(A_1 \cup \dots \cup A_n) < \infty$. Show that the *Inclusion-Exclusion Principle* holds:

$$\mu(A_1 \cup \dots \cup A_n) = \sum_{1 \le r \le n} \left((-1)^{r-1} \sum_{1 \le j_1 < \dots < j_r \le n} \mu(A_{j_1} \cap \dots \cap A_{j_r}) \right).$$

3.1.P2. When $X = \{a, b, c, d\}$ and $A = X$, $B = \{c, d\}$, $C = \{d\}$, find the canonical form of the simple function $s = \chi_A + 2\chi_B + \chi_C$.

3.1.P3. Does the data of Exercise 3.1.9 determine the sets A, B, C uniquely?

3.1.P4. Each of three junior managers in a company submits a list of three cities to the manager, because the senior manager wants them to. The manager goes through the lists and informs the senior manager that there are five cities in all the three lists taken together, and that two cities appear in two lists each and two others in three lists each. When the senior manager hears this, she is furious—Why?

3.1.P5. Let $\sum \alpha_j$ be a series of nonnegative terms. Show that there is a measure μ on $\mathbb{N}$, every subset measurable, such that $\mu(A) = \sum_{j \in A} \alpha_j$ and $\int f = \sum_{j=1}^{\infty} \alpha_j f(j)$ for every nonnegative function f on $\mathbb{N}$. (Note that when each α_j is 1, this is the counting measure.)

3.1.P6. Show that the following sequence in $\mathbb{R}$ diverges:

$$\left\{ \sum_{j=1}^{\infty} \frac{n}{j(n+j)} \right\}_{n \in \mathbb{N}}$$

3.1.P7. Let μ be a measure on X and let $A_1, \dots, A_n$, where $n \ge 5$, be measurable subsets of X such that whenever $1 \le j_1 < j_2 < j_3 < j_4 < j_5 \le n$, the intersection $A_{j_1} \cap A_{j_2} \cap A_{j_3} \cap A_{j_4} \cap A_{j_5}$ of the corresponding subsets is empty. (In symbol-free terminology, the intersection of sets with any five distinct indices is empty.) Show that

$$\sum_i \mu(A_i) \le \mu(A_1 \cup \cdots \cup A_n) + \sum_{1 \le j_1 < j_2 \le n} \mu(A_{j_1} \cap A_{j_2}).$$

3.1.P8. (a) If $A_1, \ldots, A_4$ are respectively $\{2, 3, 4\}$, $\{1, 3, 4\}$, $\{1, 2, 4\}$, $\{1, 2, 3\}$, then the intersection of all four sets is obviously empty. With the counting measure as μ, compute $\sum_{j=1}^4 \mu(A_j)$ and $\mu\left(\bigcup_{j=1}^4 A_j\right)$. Do the same when $A_1, \ldots, A_4$ are respectively $\{1,2,3,4\}$, $\{1,2,4\}$, $\{1,3\}$, $\{2,3,4\}$, which also have empty intersection.

(b) Give an example of five distinct sets $A_1, \ldots, A_5$ such that the intersection of any four among them is empty and $\sum_{j=1}^5 \mu(A_j) = 3\mu\left(\bigcup_{j=1}^5 A_j\right)$, where μ again means the counting measure.

3.1.P9. Let μ be a measure on X and let $A_1, \ldots, A_n$, where $n \ge 4$, be measurable subsets of X such that whenever $1 \le j_1 < j_2 < j_3 < j_4 \le n$, the intersection $A_{j_1} \cap A_{j_2} \cap A_{j_3} \cap A_{j_4}$ of the corresponding subsets is empty. (In symbol-free terminology, the intersection of sets with any four distinct indices is empty.) Suppose also that $\mu(A_1 \cup \cdots \cup A_n) < \infty$. Show that

$$\sum_j \mu(A_j) = \mu\left(\bigcup_j A_j\right) + \mu\left(\bigcup_{1 \le j_1 < j_2 \le n} (A_{j_1} \cap A_{j_2})\right) + \sum_{1 \le j_1 < j_2 < j_3 \le n} \mu(A_{j_1} \cap A_{j_2} \cap A_{j_3}),$$

$$\sum_{1 \le j_1 < j_2 \le n} \mu(A_{j_1} \cap A_{j_2}) = \mu\left(\bigcup_{1 \le j_1 < j_2 \le n} (A_{j_1} \cap A_{j_2})\right) + 2 \sum_{1 \le j_1 < j_2 < j_3 \le n} \mu(A_{j_1} \cap A_{j_2} \cap A_{j_3})$$

and that

$$\sum_j \mu(A_j) \le 3\mu\left(\bigcup_j A_j\right).$$

3.2 Other Measurable Functions

In order to obtain the "additivity" property $\int (f + g) = \int f + \int g$ for more general measurable functions, we first need a sort of converse for Problem 2.1.P3. We begin by showing that a nonnegative measurable function is the limit of an increasing sequence of nonnegative *simple* functions. No measure is involved and the result could logically have been presented in Sect. 2.1, but it is only now that we shall have immediate use for it.

3.2.1. Theorem. *Let f be a measurable nonnegative extended real-valued function on X. Then there exists a sequence $\{s_n\}_{n \ge 1}$ of simple functions such that, for every $x \in X$,*

(i) $0 \le s_1(x) \le s_2(x) \le \cdots \le f(x)$ (increasing),

(ii) $s_n(x) \to f(x)$ as $n \to \infty$ (converging pointwise to f).

Proof: (Figures relating the sets to be called $E_{n,k}$ and the simple functions s_n are given after this proof.) Since f is a nonnegative (extended real) valued function, its range is contained in $\mathbb{R}^{*+}$. For each positive integer n, the interval $[0, n)$ is the union of disjoint intervals

$$[\frac{k-1}{2^n}, \frac{k}{2^n}), \ 1 \le k \le n2^n.$$

Therefore $\mathbb{R}^{*+}$ is the union of the above intervals and the set $\{x \in \mathbb{R}^{*+} : x \ge n\}$, which we shall denote by $[n, \infty]$. It follows that

$$X = f^{-1}(\mathbb{R}^{*+}) = \bigcup_{k=1}^{n2^n} f^{-1}\left([\frac{k-1}{2^n}, \frac{k}{2^n})\right) \cup f^{-1}([n, \infty]).$$

For $1 \le k \le n2^n$, let

$$E_{n,k} = f^{-1}\left([\frac{k-1}{2^n}, \frac{k}{2^n})\right) \text{ and } F_n = f^{-1}([n, \infty]).$$

Then the sets $E_{n,k}$ and F_n are disjoint and measurable with union equal to X. Define the functions s_n on X as

$$s_n = \sum_{k=1}^{n2^n} \frac{k-1}{2^n} \chi_{E_{n,k}} + n\chi_{F_n} \quad \text{for each } n \in \mathbb{N}.$$

From the disjointness of the $1+n2^n$ sets $E_{n,k}$ and F_n, it follows that s_n takes the respective values $(k-1)/2^n$ on the sets $E_{n,k}$ and n on F_n.

It would be well to bear in mind during the rest of this proof that the three assertions

$$k-1 \le f(x)2^n < k, x \in E_{n,k} \text{ and } s_n(x) = (k-1)/2^n$$

are equivalent *but provided* we know that $1 \le k \le n2^n$.

Clearly, $0 \le s_n(x) \le f(x)$ for every $x \in X$. We shall show that

$$s_n(x) \le s_{n+1}(x) \quad \text{for every } x \in X, \tag{3.13}$$

whereupon (i) will have been justified.

In order to show that (3.13) holds, first consider any $x \in X$ such that $f(x) \geq n+1$. It satisfies $x \in F_{n+1} \subseteq F_n$, and hence $s_n(x) = n$ and $s_{n+1}(x) = n + 1$. Thus (3.13) holds for such x.

Next, consider an $x \in X$ for which $n \leq f(x) < n + 1$. This double inequality implies $x \in F_n$, and therefore $s_n(x) = n$. It also implies $n2^{n+1} \leq f(x)2^{n+1} < (n + 1)2^{n+1}$, which further implies that $k - 1 \leq f(x)2^{n+1} < k$, for some $k \in \mathbb{N}$ satisfying $n2^{n+1} \leq k - 1 < (n + 1)2^{n+1}$. [Choose $k = 1 + \lfloor f(x)2^{n+1} \rfloor$.] It follows that $x \in E_{n+1,k}$, where $1 \leq k \leq (n + 1)2^{n+1}$, and therefore $s_{n+1}(x) = (k - 1)/2^{n+1} \geq n$. Since $s_n(x) = n$, the statement (3.13) is found to hold when $n \leq f(x) < n + 1$.

Finally, consider the only remaining possibility, namely, $f(x) < n$. In this event, $0 \leq f(x)2^n < n2^n$ and hence $k - 1 \leq f(x)2^n < k$, for some $k \in \mathbb{N}$ [$k = 1 + \lfloor f(x)2^n \rfloor$ as before] satisfying $0 < k \leq n2^n$. It follows that $x \in E_{n,k}$, where $1 \leq k \leq n2^n$, and therefore $s_n(x) = (k - 1)/2^n$. Now, the inequality $k - 1 \leq f(x)2^n < k$ holds if and only if

$$\text{either} \quad 2(k - 1) \leq f(x)2^{n+1} < 2k - 1 \quad \text{or} \quad 2k - 1 \leq f(x)2^{n+1} < 2k.$$

This combination of double inequalities is in turn equivalent to

$$\text{either} \quad x \in E_{n+1,2k-1} \quad \text{or} \quad x \in E_{n+1,2k}$$

because $1 \leq 2k - 1 < 2k \leq (n + 1)2^{n+1}$. If $x \in E_{n+1,2k-1}$, then $s_{n+1}(x) = (2k - 2)/2^{n+1} = (k - 1)/2^n$. Since $s_n(x) = (k - 1)/2^n$ as observed above, we find that (3.13) holds if $x \in E_{n+1,2k-1}$. If $x \in E_{n+1,2k}$ instead, then $s_{n+1}(x) = (2k - 1)/2^{n+1} > (k - 1)/2^n = s_n(x)$. Thus, (3.13) turns out to hold in this remaining case too, and (i) has therefore been established.

Since the union of the disjoint sets $E_{n,k}$ and F_n is X, every $x \in F_n^c$ satisfies $x \in E_{n,k}$ for some k, and hence by the definition of s_n, satisfies

$$s_n(x) = \frac{k - 1}{2^n} \leq f(x) < \frac{k}{2^n}.$$

This shows that

$$x \in F_n^c \Rightarrow 0 \leq f(x) - s_n(x) < \frac{1}{2^n}. \tag{3.14}$$

If $f(x) < \infty$, then $x \in F_n^c$ for all sufficiently large n and it follows from (3.14) that $0 \leq f(x) - s_n(x) < 2^{-n}$ for all sufficiently large n. Therefore (ii) holds when $f(x) < \infty$. If on the other hand $f(x) = \infty$, then for all n, we have $x \in F_n$ and hence $s_n(x) = n$, so that (ii) still holds. $\qquad \square$

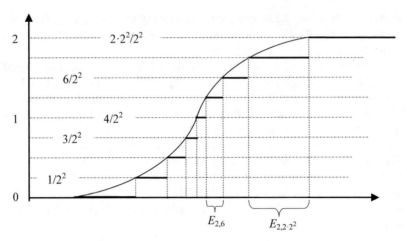

An increasing function and its s_2 (shown in thick horizontal segments)

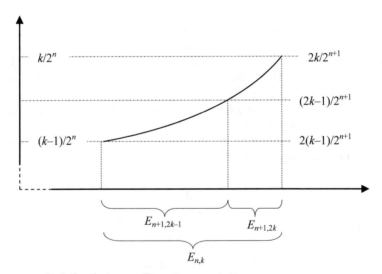

Relation between $E_{n,k}$, $E_{n+1,2k}$ and $E_{n+1,2k-1}$

We can now establish the additivity and "homogeneity" properties of the integral for nonnegative extended real-valued (measurable) functions.

3.2.2. Theorem. *If f and g are measurable nonnegative functions on a measure space X, then*

$$\int (f + g) = \int f + \int g,$$

and for $0 \le c < \infty$, we have

$$\int (cf) = c \int f.$$

Proof: The second equality is merely a restatement of Proposition 2.2.7.

As for the first equality, Theorem 3.2.1 yields increasing sequences $\{s_n\}_{n\geq 1}$ and $\{t_n\}_{n\geq 1}$ of nonnegative simple functions converging pointwise to f and g respectively. Taking $u_n = s_n + t_n$, we obtain an increasing sequence $\{u_n\}_{n\geq 1}$ of nonnegative simple functions converging pointwise to $f + g$. By the Monotone Convergence Theorem 2.3.2,

$$\int f = \lim_{n\to\infty} \int s_n, \int g = \lim_{n\to\infty} \int t_n \text{ and } \int (f+g) = \lim_{n\to\infty} \int u_n. \qquad (3.15)$$

But by Proposition 3.1.8,

$$\int u_n = \int s_n + \int t_n.$$

Combining this with (3.15), we get the required equality.   □

3.2.3. Theorem (Fatou's Lemma). *Let* $\{f_n\}_{n\geq 1}$ *be a sequence of measurable nonnegative functions on a measure space* X. *Then*

$$\int (\liminf_{n\to\infty} f_n) \leq \liminf_{n\to\infty} \int f_n.$$

Proof: Let $\{g_n\}_{n\geq 1}$ be the sequence of functions on X defined by

$$g_n(x) = \inf_{k\geq n} f_k(x) \text{ for each } x \in X \text{ and each } n \in \mathbb{N}.$$

By Problem 2.2.P3, each function g_n is measurable. Also, $g_n \leq f_n$ everywhere on X, and hence

$$\int g_n \leq \int f_n. \qquad (3.16)$$

Moreover, $\{g_n\}_{n\geq 1}$ is an increasing sequence of nonnegative measurable functions satisfying $\lim_{n\to\infty} g_n = \lim_{n\to\infty} \inf f_n$ everywhere. Therefore we may infer from the Monotone Convergence Theorem 2.3.2 that

$$\int (\liminf_{n\to\infty} f_n) = \lim_{n\to\infty} \int g_n.$$

When an increasing sequence in $\mathbb{R}^{*+}$ has a limit in $\mathbb{R}^{*+}$, the liminf agrees with the limit. Therefore

$$\lim_{n\to\infty} \int g_n = \liminf_{n\to\infty} \int g_n \leq \liminf_{n\to\infty} \int f_n$$

by (3.16).   □

The integral of the limit inferior can actually be less than the limit inferior of the integrals. (See Problem 3.2.P2.)

Before taking up the next proof, it may be useful to recount from Definition 2.2.5 that for any measurable real-valued function f on a measurable set X, the *measure space integral of f* is defined to be

$$\int f = \int f^+ - \int f^-,$$

provided that $\int f^-$ is finite (i.e. is not ∞). The function f is said to be *summable* or *integrable* if both $\int f^+$ and $\int f^-$ are finite, or equivalently, if $\int f$ is finite.

3.2.4. Proposition. (a)   *If $\int f$ exists (∞ permissible), then $\left| \int f \right| \leq \int |f|$.*
(b)   *The function f is integrable if and only if its absolute value $|f|$ is integrable.*

Proof: (a)   We need consider only the case when $\int |f| < \infty$. Since $|f| = f^+ + f^-$, we have $\int |f| = \int f^+ + \int f^-$ by Theorem 3.2.2. Therefore both $\int f^+$ and $\int f^-$ are finite. Hence by definition of the measure space integral,

$$\left| \int f \right| = \left| \int f^+ - \int f^- \right| \leq \left| \int f^+ \right| + \left| \int f^- \right| = \int f^+ + \int f^- = \int |f|.$$

(b)   By definition, f is integrable if and only if $\int f^+$ and $\int f^-$ are both finite. This is equivalent to $\int |f|$ being finite, because $\int |f| = \int f^+ + \int f^-$, as already observed.   □

Since $(f+g)^+$ is not always the same as $f^+ + g^+$, proving the analog of Theorem 3.2.2 for functions that fail to be nonnegative is not altogether straightforward.

3.2.5. Theorem. *Let f and g be integrable functions on a measure space X and α, β be real. Then $\alpha f + \beta g$ is integrable and*

$$\int (\alpha f + \beta g) = \alpha \int f + \beta \int g.$$

If $f \geq g$ everywhere, then $\int f \geq \int g$.

Proof: By Theorem 2.2.3, $\alpha f + \beta g$ is measurable. Hence by Proposition 2.2.8 and Theorem 3.2.2,

$$\int |\alpha f + \beta g| \leq \int (|\alpha||f| + |\beta||g|) = |\alpha| \int |f| + |\beta| \int |g| < \infty.$$

Thus $\alpha f + \beta g$ is integrable by Proposition 3.2.4.

To establish the equality (sometimes called *linearity of the integral for integrable functions*) it is sufficient to prove *additivity*:

$$\int (f + g) = \int f + \int g$$

and *homogeneity*:

$$\int (\alpha f) = \alpha \int f.$$

As f and g are both integrable, the same is true of f^+, f^-, g^+, g^- as well as $f^+ + g^+$ and $f^- + g^-$. As shown above, $f + g$ is integrable and hence so are $(f + g)^+$ and $(f + g)^-$. Moreover,

$$(f^+ + g^+) - (f^- + g^-) = f + g = (f + g)^+ - (f + g)^-$$

and hence

$$(f^+ + g^+) + (f + g)^- = (f + g)^+ + (f^- + g^-).$$

Upon applying Theorem 3.2.2 repeatedly, we get $\int f^+ + \int g^+ + \int (f + g)^- = \int (f + g)^+ + \int f^- + \int g^-$. Since none of the integrals here is ∞, we obtain

$$\int (f + g) = \int (f + g)^+ - \int (f + g)^- = \int f^+ - \int f^- + \int g^+ - \int g^- = \int f + \int g.$$

This proves additivity.

To prove homogeneity, first consider $\alpha \geq 0$. The required equality then follows from Definition 2.2.5 and the homogeneity of the integral for nonnegative functions (Theorem 3.2.2). The case of negative α can now be deduced from the relations $f^- = (-f)^+$ and $f^+ = (-f)^-$. This completes the proof of homogeneity.

To prove the last part, we use the consequence $\int (f - g) = \int f - \int g$ of what has just been proved to argue that

$$f \geq g \Rightarrow f - g \geq 0 \Rightarrow \int (f - g) \geq 0 \Rightarrow \int f \geq \int g. \qquad \square$$

We close this section with the second of the two fundamental convergence theorems that embody the advantage of the measure space integral over the Riemann integral (the first one was the Monotone Convergence Theorem 2.3.2).

3.2.6. Dominated Convergence Theorem. *Let $\{f_n\}_{n \geq 1}$ be a sequence of measurable functions on a measure space X, converging everywhere to a function f, and suppose there exists an integrable (therefore real-valued) function g on X such that*

$$|f_n(x)| \leq g(x) \quad for\ every\ x \in X.$$

Then the functions f and f_n are all integrable,

$$\lim_{n\to\infty} \int |f_n - f| = 0 \quad and \quad \lim_{n\to\infty} \int f_n = \int f.$$

In particular, $\lim_{n\to\infty} \int f_n$ *exists.*

Proof: Since $f = \lim_{n\to\infty} \int f_n$ and $|f_n| \le g$ everywhere, we also have $|f| \le g$ everywhere. Therefore the functions f and f_n are all real-valued and hence so is $f_n - f$. Besides, f is measurable by Problem 2.2.P4. The assumed integrability of g therefore leads to that of f and f_n upon applying Theorem 3.2.5 and Proposition 3.2.4(b).

Now it follows from the hypotheses that $|f_n - f| \le 2g$, so that the functions $2g - |f_n - f|$ are nonnegative, and also that $\lim_{n\to\infty} \inf(2g - |f_n - f|) = 2g$. Moreover, the functions are bounded above at each $x \in X$ by $2\,g(x)$ and their integrals by $\int (2g)$, which is assumed to be finite. Therefore by Fatou's Lemma 3.2.3 and Theorem 3.2.5, we have

$$\int (2g) \le \lim_{n\to\infty} \inf \int (2g - |f_n - f|) = \int (2g) + \lim_{n\to\infty} \inf \left(-\int |f_n - f| \right)$$

$$= \int (2g) - \lim_{n\to\infty} \sup \int |f_n - f|.$$

Since $\int (2g) < \infty$, it follows from here that

$$\lim_{n\to\infty} \sup \int |f_n - f| \le 0.$$

This establishes the first of the two equalities sought to be proved. The second equality and the existence of the limit therein follow from the first equality upon applying Theorem 3.2.5 to f_n and f and then applying Proposition 3.2.4(a) to $f_n - f$. $\square$

Problem Set 3.2
3.2.P1. Let μ be a measure on X and f, g be integrable functions such that $f \ge g$ everywhere and $\int f = \int g$. Prove $\mu(X(f > g)) = 0$.

3.2.P2. Let $X = \{a, b\}$ with the counting measure. Let $\{f_n\}_{n\ge 1}$ be the sequence of functions on X defined by $f_n = \chi_{\{a\}}$ if n is even and $\chi_{\{b\}}$ if n is odd. Find

$$\int (\lim_{n\to\infty} \inf f_n) \text{ and } \lim_{n\to\infty} \inf \int f_n.$$

3.2.P3. In this problem, $(X, \mathcal{F}, \mu)$ is a measure space and f, g are measurable real-valued functions on X. Then the nonnegative real-valued function $|fg|$ is measurable by Theorem 2.2.3(c) and Proposition 2.2.2(b); also, for any real $p > 0$ and $q > 0$, the real-valued functions $|f + g|$, $|f + g|^p$, $|f|^p$ and $|g|^q$ are measurable by Theorem 2.2.3(b), Proposition 2.2.2(b) and Problem 2.2.P6(a). If $p > 1$, then $|f + g|^{p-1}$ is also measurable.

(a) Suppose p, q are positive real numbers such that $\frac{1}{p} + \frac{1}{q} = 1$. Show that

$$\int |fg| \le \left(\int |f|^p\right)^{\frac{1}{p}} \left(\int |g|^q\right)^{\frac{1}{q}}.$$

This is known as **Hölder's Inequality**.

(b) Suppose p is a real number such that $p \ge 1$. Show that

$$\left(\int |f + g|^p\right)^{\frac{1}{p}} \le \left(\int |f|^p\right)^{\frac{1}{p}} + \left(\int |g|^p\right)^{\frac{1}{p}}.$$

This is known as **Minkowski's Inequality**.

3.2.P4. (Needed in Proposition 7.7.5) Let f be an integrable function on X, where $(X, \mathcal{F}, \mu)$ is a measure space. Given any $\varepsilon > 0$, show that there exists a $\delta > 0$ such that every $A \in \mathcal{F}$ with $\mu(A) < \delta$ satisfies $\left|\int (\chi_A f)\right| < \varepsilon$.

3.2.P5. Show that the "dominated" hypothesis of the Dominated Convergence Theorem 3.2.6, namely that, for some integrable g, the inequality $|f_n(x)| \le g(x)$ holds for every $x \in X$, cannot be dropped.

3.2.P6. For each fixed $n \in \mathbb{N}$, the series $\sum_{j=1}^{\infty} \frac{1}{n}\left(\frac{n}{n+1}\right)^{j-1}$ converges to $\frac{n+1}{n}$. Show that any series $\sum_{j=1}^{\infty} a_j$ satisfying $a_j > \frac{1}{n}\left(\frac{n}{n+1}\right)^{j-1}$ for every $n \in \mathbb{N}$ and every $j \in \mathbb{N}$ diverges.

3.2.P7. Show that the following sequence converges to 0:

$$\left\{\sum_{j=1}^{\infty} \frac{n}{j(n^2 + j^2)}\right\}_{n \in \mathbb{N}}$$

3.2.P8. Given any measure space, show that a nonnegative measurable function f satisfies

$$\int f = \sup\left\{\int s : 0 \le s \le f, \text{ where } s \text{ is simple}\right\}.$$

Hint: Use Theorem 3.2.1 and the Monotone Convergence Theorem 2.3.2.

3.2.P9. Given any measure space $(X, \mathcal{F}, \mu)$, let $\mathcal{S}$ denote the class of nonnegative simple functions s vanishing outside a set of finite measure, i.e. $\mu(X(s > 0)) < \infty$. This is obviously equivalent to $\int s < \infty$.

(a) If every set of infinite measure contains a subset of finite positive measure, show for any nonnegative measurable function f that

$$\int f = \sup\left\{\int s : 0 \le s \le f, s \in \mathcal{S}\right\}$$

(b) Prove the converse of (a).

Hint: (a) Use Problem 2.3.P3.

(b) Take f to be the characteristic function of any given set of infinite measure; the condition yields $s \in S$ with $\int s > 0$. Now consider any set of positive measure on which s takes a positive value.

3.2.P10. (a) Let μ be a measure on a set X. Suppose f is a nonnegative real-valued function on X and $\{f_n\}_{n\geq 1}$ a sequence of measurable nonnegative real-valued functions such that

$$\text{(i) } f_n(x) \geq f_{n+1}(x) \quad \text{and} \quad \text{(ii) } f(x) = \lim_{n\to\infty} f_n(x) \text{ for every } x \in X.$$

Then f is measurable, and if $\int f_N < \infty$ for some N, then $\int f = \lim_{n\to\infty} f_n$.

(b) Show that in part (a), the hypothesis that $\int f_N < \infty$ for some N cannot be dropped even if $\mu(X)$ is finite and the only subset of measure 0 is the empty set.

3.2.P11. Let the function $f : [0, 1] \to \mathbb{R}$ be given by $f(x) = x^{\frac{1}{2}}$. In the notation of the proof of Theorem 3.2.1, find the set $E_{3,9}$ and the number $s_3(\frac{3}{4})$.

3.3 Subadditive Fuzzy Measures

This section can be omitted without loss of continuity.

Example 2.1.3(b) exhibits a nonnegative-valued function μ with a σ-algebra $\mathcal{F}$ as its domain, vanishing at the empty set and satisfying monotonicity: $A \subseteq B \Rightarrow \mu(A) \leq \mu(B)$, where it is understood of course that $A, B \in \mathcal{F}$. Recall that μ is not a measure because it is not even finitely additive. It was noted just before Definition 2.2.1 that the concept of the integral of a nonnegative measurable function makes sense when μ is merely nonnegative and monotone. In this section, we shall mostly be concerned with a monotone nonnegative-valued function with a σ-algebra $\mathcal{F}$ as its domain and vanishing at the empty set; such an object will be referred to as a **fuzzy measure** on X, it being understood that the domain is a σ-algebra $\mathcal{F}$ of subsets of X. Thus what was called "μ" in Example 2.1.3(b) is a fuzzy measure. To avoid confusion, we shall denote a fuzzy measure by $\widetilde{\mu}$ and integration with respect to it by $\int_\sim$ or by $\int_\sim d\widetilde{\mu}$, the latter being more useful when there are two or more fuzzy measures in the discussion.

Observe that Example 2.1.8(c) computes $\int_\sim (f + g)$, $\int_\sim f$ and $\int_\sim g$ for certain functions f and g, and demonstrates that the analog of Minkowski's Inequality of Problem 3.2.P3(b) need not hold for a fuzzy measure when $p = 1$. Similarly, Problem 2.1.P6 shows that it need not hold when $p = 2$. A suitably modified version of Minkowski's Inequality will be proved below in Theorem 3.3.5 for fuzzy measures that are finitely subadditive in the sense of Problem 2.1.P5. In this context, we shall drop the qualification "finitely" in the present section.

The fuzzy measure of Example 2.1.3(b) is subadditive. Besides, the outer measures to be studied in Chap. 4 will be subadditive fuzzy measures. From a given subadditive fuzzy measure $\widetilde{\mu}$, one can easily generate other subadditive fuzzy measures $\widetilde{\nu}$ by choosing any increasing function $\phi : \mathbb{R}^{*+} \to \mathbb{R}^{*+}$ such that $\phi(0) = 0$, $\phi(u + v)) \leq \phi(u) + \phi(v)$ and setting $\widetilde{\nu}(E) = \phi(\widetilde{\mu}(E))$.

3.3.1. Exercise Let $\widetilde{\mu}$ be a subadditive fuzzy measure with domain a σ-algebra $\mathcal{F}$. If $A, B \in \mathcal{F}$ and $\widetilde{\mu}(B) = 0$, show that $\widetilde{\mu}(A \backslash B) = \widetilde{\mu}(A)$.

Solution: On the one hand, we have

$$\widetilde{\mu}(A) \geq \widetilde{\mu}(A \backslash B) \text{ and } \widetilde{\mu}(A \cap B) = 0 \text{ by monotonicity,}$$

while on the other hand,

$$\widetilde{\mu}(A) \leq \widetilde{\mu}(A \backslash B) + \widetilde{\mu}(A \cap B) \text{ by subadditivity.}$$

Proposition 2.2.7 and Proposition 2.2.8 about the integral with respect to a measure μ do not use the full force of the hypothesis that μ is a measure. The reader can check that their analogs for any subadditive fuzzy measure $\widetilde{\mu}$ are valid. We record the analogs here for convenience.

3.3.2. Proposition. *Suppose $\widetilde{\mu}$ is a subadditive fuzzy measure on a set X. For a measurable $f : X \to \mathbb{R}^{*+}$ and for $0 \leq c < \infty$, we have*

$$\int_{\sim} (cf) = c \int_{\sim} f.$$

Proof: The proof of Proposition 2.2.7 carries over. $\qquad\qquad\qquad\qquad\qquad \square$

3.3.3. Proposition. *Suppose $\widetilde{\mu}$ is a subadditive fuzzy measure on a set X and the nonnegative measurable functions f and g on X satisfy $f(x) \leq g(x)$ for every $x \in X$. Then $\int_{\sim} f \leq \int_{\sim} g$.*

Proof: The proof of Proposition 2.2.8 carries over. $\qquad\qquad\qquad\qquad\qquad \square$

3.3.4. Proposition. *Suppose $\widetilde{\mu}$ is a subadditive fuzzy measure on a set X. Then*

(a) *For a measurable function $f : X \to \mathbb{R}^{*+}$ and any real $p > 0$, the nonnegative function $f^p : X \to \mathbb{R}^{*+}$ is measurable; moreover,*

$$\int_{\sim} f = 0 \text{ if and only if } \tilde{f}(t) = 0 \text{ for every } t > 0,$$

which, in turn, is equivalent to $\int_{\sim} f^p = 0$ for any $p > 0$.

(b) *If $f : X \to \mathbb{R}^{*+}$ and $g : X \to \mathbb{R}^{*+}$ are measurable functions satisfying $\int_{\sim} f = 0 = \int_{\sim} g$, then $\int_{\sim} (f + g) = 0$.*

Proof: The solution of Problem 2.2.P6 carries over.   □

The result of the next theorem has already been illustrated in Example 2.1.8(c) with $p=1$ (but cf. Problem 3.3.P2 and Problem 3.3.P6) and in Problem 2.1.P6 with $p=2$. In both illustrations, the subadditive fuzzy measure was not a measure.

3.3.5. Theorem. *Suppose $\tilde{\mu}$ is a subadditive fuzzy measure on a set X and $f : X \to \mathbb{R}^{*+}$, $g : X \to \mathbb{R}^{*+}$ are measurable functions. Then for any real $p > 0$,*

$$\left(\int_{\sim} (f+g)^p\right)^{\frac{1}{p+1}} \leq \left(\int_{\sim} f^p\right)^{\frac{1}{p+1}} + \left(\int_{\sim} g^p\right)^{\frac{1}{p+1}}$$

and

$$\int_{\sim} (f+g)^p \leq 2^p \left(\int_{\sim} f^p + \int_{\sim} g^p\right).$$

Proof: If either $\int_{\sim} f^p$ or $\int_{\sim} g^p$ is ∞, there is nothing to prove. So, we may assume both to be finite. If both are 0, then $\int_{\sim} (f+g)^p = 0$ by Proposition 3.3.4 and the inequalities hold. So, we proceed only with the case $\int_{\sim} f^p > 0$.

Consider an arbitrary λ satisfying $0 < \lambda < 1$. For any such λ, we have

$$X\left((f+g)^p > t\right) = X\left(f+g > t^{\frac{1}{p}}\right) \subseteq X\left(f > \lambda t^{\frac{1}{p}}\right) \cup X\left(g > (1-\lambda)t^{\frac{1}{p}}\right).$$

Therefore the monotonicity and subadditivity of $\tilde{\mu}$ imply that

$$\widetilde{(f+g)^p}(t) \leq \widetilde{f^p}\left(\lambda^p t\right) + \widetilde{g^p}\left((1-\lambda)^p t\right).$$

It follows from here that

$$\int_{\sim} (f+g)^p \leq \int_0^\infty \widetilde{f^p}(\lambda^p t)dt + \int_0^\infty \widetilde{g^p}\left((1-\lambda)^p t\right)dt.$$

On the basis of Proposition 2.2.6, we know that

$$\int_0^\infty \widetilde{f^p}(\lambda^p t)dt = \frac{1}{\lambda^p} \int_0^\infty \widetilde{f^p}(t)dt = \frac{1}{\lambda^p} \int_{\sim} f^p$$

and

$$\int_0^\infty \widetilde{g^p}((1-\lambda)^p t)dt = \frac{1}{(1-\lambda)^p} \int_0^\infty \widetilde{g^p}(t)dt = \frac{1}{(1-\lambda)^p} \int_{\sim} g^p.$$

Therefore

$$\int_{\sim} (f+g)^p \leq \frac{1}{\lambda^p} \int_{\sim} f^p + \frac{1}{(1-\lambda)^p} \int_{\sim} g^p. \tag{3.17}$$

Since this holds for all λ satisfying $0 < \lambda < 1$, we may take the limit as $\lambda \to 1$. If $\int_{\sim} g^p = 0$, taking the limit yields the first one of the desired inequalities. However, if $\int_{\sim} g^p > 0$, it is permissible to select

$$\lambda = \frac{\left(\int_{\sim} f^p\right)^{\frac{1}{p+1}}}{\left(\int_{\sim} f^p\right)^{\frac{1}{p+1}} + \left(\int_{\sim} g^p\right)^{\frac{1}{p+1}}}$$

because this number λ satisfies $0 < \lambda < 1$, considering that we now have $\int_{\sim} f^p > 0$ as well as $\int_{\sim} g^p > 0$. Upon using it in (3.17), we obtain the first of the desired inequalities after a brief but elementary computation.

The second inequality follows from (3.17) by taking $\lambda = \frac{1}{2}$. □

3.3.6. Exercise. Let μ be the counting measure on the two-element set $X = \{a, b\}$. Choose f to be the characteristic function of $\{a\}$ and g to be the characteristic function of $\{b\}$. Show that for $p = \frac{1}{2}$, Minkowski's Inequality does not hold but the inequality of Theorem 3.3.5 does.

Solution: Being a measure, μ is also a subadditive fuzzy measure and $\int$ means the same thing as $\int_{\sim}$.

Now, $\int f^p = 1 = \int g^p$ and $\int (f+g)^p = 2$. Therefore $\left(\int f^p\right)^{\frac{1}{p}} = 1 = \left(\int g^p\right)^{\frac{1}{p}}$ and $\left(\int (f+g)^p\right)^{\frac{1}{p}} = 4$. Since $4 > 1 + 1$, Minkowski's Inequality does not hold. However, $p + 1 = \frac{3}{2}$ and therefore, $\left(\int f^p\right)^{\frac{1}{p+1}} = 1 = \left(\int g^p\right)^{\frac{1}{p+1}}$, $\left(\int (f+g)^p\right)^{\frac{1}{p+1}} = 2^{\frac{2}{3}}$. Since $2^{\frac{2}{3}} < 1 + 1$, the inequality of Theorem 3.3.5 does hold.

3.3.7. Exercise. Suppose $\tilde{\mu}$ is a subadditive fuzzy measure on a set X and $\phi_k : X \to \mathbb{R}^{*+}$ are measurable functions for $1 \leq k \leq n$, where $n \geq 2$. For $p > 0$, show that

$$\int_{\sim} \left(\sum_{k=1}^{n} \phi_k\right)^p \leq \sum_{k=1}^{n-1} 2^{pk} \int_{\sim} \phi_k^p + 2^{p(n-1)} \int_{\sim} \phi_n^p.$$

Solution: Use induction on n. For $n = 2$, the assertion is that $\int_{\sim} (\phi_1 + \phi_2)^p \leq 2^p \int_{\sim} \phi_2^p + 2^p \int_{\sim} \phi_1^p$, which is true by virtue of the second inequality of Theorem 3.3.5. Assume as induction hypothesis that the result is true for some $n \geq 2$. Then by virtue of the same inequality again,

$$\int_{\sim} \left(\sum_{k=1}^{n+1} \phi_k\right)^p \leq 2^p \int_{\sim} \phi_1^p + 2^p \int_{\sim} \left(\sum_{k=2}^{n+1} \phi_k\right)^p.$$

The induction hypothesis further yields

$$\int_{\sim} \left(\sum_{k=1}^{n+1} \phi_k\right)^p \leq 2^p \int_{\sim} \phi_1^p + 2^p \left(\sum_{k=2}^{n} 2^{p(k-1)} \int_{\sim} \phi_k^p + 2^{p(n-1)} \int_{\sim} \phi_{n+1}^p\right)$$

$$\leq 2^p \int_{\sim} \phi_1^p + \left(\sum_{k=2}^{n} 2^{pk} \int_{\sim} \phi_k^p + 2^{pn} \int_{\sim} \phi_{n+1}^p\right)$$

$$\leq \sum_{k=1}^{n} 2^{pk} \int_{\sim} \phi_k^p + 2^{pn} \int_{\sim} \phi_{n+1}^p,$$

which shows that the result is true for $n+1$.

Readers who are aware of pseudometric spaces and completeness will note that setting $\rho(f, g) = \left(\int |f - g|^p\right)^{\frac{1}{p}}$, where $p \geq 1$ and the integral is taken with respect to a measure, gives rise to a pseudometric on the space of all real-valued measurable functions f with $\int |f|^p < \infty$, the crucial triangle inequality being a consequence of Minkowski's Inequality. It is well known that the corresponding metric space is complete (see 5.6.6 of Craven [1, p. 157]). If we take a subadditive fuzzy measure $\widetilde{\mu}$ instead of a measure, then setting $\rho(f, g) = \left(\int_{\sim} |f - g|^p\right)^{\frac{1}{p+1}}$, where $p > 0$, can be seen to give rise to a pseudometric on the space of all real-valued measurable functions f with $\int_{\sim} |f|^p < \infty$, the crucial triangle inequality now being a consequence of Theorem 3.3.5. Completeness of the corresponding metric space can be proved with the additional hypothesis that inner continuity as described in Sect. 2.3 holds for $\widetilde{\mu}$. By Proposition 2.3.1, inner continuity holds for any arbitrary measure, but this is not so for subadditive fuzzy measures. For a counterexample, consider the set $\mathbb{N}$ with $\widetilde{\mu}$ defined on the σ-algebra of all subsets by putting $\widetilde{\mu}(\emptyset) = 0$, $\widetilde{\mu}(\mathbb{N}) = 2$ and $\widetilde{\mu}(E) = 1$ whenever $\emptyset \neq E \neq \mathbb{N}$; it is easy to check that $\widetilde{\mu}$ is a subadditive fuzzy measure but is not inner continuous. Given inner continuity for a subadditive fuzzy measure, the proof of completeness can be carried out by taking advantage of Exercise 3.3.7. Details may be found in Shirali [8].

Theorem 3.3.5 is an analog of Minkowski's Inequality, valid for subadditive fuzzy measures. The next result is an analog of Hölder's Inequality of Problem 3.2.P3(a), valid for subadditive fuzzy measures.

3.3.8. Theorem. *Suppose $\widetilde{\mu}$ is a subadditive fuzzy measure on a set X and $f : X \to \mathbb{R}^{**}$, $g : X \to \mathbb{R}^{**}$ are measurable functions. Suppose p, q are positive real numbers such that $\frac{1}{p} + \frac{1}{q} = 1$. Then*

$$\int_{\sim} fg \leq \left(\frac{1}{\sqrt{p}} + \frac{1}{\sqrt{q}}\right)^2 \left(\int_{\sim} f^p\right)^{\frac{1}{p}} \left(\int_{\sim} g^q\right)^{\frac{1}{q}}.$$

Proof: First we dispose of the case when $\int_{\sim} f^p = 0$. Proposition 3.3.4 implies that $\widetilde{f}(t) = 0$ for every $t > 0$. Take an arbitrary $t > 0$. Then for any $x \in X$,

$$f(x)g(x) > t \Rightarrow f(x) > 0 \Rightarrow f(x) > s \text{ for some } s > 0.$$

In other words, $X(fg > t) \subseteq X(f > s)$ for some $s > 0$. By monotonicity of $\widetilde{\mu}$, it follows that $\widetilde{fg}(t) = 0$. Since $t > 0$ is arbitrary, this implies $\int_{\sim} fg = 0$ by Proposition 3.3.4. Thus the inequality holds when $\int_{\sim} f^p = 0$. Similarly, it holds when $\int_{\sim} g^q = 0$.

It remains to establish the inequality when $\int_{\sim} f^p > 0$ and $\int_{\sim} g^q > 0$. If one of them is ∞, the right side is ∞, and there is nothing to prove. So, we also assume both of them to be finite.

Set $F = f/(\int_{\sim} f^p)^{1/p}$ and $G = g/(\int_{\sim} g^q)^{1/q}$. Then by Proposition 3.3.2, $\int_{\sim} F^p = 1 = \int_{\sim} G^q$. By the inequality

$$ab \le \frac{1}{p}a^p + \frac{1}{q}b^q \text{ for any nonnegative } a \text{ and } b$$

(easily proved by considering the derivative of $\phi(\xi) = 1/q + \xi/p - \xi^{1/p}$ for $\xi \in (0, \infty)$), we have $FG \le F^p/p + G^q/q$. Employing Proposition 3.3.2, Proposition 3.3.3 and Theorem 3.3.5 (wherein we select 'p' to be 1), we obtain $\int_{\sim} FG \le \left(\frac{1}{\sqrt{p}} + \frac{1}{\sqrt{q}}\right)^2$. This leads to the required inequality upon using Proposition 3.3.2 once again. $\qquad\square$

3.3.9. Example. We illustrate the set of nonnegative functions with $\int_{\sim} f = 1$ in the particular case when X consists of two distinct elements a and b, the σ-algebra $\mathcal{F}$ consists of all subsets, and $\widetilde{\mu}$ is defined by setting $\widetilde{\mu}\emptyset = 0$, $\widetilde{\mu}\{a\} = 2$, $\widetilde{\mu}\{b\} = 3$ and $\widetilde{\mu}\{a, b\} = 4$. It is easy to check that $\widetilde{\mu}$ is a subadditive fuzzy measure. Since X consists of two points, a nonnegative real-valued function on it can be represented by a point in the first quadrant of the coordinate plane. Let $f(a) = x$ and $f(b) = y$, where x and y are both positive. Its representation in the coordinate plane is the point with coordinates (x, y). Now,

$$\widetilde{f}(t) = \begin{cases} 4 & 0 \le t < x \\ 3 & x \le t < y \\ 0 & y \le t < \infty \end{cases} \text{ if } x < y \text{ and } \widetilde{f}(t) = \begin{cases} 4 & 0 \le t < y \\ 2 & y \le t < x \\ 0 & x \le t < \infty \end{cases} \text{ if } y < x.$$

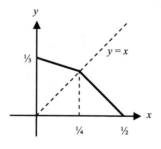

In the remaining case, i.e. $x = y$, the function $\tilde{f}$ equals 4 when $0 < t < x$ and 0 when $x \leq t < \infty$. Therefore, depending on whether $x < y$ or $y < x$ or $x = y$, we have $\int_{\sim} f = 4x + 3(y - x) = x + 3y$ or $4y + 2(x - y) = 2x + 2y$ or $4x$.

With each such f represented by the point (x, y) in the coordinate plane, the set of functions with $\int_{\sim} f = 1$ has the graph shown in the accompanying Figure.

3.3.10. Exercise. Let $\alpha, \beta > 0$ and $0 < s < 1$. For any x such that $\max\{\alpha, \beta\} \leq x \leq \alpha + \beta$, show that

$$\left((1 + s)^2 x\right)^{\frac{1}{3}} \leq \left(s^2 x + (1 - s^2)\alpha\right)^{\frac{1}{3}} + \left(s^2 x + (1 - s^2)\beta\right)^{\frac{1}{3}}.$$

Solution: Define $\tilde{\mu}$ on a two element set $\{a, b\}$ as $\tilde{\mu}\emptyset = 0$, $\tilde{\mu}\{a\} = \alpha$, $\tilde{\mu}\{b\} = \beta$ and $\tilde{\mu}\{a, b\} = x$. Then $\tilde{\mu}$ is a subadditive fuzzy measure. Consider the functions f and g on $\{a,b\}$ such that $f(a) = 1, f(b) = s$ and $g(a) = s, g(b) = 1$. The sum $f + g$ takes the value $1 + s$ at both points and hence $\int_{\sim} (f + g)^2 = (1 + s)^2 x$. Also,

$$(\widetilde{f^2})(t) = \begin{cases} x & 0 \leq t < s^2 \\ \alpha & s^2 \leq t < 1 \\ 0 & 1 \leq t < \infty \end{cases} \text{ and } (\widetilde{g^2})(t) = \begin{cases} x & 0 \leq t < s^2 \\ \beta & s^2 \leq t < 1 \\ 0 & 1 \leq t < \infty. \end{cases}$$

Therefore $\int_{\sim} f^2 = s^2 x + (1 - s^2)\alpha$ and $\int_{\sim} g^2 = s^2 x + (1 - s^2)\beta$. On applying Theorem 3.3.5 with $p = 2$, we arrive at the desired inequality.

Problem Set 3.3

3.3.P1. Let $X = \mathbb{N}$ and $\mathcal{F}$ be the σ-algebra of all subsets of X. Take $\tilde{\nu}$ to be defined by $\tilde{\nu}(E) = \sum_{i=1}^{k} \beta^i$, where $\beta \in (0, 1)$ is fixed and k is the number of elements in E. (If $E = \emptyset$, then $k = 0$ and the summation becomes "empty", in which case its value is understood to be 0 as usual.) Show that $\tilde{\nu}$ is a subadditive fuzzy measure that is not a measure. Is it inner continuous?

3.3.P2. (Cf. Problem 3.3.P6) Suppose $\tilde{\mu}$ is a subadditive fuzzy measure on a set X and $f : X \to \mathbb{R}^{*+}$, $g : X \to \mathbb{R}^{*+}$ are measurable functions. If $fg = 0$ everywhere, show that $\int_{\sim} (f + g) \leq \int_{\sim} f + \int_{\sim} g$.

3.3.P3. Given a subadditive fuzzy measure $\tilde{\mu}$ on a σ-algebra $\mathcal{F}$ of subsets of a set X and a nonnegative measurable function $f : X \to \mathbb{R}^{*+}$, prove that the function $\tilde{\nu}$ defined on all sets $E \in \mathcal{F}$ by setting

$$\tilde{\nu}(E) = \int_{\sim} f \chi_E d\tilde{\mu}$$

is a subadditive fuzzy measure. Prove also that (i) $\tilde{\nu}(E) = 0$ whenever $\tilde{\mu}(E) = 0$ and (ii) $\tilde{\nu}$ is a measure (i.e. is countably additive) whenever $\tilde{\mu}$ is.

3.3.P4. Let $\widetilde{\mu}$ and $\widetilde{\nu}$ be finite subadditive fuzzy measures on a σ-algebra $\mathcal{F}$ of subsets of a set X such that $\widetilde{\nu}(E) = 0$ whenever $\widetilde{\mu}(E) = 0$. Must there exist a nonnegative measurable function $f : X \to \mathbb{R}^{*+}$ satisfying $\widetilde{\nu}(E) = \int_{\sim} f \chi_E d\widetilde{\mu}$ for each $E \in \mathcal{F}$?

3.3.P5. Let $\alpha, \beta > 0$ and $s > 1$. For any x such that $\max \{\alpha, \beta\} \le x \le \alpha + \beta$, show that

$$\left((1+s)^3 x\right)^{\frac{1}{4}} \le \left(x + (s^3 - 1)\alpha\right)^{\frac{1}{4}} + \left(x + (s^3 - 1)\beta\right)^{\frac{1}{4}}.$$

3.3.P6. (Cf. Problem 3.3.P2) Let $\widetilde{\mu}$ be a subadditive fuzzy measure on a two-element set $X = \{a,b\}$, the σ-algebra $\mathcal{F}$ being the family of all four subsets of X. For any functions $f : X \to \mathbb{R}^{*+}$, $g : X \to \mathbb{R}^{*+}$, show that $\int_{\sim} (f + g) \le \int_{\sim} f + \int_{\sim} g$. (Consider only the nontrivial case when $\widetilde{\mu}\{a\} > 0$, $\widetilde{\mu}\{b\} > 0$ and $\widetilde{\mu}\{a, b\} < \infty$.)

Chapter 4
Construction of a Measure

4.1 Lebesgue Outer Measure

Now that we have seen some of the benefits that will accrue from extending the total length concept so as to have the properties described in Remark 1.1.6, we are about to take a preliminary step towards actually obtaining the extension. We emphasize that we do not begin by taking an unambiguously determined total length of a union of finitely many disjoint intervals as being known (the result of Problem 1.1.P9), although the motivation stated in Remark 1.1.6 does. We shall instead deduce it from Proposition 4.2.8 and Theorem 4.3.1 in the next section.

Towards an extension to more general subsets of an interval $[a, b]$, we first introduce the following concept.

4.1.1. Definition. *A subset A of* $\mathbb{R}$ *is said to be* **covered by a family** $\mathcal{B}$ *of subsets of* $\mathbb{R}$ *if* $A \subseteq \cup_{I \in \mathcal{B}} I$. *It is said to be* **covered by a sequence** $\{I_n\}_{n \geq 1}$ *of intervals if* $A \subseteq \cup_{n=1}^{\infty} I_n$. *The* **total length of a sequence** $\{I_n\}_{n \geq 1}$ *of intervals is* $\sum_{n=1}^{\infty} \ell(I_n)$.

Any subset of $\mathbb{R}$ is covered by the sequence of open intervals $\mathbb{R}, \emptyset, \emptyset, \ldots$ or by $(-1, 1), (-2, 2), \ldots$.

It is to be borne in mind that $\sum_{n=1}^{\infty} \ell(I_n)$ can be ∞ and general statements involving such sums may sometimes have to be validated separately in order to account for that possibility.

4.1.2. Remarks. (a) The total length of a sequence of intervals as just defined is not to be taken as the total length of a union of finitely many disjoint intervals, because the former allows for repetition of terms in the sum.

Note that there is no ambiguity about the total length of a finite family (as distinct from the union thereof) of intervals. It is simply the sum of their lengths, each interval being taken into account once.

(b) Suppose that for each $n \in \mathbb{N}$, we have a series $\sum_{k=1}^{\infty} a_{n,k}$ of nonnegative terms in $\mathbb{R}^{*+}$. Let $\phi \colon \mathbb{N} \to \mathbb{N} \times \mathbb{N}$ be a bijection. Then we have the related series

© Springer Nature Switzerland AG 2018
S. Shirali, *A Concise Introduction to Measure Theory*,
https://doi.org/10.1007/978-3-030-03241-8_4

$\sum_{p=1}^{\infty} a_{\phi(p)}$. Ostensibly, the terms in this series are the same as in the repeated sum $\sum_{n=1}^{\infty} \sum_{k=1}^{\infty} a_{n,k}$. To conclude from this that they have the same sum, we must be clear about how the nonnegativity of $a_{n,k}$ comes into the picture. We shall argue only that

$$\sum_{p=1}^{\infty} a_{\phi(p)} \leq \sum_{n=1}^{\infty} \sum_{k=1}^{\infty} a_{n,k},$$

as this is all that we shall need for now. A proof of the reverse inequality is left as Problem 4.1.P5.

Since each $\phi(p)$ is an element of $\mathbb{N} \times \mathbb{N}$, it is of the form (n_p, k_p), where $n_p \in \mathbb{N}$ and $k_p \in \mathbb{N}$. Given any natural number P, let

$$N = \max\{n_p : 1 \leq p \leq P\} \text{ and } K = \max\{k_p : 1 \leq p \leq P\}.$$

Then $1 \leq p \leq P$ implies $\phi(p) = (n, k)$ with $n \leq N$ and $k \leq K$. In view of the hypothesis that each $a_{n,k}$ is nonnegative, it now follows that

$$\sum_{p=1}^{P} a_{\phi(p)} \leq \sum_{n=1}^{N} \sum_{k=1}^{K} a_{n,k} \leq \sum_{n=1}^{\infty} \sum_{k=1}^{K} a_{n,k} \leq \sum_{n=1}^{\infty} \sum_{k=1}^{\infty} a_{n,k}.$$

Since this true for every natural number P and each $a_{\phi(p)}$ is nonnegative, the inequality in question must hold.

4.1.3. Definition. *For any subset A of $\mathbb{R}$, we define the* **Lebesgue outer measure** $m^*(A)$ *to be the infimum of the set of all sums*

$$\sum_{n=1}^{\infty} \ell(I_n),$$

where $\{I_n\}_{n\geq 1}$ is a sequence of open intervals that covers A; the infimum is taken over all such sequences. If all such sums are ∞, then the infimum is taken to be ∞.

4.1.4. Remark. It is a consequence of this definition that $m^*(A)$ is a nonnegative extended real number and has the property that, for any $\varepsilon > 0$, there exists a sequence $\{I_n\}_{n\geq 1}$ of open intervals I_n such that $A \subseteq \cup_{n=1}^{\infty} I_n$ and

$$\sum_{n=1}^{\infty} \ell(I_n) \leq m^*(A) + \varepsilon.$$

When it is finite, i.e. when $m^*(A) < \infty$, one may replace "$\leq$" in the above inequality by "$<$".

The following properties of Lebesgue outer measure are almost immediate from the definition.

4.1.5. Proposition. (a) $0 \leq m^*(A) \leq b - a$ *for all subsets A of* $[a, b]$.
(b) $m^*(\emptyset) = 0$.
(c) *If* $A \subseteq B$, *then* $m^*(A) \leq m^*(B)$.
(d) *For* $x \in \mathbb{R}$, *we have* $m^*(\{x\}) = 0$.
(e) *For* $x \in \mathbb{R}$ *and* $A \subseteq \mathbb{R}$, *we have* $m^*(A + x) = m^*(A)$, *where* $A + x = \{a + x : a \in A\}$.

Proof: (a) For any $\varepsilon > 0$, the sequence $\{I_n\}_{n \geq 1}$ of open intervals with $I_1 = (a - \varepsilon, b + \varepsilon)$ and $I_n = \emptyset$ for $n > 1$ covers any subset of $[a, b]$, and satisfies $0 \leq \sum_{n=1}^{\infty} \ell(I_n) = b - a + 2\varepsilon$. So by Definition 4.1.3, $0 \leq m^*(A) \leq b - a + 2\varepsilon$ for every $\varepsilon > 0$.

(b) The empty set $\emptyset$ is covered by the sequence $\{I_n\}_{n \geq 1}$ of open intervals with every $I_n = \emptyset$. Since $\sum_{n=1}^{\infty} \ell(I_n) = 0$, it follows that $m^*(\emptyset) = 0$.

(c) Any sequence of open intervals that covers B also covers A and hence the set of numbers of which $m^*(A)$ is the infimum contains the corresponding set for $m^*(B)$. Consequently, $m^*(A) \leq m^*(B)$.

(d) For every $\varepsilon > 0$, the sequence $\{I_n\}_{n \geq 1}$ of open intervals with $I_1 = (x - \varepsilon, x + \varepsilon)$ and $I_n = \emptyset$ for $n > 1$ covers $\{x\}$ and satisfies $\sum_{n=1}^{\infty} \ell(I_n) = 2\varepsilon$.

(e) Whenever $\{I_n\}_{n \geq 1}$ is a sequence of open intervals that covers A, $\{I_n + x\}_{n \geq 1}$ is a sequence of open intervals that covers $A + x$ and $\ell(I_n) = \ell(I_n + x)$ for each n. It follows that $m^*(A) \geq m^*(A + x)$. Similarly, $m^*(A + x) \geq m^*((A + x) + (-x)) = m^*(A)$. □

The above proposition says essentially that Lebesgue outer measure m^* is defined on the σ-algebra of all subsets of $\mathbb{R}$ and it has two special properties: (i) m^* of any set containing a single element is 0; and (ii) no set has m^* greater than the length of any interval it may be contained in. The same is true of the restriction of m^* to all subsets of an interval $[a, b]$. The rest of this section is devoted primarily to establishing the two further properties that (iii) m^* of an interval agrees with its length (Theorem 4.1.8) and (iv) m^* is "countably subadditive" (Proposition 4.1.9).

4.1.6. Example. Suppose that $A \subseteq \mathbb{R}$ is countable and infinite. Then $m^*(A)$ can be shown to be 0 as follows: The supposition means that the points of A can be arranged in a sequence $\{x_k\}_{k \geq 1}$. Given $\varepsilon > 0$, let I_k be an open interval of length $\varepsilon/2^k$ with midpoint x_k. The sequence $\{I_k\}_{k \geq 1}$ of open intervals then covers A and satisfies $\sum_{k=1}^{\infty} \ell(I_k) = \sum_{k=1}^{\infty} \varepsilon/2^k = \varepsilon$. So, $m^*(A) \leq \varepsilon$. A particular case is that $m^*(\mathbb{Q}) = 0$. It is of course trivial to show that $m^*(A) = 0$ if A is finite.

It follows from what has just been noted that a set $A \subseteq \mathbb{R}$ for which $m^*(A) > 0$ must be uncountable.

4.1.7. Proposition. *When a nonempty closed bounded interval is covered by a finite family of open intervals, the total length of the intervals in the family exceeds that of the closed interval.*

Proof: When there is only a single interval in the family, there is nothing to prove. Assume as induction hypothesis that whenever a nonempty closed bounded interval I is covered by a family of n open intervals, their total length exceeds that of I.

Now consider a family of $n+1$ open intervals covering a nonempty closed bounded interval $[a, b]$. One of them, call it (α, β), must contain a. Then $\alpha < a < \beta$ and it follows that $b - \alpha > b - a$. Consequently, it is sufficient to argue that the total length of all the $n+1$ intervals exceeds or equals $b - \alpha$.

If $\beta \geq b$, it follows that $\beta - \alpha \geq b - \alpha$, which means the length of the interval (α, β) alone exceeds or equals $b - \alpha$.

If $\beta < b$, we have $[\beta, b] \subseteq [a, b]$ but $[\beta, b] \cap (\alpha, \beta) = \emptyset$. It follows that the other n intervals in the family must cover $I = [\beta, b]$, which is a nonempty closed bounded interval. Therefore by the induction hypothesis, their total length exceeds $b - \beta$ and hence the total length of all the $n+1$ intervals exceeds $(b - \beta) + (\beta - \alpha) = b - \alpha$. □

4.1.8. Theorem. *If A is an interval, then its Lebesgue outer measure is the same as its length: $m^*(A) = \ell(A)$.*

Proof: As there is nothing to prove when $A = \emptyset$, we shall consider only nonempty A.

First suppose that A is a bounded interval with endpoints a and b, where $a \leq b$. Then $A \subseteq [a, b]$ and $\ell(A) = b - a$. Therefore it follows from Proposition 4.1.5(a) that $m^*(A) \leq \ell(A)$. We shall show that $m^*(A) \leq \ell(A)$ by considering the separate cases when A is closed and when A is not closed.

Consider the case when the nonempty bounded interval A is closed; call it $[\alpha, \beta]$. Let $\{I_n\}_{n \geq 1}$ be a sequence of open intervals that covers A and the family $\mathcal{B}$ be the range of the sequence. Since A is closed and bounded, the Heine–Borel Theorem provides a finite subfamily $\mathcal{A} \subseteq \mathcal{B}$ that covers A. By Proposition 4.1.7, the total length of the intervals in $\mathcal{A}$ is greater than $\beta - \alpha$. Now the length of each interval of $\mathcal{A}$ appears as a term at least once in the series $\sum_{n=1}^{\infty} \ell(I_n)$. It follows that the sum of this series is greater than or equal to the total length of the intervals in $\mathcal{A}$, and is therefore greater than $\beta - \alpha$. This establishes that

$$m^*(A) \geq \beta - \alpha = \ell(A)$$

in the case when A is closed.

Next, consider the case when the nonempty bounded interval A is not closed. When this is so, A must contain more than one point and, in fact, for every $\varepsilon > 0$, there must exist a closed interval $I \subseteq A$ such that

$$\ell(I) > \ell(A) - \varepsilon.$$

Since I is closed, we know from the case already established above that $m^*(I) = \ell(I)$, and since $I \subseteq A$, we also know that $m^*(A) \geq m^*(I)$. It follows that $m^*(A) > \ell(A) - \varepsilon$.

This inequality has been shown to hold for any $\varepsilon > 0$, and therefore we have $m^*(A) \geq \ell(A)$.

Finally, suppose A is not a bounded interval, which is to say, it is an unbounded interval and $\ell(A) = \infty$. Then it contains bounded intervals of arbitrarily great length and hence of arbitrarily great Lebesgue outer measure. From Proposition 4.1.5(c), it follows that $m^*(A) = \infty$. $\qquad\qquad\square$

The next proposition provides finite subadditivity of m^* upon taking $A_j = \emptyset$ for $j \geq 3$. In presenting its proof, it is customary to dispense with the song and dance of Remark 4.1.2(b), regarding it as obvious that the repeated sum in (4.2) below is equal to the sum of lengths of a suitable sequence of open intervals.

4.1.9. Proposition. *The Lebesgue outer measure m^* is* **countably subadditive***; that is, for every sequence $\{A_n\}_{n \geq 1}$ of subsets of $\mathbb{R}$,*

$$m^*\left(\bigcup_{n=1}^{\infty} A_n\right) \leq \sum_{n=1}^{\infty} m^*(A_n). \tag{4.1}$$

Proof: Consider any $\varepsilon > 0$. In accordance with Remark 4.1.4, corresponding to each $n \in \mathbb{N}$, there is a sequence $\{I_{n,k}\}_{k \geq 1}$ of open intervals $I_{n,k}$ such that $A_n \subseteq \bigcup_{k=1}^{\infty} I_{n,k}$ and

$$\sum_{k=1}^{\infty} \ell(I_{n,k}) \leq m^*(A_n) + \frac{\varepsilon}{2^n}.$$

Then the family $\{I_{n,k} : (n, k) \in \mathbb{N} \times \mathbb{N}\}$ covers $\bigcup_{n=1}^{\infty} A_n$ and

$$\sum_{n=1}^{\infty} \sum_{k=1}^{\infty} \ell(I_{n,k}) \leq \sum_{n=1}^{\infty} m^*(A_n) + \varepsilon. \tag{4.2}$$

Next, let $\phi : \mathbb{N} \to \mathbb{N} \times \mathbb{N}$ be a bijection. Then the sequence $\{I_{\phi(p)}\}_{p \geq 1}$ of open intervals covers $\bigcup_{n=1}^{\infty} A_n$ and hence

$$m^*\left(\bigcup_{n=1}^{\infty} A_n\right) \leq \sum_{p=1}^{\infty} \ell(I_{\phi(p)}).$$

It follows from Remark 4.1.2(b) and from (4.2) that

$$m^*\left(\bigcup_{n=1}^{\infty} A_n\right) \leq \sum_{n=1}^{\infty} m^*(A_n) + \varepsilon.$$

Since this holds for every $\varepsilon > 0$, the desired inequality (4.1) follows. $\qquad\qquad\square$

4.1.10. Exercise. If B is the set of all irrational numbers in $[a, b]$, show that $m^*(B) = b - a$.

Solution: By Proposition 4.1.5(a), $m^*(B) \leq b - a$. Let A be the set of all rational numbers in $[a, b]$. Then A is countable and hence $m^*(A) = 0$, as shown in Example 4.1.6. By Proposition 4.1.9, we have

$$m^*(A \cup B) \leq m^*(A) + m^*(B) = m^*(B) \leq b - a.$$

On the other hand, $A \cup B = [a, b]$, so that $m^*(A \cup B) = b - a$ by Theorem 4.1.8. Therefore the inequality displayed above yields $b - a \leq m^*(B) \leq b - a$.

Problem Set 4.1

4.1.P1. The outer measure m^* on $[a, b]$ is defined on the σ-algebra $\mathcal{F}$ of all subsets of $[a, b]$ and therefore all functions are measurable and have a distribution. Find the distribution $\tilde{f}$ of the function $f : [a, b] \to \mathbb{R}$ such that $f(x) = 1$ if x is irrational and 0 otherwise and hence find $\int_0^{\sup f} \tilde{f}$. (The reader is reminded that the Riemann integral $\int_a^b f$ does not exist.)

4.1.P2. For nonempty $A \subseteq \mathbb{R}$, the intersection I_A of all intervals containing A is a nonempty interval. Show that $m^*(A) \leq \sup\{|x - y| : x, y \in A\} = \ell(I_A)$.

4.1.P3. For any subset A of $[a, b]$ and any $\varepsilon > 0$, show that there exists a subset B of $[a, b]$ such that $A \subseteq B = \cup_{j=1}^{\infty} J_j$, $m^*(B) < m^*(A) + \varepsilon$ and each J_j is the intersection of an open interval with $[a, b]$.

4.1.P4. Consider the set C, known as the **Cantor set**, defined as $\cap_{n=1}^{\infty} C_n$, where $C_1 = \left[0, \frac{1}{3}\right] \cup \left[\frac{2}{3}, 1\right]$, and C_{n+1} is obtained from C_n by removing the open middle thirds of its closed intervals. Thus, $C_2 = \left[0, \frac{1}{9}\right] \cup \left[\frac{2}{9}, \frac{1}{3}\right] \cup \left[\frac{2}{3}, \frac{7}{9}\right] \cup \left[\frac{8}{9}, 1\right]$, and so on.

Show that $m^*(C) = 0$.

4.1.P5. Prove the reverse of the inequality established in Remark 4.1.2(b): Suppose that for each $n \in \mathbb{N}$, we have a series $\sum_{k=1}^{\infty} a_{n,k}$ of nonnegative terms in $\mathbb{R}^{*+}$. Let $\phi : \mathbb{N} \to \mathbb{N} \times \mathbb{N}$ be a bijection. Prove that

$$\sum_{p=1}^{\infty} a_{\phi(p)} \geq \sum_{n=1}^{\infty} \sum_{k=1}^{\infty} a_{n,k}.$$

Hint: First show that, given any $N, K \in \mathbb{N}$, there exists a $P \in \mathbb{N}$ such that $\sum_{n=1}^{N} \sum_{k=1}^{N} a_{n,k} \leq \sum_{p=1}^{P} a_{\phi(p)}$. Also use the fact that the limit of a finite sum of sequences is the sum of their separate limits.

4.2 Measure from Outer Measure

The outer measure m^* as defined in Sect. 4.1 is an example of a nonnegative, monotone, countably subadditive set function that is defined on all subsets of $\mathbb{R}$ and satisfies
$m^*(\emptyset) = 0$. However, it does not provide the extension we wanted in Remark 1.1.6
because we have not proved that m^* satisfies requirement (3) recorded there. It may
be noted that we did not necessarily want the extended total length to be defined
on *all* subsets. What we shall do now is select a σ-algebra of subsets such that the
restriction of m^* to the σ-algebra is a measure satisfying (1), (2) as well as (3) of that
remark. Of course, the selected σ-algebra will be shown to contain intervals, so that
the restriction is indeed an extension of total length.

The process of obtaining a σ-algebra such that the restriction of m^* to it is a
measure is actually quite general and can be carried out with a general set X instead
of $\mathbb{R}$ or $[a, b]$ as the underlying set. The reader may recall that before Sect. 4.1, our
σ-algebras consisted of subsets of a general set X. We return to that wider context.
Instead of m^*, we consider a nonnegative countably subadditive function defined on
the σ-algebra of *all* subsets of some (nonempty) set X, vanishing at $\emptyset$ and taking
"bigger values on bigger sets". Such an entity is called an *outer measure*.

4.2.1. Definition. *An **outer measure** μ^* on a nonempty set X is an $\mathbb{R}^{*+}$-valued
function defined on every subset of X and having the properties that*

(a) $\mu^*(\emptyset) = 0$ (vanishes at the empty set);
(b) $A \subseteq B \Rightarrow \mu^*(A) \leq \mu^*(B)$(monotone);
(c) *for every sequence* $\{A_n\}_{n \geq 1}$ *of subsets of* X,

$$\mu^*\left(\bigcup_{n=1}^{\infty} A_n\right) \leq \sum_{n=1}^{\infty} \mu^*(A_n) \quad \text{(countably subadditive)}.$$

Finite subadditivity follows exactly as it does for Lebesgue outer measure.

In the preceding section, we have shown that m^* has all the properties of Definition
4.2.1 and it is therefore an outer measure on $\mathbb{R}$. When X is finite, any sequence of
subsets must eventually repeat terms. So, a finitely subadditive nonnegative extended
real-valued function defined on all subsets is perforce countably subadditive and is
therefore an outer measure if it is monotone and vanishes at $\emptyset$. Examples of these
have been presented before and can now be marshaled to serve as illustrations of
outer measures.

It is trivial to see that when an outer measure on a set X is restricted to the subsets
of some $Y \subseteq X$, it becomes an outer measure on Y. Usually it is convenient to avoid
introducing a new name or symbol for a restriction. Thus we have Lebesgue measure
m^* on $\mathbb{R}$ as well as on its subsets, particularly intervals.

If there is a σ-algebra of subsets such that the restriction of μ^* to the σ-algebra
is countably additive, then it is a trivial deduction that the restriction is a measure.
One such σ-algebra is the one consisting of only the two subsets $\emptyset$ and X. We shall

discuss how to construct one that has a chance of being larger, and in the case of Lebesgue outer measure on an interval, is large enough to include all subintervals.

4.2.2. Definition. *When μ^* is an outer measure on X, a subset $E \subseteq X$ is said to be* μ^*-**measurable** *if*

$$\mu^*(A) = \mu^*(A \cap E) + \mu^*(A \cap E^c) \quad for\ every\ A \subseteq X.$$

An informal description of μ^*-measurability of a set E is that E breaks up every subset A of X in such a way as to make μ^* look like it were additive.

A set which is m^*-measurable, where m^* is Lebesgue outer measure, is said to be **Lebesgue measurable**.

4.2.3 Remarks. (a) By finite subadditivity, we have

$$\mu^*(A) \leq \mu^*(A \cap E) + \mu^*(A \cap E^c) \quad \text{for every } A \subseteq X.$$

Thus, to prove μ^*-measurability of E, it is enough to show the reverse of this inequality, which is known as the **Carathéodory condition**. It holds trivially when $A \subseteq E$ or $A \subseteq E^c$ and also when $\mu^*(A) = 0$ or ∞.

(b) If E is μ^*-measurable, so is its complement E^c.

(c) If $\mu^*(E) = 0$, then E is μ^*-measurable. This is because, for any $A \subseteq X$, nonnegativity and monotonicity of μ^* yield $0 \leq \mu^*(A \cap E) \leq \mu^*(E) = 0$ and $\mu^*(A \cap E^c) \leq \mu^*(A)$. So, $\mu^*(A) \geq \mu^*(A \cap E^c) + 0 = \mu^*(A \cap E^c) + \mu^*(A \cap E)$, which implies E is μ^*-measurable by (a) above. By monotonicity, every subset of E also has outer measure 0 and is therefore μ^*-measurable. Moreover by (b), it follows that the complement is μ^*-measurable. In particular, $\emptyset$ and X are always μ^*-measurable.

(d) A countable subset of $\mathbb{R}$ is m^*-measurable. This follows from (c) above and Example 4.1.6.

4.2.4. Example. Let X be the set consisting of two elements a and b, and μ^* be the outer measure given by $\mu^*(\emptyset) = 0$, $\mu^*(\{a\}) = 2$, $\mu^*(\{b\}) = 3$ and $\mu^*(X) = 4$. The subset $E = \{a\}$ is not μ^*-measurable, because when $A = X$, we find that $\mu^*(A \cap E) = \mu^*(E) = 2$, $\mu^*(A \cap E^c) = \mu^*(\{b\}) = 3$ and $\mu^*(A) = \mu^*(X) = 4 \neq 2 + 3$. It follows that the complement $E^c = \{b\}$ is also not μ^*-measurable. Thus the only μ^*-measurable subsets are $\emptyset$ and X.

4.2.5. Definition. *The collection of all μ^*-measurable subsets of X will be denoted by* $\mathcal{M}$.

We propose to show that $\mathcal{M}$ is a σ-algebra and that the restriction of μ^* to $\mathcal{M}$ is countably additive and hence a measure. This will be achieved in several stages.

4.2.6. Proposition. $\mathcal{M}$ *is closed under finite unions and finite intersections.*

Proof: It is sufficient to prove that $E_1, E_2 \in \mathcal{M} \Rightarrow E_1 \cup E_2 \in \mathcal{M}$. The rest follows by induction and the fact that $\mathcal{M}$ is closed under complements.

So, let E_1, $E_2 \in \mathcal{M}$. In order to prove that $E_1 \cup E_2 \in \mathcal{M}$, consider any subset A $\subseteq X$. Since $E_1 \in \mathcal{M}$ (which means it is μ^*-measurable), we have

$$\mu^*(A) = \mu^*(A \cap E_1) + \mu^*(A \cap E_1^c). \tag{4.3}$$

Since $E_2 \in \mathcal{M}$, applying Definition 4.2.2 with $A \cap E_1$ and $A \cap E_1^c$ in place of A, we get

$$\mu^*(A \cap E_1) = \mu^*(A \cap E_1 \cap E_2) + \mu^*(A \cap E_1 \cap E_2^c) \tag{4.4}$$

and

$$\mu^*(A \cap E_1^c) = \mu^*(A \cap E_1^c \cap E_2) + \mu^*(A \cap E_1^c \cap E_2^c). \tag{4.5}$$

Now,

$$(E_1 \cap E_2) \cup (E_1 \cap E_2^c) \cup (E_1^c \cap E_2) = E_1 \cup E_2$$

and hence

$$(A \cap E_1 \cap E_2) \cup (A \cap E_1 \cap E_2^c) \cup (A \cap E_1^c \cap E_2) = A \cap (E_1 \cup E_2). \tag{4.6}$$

Also,

$$(A \cap E_1^c \cap E_2^c) = A \cap (E_1 \cup E_2)^c. \tag{4.7}$$

On using (4.6), (4.7) and subadditivity of outer measure, we obtain

$$\mu^*(A \cap (E_1 \cup E_2)) + \mu^*(A \cap (E_1 \cup E_2)^c) \leq$$
$$\mu^*(A \cap E_1 \cap E_2) + \mu^*(A \cap E_1 \cap E_2^c) + \mu^*(A \cap E_1^c \cap E_2) + \mu^*(A \cap E_1^c \cap E_2^c).$$

On using (4.4), (4.5) and then (4.3), we further obtain

$$\mu^*(A \cap (E_1 \cup E_2)) + \mu^*(A \cap (E_1 \cup E_2)^c) \leq \mu^*(A \cap E_1) + \mu^*(A \cap E_1^c) = \mu^*(A).$$

In view of Remark 4.2.3(a), this completes the proof that $E_1 \cup E_2 \in \mathcal{M}$. $\qquad\square$

4.2.7. Exercise. (a) Let $B \subseteq A \cup E$ and $\mu^*(E) = 0$. Show that $A^c \cap B \in \mathcal{M}$.
(b) Let $B \subseteq A \cup E$, $B \in \mathcal{M}$ and $\mu^*(E) = 0$. Show that $A \cap B \in \mathcal{M}$.
(c) Let $B \subseteq A \cup E$, $A \subseteq B \cup E$, $B \in \mathcal{M}$ and $\mu^*(E) = 0$. Show that $A \in \mathcal{M}$.

Solutions: (a) $A^c \cap B \subseteq A^c \cap (A \cup E) = (A^c \cap A) \cup (A^c \cap E) = \emptyset \cup (A^c \cap E) \subseteq E$.
Since $\mu^*(E) = 0$, monotonicity implies $\mu^*(A^c \cap B) = 0$ and hence $A^c \cap B \in \mathcal{M}$ by Remark 4.2.3(c).

(b) $A \cap B = (A \cap B) \cup (B^c \cap B) = (A \cup B^c) \cap B = (A^c \cap B)^c \cap B$. Since $A^c \cap B \in \mathcal{M}$
by part (a) and $B \in \mathcal{M}$ by hypothesis, we have $A \cap B \in \mathcal{M}$ by Proposition
4.2.6.
(c) $A = A \cap (B^c \cup B) = (A \cap B^c) \cup (A \cap B)$. Since $A \cap B^c \in \mathcal{M}$ by part (a) and
$A \cap B \in \mathcal{M}$ by part (b), we get $A \in \mathcal{M}$ by Proposition 4.2.6.

To prove closure under countable unions, we shall use one more result, namely,
that the restriction of μ^* to $\mathcal{M}$ is finitely additive.

4.2.8. Proposition. *If $E_1, E_2, \ldots, E_n$ is a finite sequence of disjoint μ^*-measurable
subsets of X and $A \subseteq X$ is any subset whatsoever, then*

$$\mu^*(A \cap \bigcup_{j=1}^{n} E_j) = \sum_{j=1}^{n} \mu^*(A \cap E_j). \tag{4.8}$$

In particular, if $A = X$, then

$$\mu^*(\bigcup_{j=1}^{n} E_j) = \sum_{j=1}^{n} \mu^*(E_j).$$

Proof: We shall prove this by induction on n. For $n = 1$, the equality is obvious.
Suppose for some $n \in \mathbb{N}$ that (4.8) holds whenever $E_1, E_2, \ldots, E_n$ is a finite sequence
of disjoint μ^*-measurable subsets of X and $A \subseteq X$ is arbitrary (induction hypothesis).
In order to arrive at the validity of (4.8) with $n+1$ in place of n, first note that $\mu^*(E)$-
measurability of E_{n+1} entails that

$$\mu^*(A \cap \bigcup_{j=1}^{n+1} E_j) = \mu^*((A \cap \bigcup_{j=1}^{n+1} E_j) \cap E_{n+1}) + \mu^*((A \cap \bigcup_{j=1}^{n+1} E_j) \cap E_{n+1}^c). \tag{4.9}$$

On the basis of the disjointness of $E_1, E_2, \ldots, E_{n+1}$, we have

$$(A \cap \bigcup_{j=1}^{n+1} E_j) \cap E_{n+1} = A \cap E_{n+1} \text{ and } (A \cap \bigcup_{j=1}^{n+1} E_j) \cap E_{n+1}^c = A \cap \bigcup_{j=1}^{n} E_j.$$

In conjunction with (4.9), this implies

$$\mu^*(A \cap \bigcup_{j=1}^{n+1} E_j) = \mu^*(A \cap E_{n+1}) + \mu^*(A \cap \bigcup_{j=1}^{n} E_j).$$

The induction hypothesis now yields equality (4.8) with $n+1$ in place of n.   □

4.2.9. Theorem. $\mathcal{M}$ *is a σ-algebra.*

Proof: From Remark 4.2.3(b) and (c), we know that $\mathcal{M}$ is closed under complements and that $\emptyset$ is an element of $\mathcal{M}$. It remains only to show that, if $E_j \subseteq X$ are all $\mu^*(E)$-measurable ($j \in \mathbb{N}$), then $\cup_{j=1}^{\infty} E_j$ is also μ^*-measurable.

Let $F_1 = E_1$ and $F_k = E_k \cap \left(\cup_{j=1}^{k-1} E_j \right)^c = E_k \backslash (E_1 \cup \cdots \cup E_{k-1})$ for $k > 1$. Then each $F_k \subseteq E_k$ and hence $\mu^*(F_k) \le \mu^*(E_k)$. Furthermore, the sets F_k are all disjoint and μ^*-measurable.

The definition of F_k also implies that $\cup_{j=1}^{n} E_j = \cup_{j=1}^{n} F_j$ for all $n \in \mathbb{N}$ and that $\cup_{j=1}^{\infty} E_j = \cup_{j=1}^{\infty} F_j$. Therefore, if we show

$$\mu^*(A) \ge \mu^*(A \cap (\bigcup_{j=1}^{\infty} F_j)) + \mu^*(A \cap (\bigcup_{j=1}^{\infty} F_j)^c) \tag{4.10}$$

for every $A \subseteq X$, the desired μ^*-measurability of $\cup_{j=1}^{\infty} E_j$ will follow.

For any positive integer p, the finite union $\cup_{j=1}^{p} F_j$ is known to be μ^*-measurable by Proposition 4.2.6. Therefore we know that

$$\mu^*(A) = \mu^*(A \cap (\bigcup_{j=1}^{p} F_j)) + \mu^*(A \cap (\bigcup_{j=1}^{p} F_j)^c). \tag{4.11}$$

Since the sets F_j are disjoint and μ^*-measurable,

$$\mu^*(A \cap (\bigcup_{j=1}^{p} F_j)) = \sum_{j=1}^{p} \mu^*(A \cap F_j) \tag{4.12}$$

by finite additivity of μ^* on $\mathcal{M}$ (Proposition 4.2.8). Also, since $\left(\cup_{j=1}^{p} F_j \right)^c \supseteq \left(\cup_{j=1}^{\infty} F_j \right)^c$ and μ^* is monotone, we have

$$\mu^*(A \cap (\bigcup_{j=1}^{p} F_j)^c) \ge \mu^*(A \cap (\bigcup_{j=1}^{\infty} F_j)^c). \tag{4.13}$$

Now, (4.11), (4.12) and (4.13) together lead to

$$\mu^*(A) \ge \sum_{j=1}^{p} \mu^*(A \cap F_j) + \mu^*(A \cap (\bigcup_{j=1}^{\infty} F_j)^c).$$

This is valid for all p while the sequence $\{\sum_{j=1}^{\infty} \mu^*(A \cap F_j)\}_{p \ge 1}$ has supremum $\sum_{j=1}^{\infty} \mu^*(A \cap F_j)$. Therefore

$$\mu^*(A) \geq \sum_{j=1}^{\infty} \mu^*(A \cap F_j) + \mu^*(A \cap (\bigcup_{j=1}^{\infty} F_j)^c)$$

$$\geq \mu^*(\bigcup_{j=1}^{\infty} (A \cap F_j)) + \mu^*(A \cap (\bigcup_{j=1}^{\infty} F_j)^c)$$

by countable subadditivity of μ^*. Since $\cup_{j=1}^{\infty} (A \cap F_j) = A \cap (\cup_{j=1}^{\infty} F_j)$, the inequality just derived asserts the same thing as (4.10). □

As was mentioned in Sect. 2.1, a set is said to be $\mathcal{F}$-measurable in the context of a σ-algebra $\mathcal{F}$ if it belongs to $\mathcal{F}$. Being an $\mathcal{M}$-measurable set is the same as being a μ^*-measurable set.

4.2.10. Exercise. Let $g:X \to \mathbb{R}$ be an $\mathcal{M}$-measurable function (which means $X(g > t) \in \mathcal{M}$ for every real t) and suppose $f:X \to \mathbb{R}$ satisfies $\mu^*(X(g \neq f)) = 0$. Show that f is $\mathcal{M}$-measurable.

Solution: Let $E = X(g \neq f)$. Then $\mu^*(E) = 0$ by hypothesis. For any t, let $A = X(f > t)$ and $B = X(g > t)$. Then $B \in \mathcal{M}$ by hypothesis. Also,

$$x \in A \quad \Rightarrow \quad f(x) > t \quad \Rightarrow \quad g(x) > t \text{ or } g(x) \neq f(x) \quad \Rightarrow \quad x \in B \text{ or } x \in E,$$

so that $A \subseteq B \cup E$. A similar argument shows that $B \subseteq A \cup E$. It follows by Exercise 4.2.7(c) that $A \in \mathcal{M}$, i.e. $X(f > t) \in \mathcal{M}$. Thus f is $\mathcal{M}$-measurable.

Recall that when μ^* is m^*, whether on all subsets of $\mathbb{R}$ or of $[a, b]$, an $\mathcal{M}$-measurable function is said to be Lebesgue measurable. Thus $f:[a, b] \to \mathbb{R}$ is Lebesgue measurable if and only if, for every real t, the set $\{x \in [a, b] : f(x) > t\}$ is m^*-measurable.

Theorem. 4.2.11. *The outer measure μ^* is countably additive on the σ-algebra $\mathcal{M}$ of μ^*-measurable subsets of X and its restriction to $\mathcal{M}$ is therefore a measure.*

Proof: Let $\{E_j\}_{j \geq 1}$ be a sequence of disjoint μ^*-measurable subsets. By monotonicity of μ^* and its finite additivity proved in Proposition 4.2.8, it is true for an arbitrary positive integer p that

$$\mu^*(\bigcup_{j=1}^{\infty} E_j) \geq \mu^*(\bigcup_{j=1}^{p} E_j) = \sum_{j=1}^{p} \mu^*(E_j).$$

Therefore

$$\mu^*(\bigcup_{j=1}^{\infty} E_j) \geq \sum_{j=1}^{\infty} \mu^*(E_j).$$

The reverse inequality follows from the countable subadditivity of μ^*. □

Problem Set 4.2

4.2.P1. Let μ^* be a finite outer measure on a set X and $A \subseteq B$ be subsets of X. Show that

(a) if $\mu^*(B \backslash A) = 0$, then $\mu^*(B) = \mu^*(A)$;
(b) if $\mu^*(B) = \mu^*(A)$ and $A \in \mathcal{M}$, then $\mu^*(B \backslash A) = 0$.

State the analog for three sets.

4.2.P2. Show that, if E_1 and E_2 are μ^*-measurable, where μ^* is an outer measure, then

$$\mu^*(E_1 \cup E_2) + \mu^*(E_1 \cap E_2) = \mu^*(E_1) + \mu^*(E_2).$$

4.2.P3. What has been called "μ" in Example 2.1.3(b) is easily seen to be an outer measure μ^*. Which subsets are μ^*-measurable?

4.2.P4. Let $X = \{a, b, c\}$ and $\mathcal{F}$ be the σ-algebra consisting of the subsets $\{a, b\}, \{c\}$ besides $\emptyset$ and X. Then $\mu \colon \mathcal{F} \to R$, where $\mu\emptyset = \mu\{a, b\} = 0, \mu\{c\} = \mu X = 1$, is clearly a measure on X. For an arbitrary subset $A \subseteq X$, define $\mu^*(A)$ to be the infimum of all numbers $\sum_{n=1}^{\infty} \mu(I_n)$, where $\{I_n\}_{n \geq 1}$ is a sequence of sets in $\mathcal{F}$ that covers A (the infimum is taken over all such sequences, but only finitely many need be used). Find $\mu^*(A)$ for each of the eight subsets of X. It can be checked easily that μ^* is an outer measure; find the μ^*-measurable subsets.

4.2.P5. Let $\mathcal{H}$ be a family of subsets of a nonempty set X such that $\emptyset \in \mathcal{H}$ and suppose $\alpha \colon \mathcal{H} \to \mathbb{R}$ is nonnegative-valued, satisfying $\alpha\emptyset = 0$. Assume that some sequence $\{H_n\}_{n \geq 1}$ of sets belonging to $\mathcal{H}$ covers X, hence also every subset of X. Show that

$$\alpha^* A = \inf\{\sum_{n=1}^{\infty} \alpha(I_n) \colon \text{every } I_n \in \mathcal{H} \text{ and } A \subseteq \bigcup_{n \geq 1} I_n\}$$

defines an outer measure on X.

4.2.P6. For $X = \{a, b, c, d\}$, a set of four elements, let

$$\mathcal{H} = \{\emptyset, \{d\}, \{a, b\}, \{a, c\}, \{a, b, c\}, X\}$$

and $\alpha \colon \mathcal{H} \to \mathbb{R}$ be defined as

$$\alpha\emptyset = 0, \quad \alpha\{d\} = 3, \quad \alpha\{a, b\} = 0, \quad \alpha\{a, c\} = 3, \quad \alpha\{a, b, c\} = 2 \quad \text{and} \quad \alpha X = 2.$$

Find $\alpha^*(A)$ for all the sixteen subsets of X and determine which ones are α^*-measurable.

4.2.P7. Let $\mathcal{H}$ be a σ-algebra of subsets of X and $(X, \mathcal{H}, \alpha)$ be a measure space. Any sequence of sets in $\mathcal{H}$ having X as one of its terms covers X. Show that for the outer measure α^* obtained as in Problem 4.2.P5, every set in $\mathcal{H}$ is α^*-measurable. Is the converse true?

4.2.P8. For infinite $A \subseteq \mathbb{N}$, let $\alpha(A)=1$ and for finite $A \subseteq \mathbb{N}$, let $\alpha(A) = \sum_{j=1}^{n} \left(1/2^j\right)$, where n is the number of elements in A. (If $A = \emptyset$, the sum becomes "empty" and is to be interpreted as usual to be 0.) Show that α is an outer measure on $\mathbb{N}$ and find all α-measurable subsets of $\mathbb{N}$.

4.2.P9. (Cf. Problem 4.3.P7) Let $\mathcal{F}$ be a σ-algebra of subsets of a nonempty set X and suppose the set function $\alpha: \mathcal{F} \to \mathbb{R}^{*+}$ vanishes at the empty set, is monotone and countably subadditive. (In other words α has all the properties of an outer measure except that its domain is $\mathcal{F}$.) For any $A \in \mathcal{F}$, put

$$\mu(A) = \sup\{\sum_{j=1}^{\infty} \alpha(E_j): \bigcup_{j=1}^{\infty} E_j = A, E_j \cap E_k = \emptyset \text{ whenever } j \neq k, \text{ every } E_j \in \mathcal{F}\}.$$

(a) Show that μ is a measure and that, for all $A \in \mathcal{F}$, we have (i) $\mu(A) \geq \alpha(A)$ (ii) $\mu(A) = 0 \Leftrightarrow \alpha(A) = 0$.
(b) If α is as in Problem 4.2.P8, what is μ?

4.3 Lebesgue Measure

In the light of what has been shown so far, the restriction of Lebesgue outer measure m^* on an interval I to the σ-algebra $\mathcal{M}$ of m^*-measurable sets of the interval is a measure on that σ-algebra. It is known as **Lebesgue measure (on I)** and we shall denote it by the letter m. Thus the triplet $(I, \mathcal{M}, m)$ is a measure space in accordance with Definition 2.1.2. The measure space integral with reference to Lebesgue measure is called the **Lebesgue integral**. We shall mostly be concerned with the cases when I is either $\mathbb{R}$ or a closed bounded interval $[a, b]$.

We have yet to prove that the σ-algebra of Lebesgue measurable sets contains all subintervals of $[a, b]$, failing which, Lebesgue measure m is not an extension of the total length concept. This is all that remains to be done for achieving what was envisaged in Remark 1.1.6 and we are now in a position to bestow the idea with entelechy.

4.3.1. Theorem. *When m^* is Lebesgue outer measure on $\mathbb{R}$ or on an interval $[a, b]$, every subinterval is m^*-measurable. Furthermore, it has Lebesgue measure equal to its length.*

Proof: It is sufficient to prove that any subinterval I of the form $(c, b]$ is m^*-measurable. Let A be any subset of $[a, b]$ and let $A_1 =A \cap I$ and $A_2 =A \cap I^c = A \cap [a, c]$. We must show that

$$m^*(A) \geq m^*(A_1) + m^*(A_2). \tag{4.14}$$

We need consider only the case when $m^*(A)$ is finite. The definition of m^* guarantees that, for every $\varepsilon > 0$, there exists a sequence $\{I_n\}_{n \geq 1}$ of open intervals such that $A \subseteq \cup_{n=1}^{\infty} I_n$ and

$$\sum_{n=1}^{\infty} \ell(I_n) \leq m^*(A) + \varepsilon.$$

Let $I'_n = I_n \cap I$ and $I''_n = I_n \cap I^c$. Then I'_n and I''_n are subintervals of $[a, b]$ (some of which may be empty) and by Problem 1.1.P6, $\ell(I_n) = \ell(I'_n) + \ell(I''_n)$. Now, the length of a subinterval is the same as its Lebesgue outer measure according to Theorem 4.1.8. Therefore

$$\ell(I_n) = m^*(I'_n) + m^*(I''_n).$$

Since $A \subseteq \cup_{n=1}^{\infty} I_n$, we have $A_1 \subseteq \cup_{n=1}^{\infty} I'_n$ and $A_2 \subseteq \cup_{n=1}^{\infty} I''_n$. Therefore

$$m^*(A_1) \leq m^*(\bigcup_{n=1}^{\infty} I'_n) \leq \sum_{n=1}^{\infty} m^*(I'_n)$$

and

$$m^*(A_2) \leq m^*(\bigcup_{n=1}^{\infty} I''_n) \leq \sum_{n=1}^{\infty} m^*(I''_n).$$

Hence

$$m^*(A_1) + m^*(A_2) \leq \sum_{n=1}^{\infty} m^*(I'_n) + \sum_{n=1}^{\infty} m^*(I''_n) = \sum_{n=1}^{\infty} [m^*(I'_n) + m^*(I''_n)]$$

$$= \sum_{n=1}^{\infty} \ell(I_n) \leq m^*(A) + \varepsilon.$$

Since this is true for every $\varepsilon > 0$, we conclude that (4.14) holds.

It follows by Theorem 4.1.8 that the Lebesgue measure of an interval equals its length. $\qquad \square$

Note that it is a consequence of the preceding Theorem and Proposition 4.2.8 that there is an unambiguously determined total length of a union E of a finite family of disjoint bounded intervals, namely, $m(E)$. Thus, we have another proof of the result of Problem 1.1.P9.

4.3.2. Corollary. *A nonnegative-valued step function on* $[a, b]$ *is Lebesgue measurable and its Riemann integral equals its Lebesgue integral.*

Proof: The Lebesgue measurability follows from the first assertion of Theorem 4.3.1. The second assertion thereof implies that the distribution function as understood in the context of the Lebesgue measure space (see the paragraph immediately after Definition 2.1.4) is the same as the distribution as understood for the purpose of Proposition 1.1.3. Therefore the equality claimed in the present corollary is just a restatement of Proposition 1.1.3.   □

4.3.3. Exercise. For the functions f_n of Exercise 1.1.5 and their limit f, show that $\int f$ exists and equals $\lim_{n \to \infty} \int f_n$, although the Riemann integral $\int_0^1 f$ does not exist.

Solution: Consider any n. As noted in Exercise 1.1.5, when $0 \le t < 1$, the t-cut $S_n = \{x \in [0,1] : f_n(x) > t\}$ is a finite union of disjoint intervals with total length 1; obviously, it is empty when $1 \le t$. So, $m(S_n)$ equals 1 for all $t \in [0, 1)$ and equals 0 for all $t \in [1, \infty)$. Hence $\tilde{f}_n$ equals 1 on $[0, 1)$ and equals 0 on $[1, \infty)$ for every n, which implies $\int f_n = 1$ for every n. It was also noted that the t-cut $S = \{x \in [0, 1] : f(x) > t\}$ is the set of all irrational numbers in $[0, 1]$ when $0 \le t < 1$; obviously, it is empty when $1 \le t$. By finite additivity of m and Example 4.1.6, it follows that $m(S)$ equals 1 for all $t \in [0, 1)$ and equals 0 for all $t \in [1, \infty)$. Thus $\tilde{f} = \tilde{f}_n$ for all n, which implies $\int f = 1$. Therefore $\int f = \lim_{n \to \infty} \int f_n$ trivially. Since $f(x)$ equals 1 if x is irrational and equals 0 otherwise, it follows that the Riemann integral $\int_0^1 f$ does not exist.

In Example 1.1.4, we considered two Riemann integrable functions $f : [a, b] \to \mathbb{R}$ and computed $\int_a^b f$ as well as the integral $\int_0^{\sup f} \tilde{f}$, which was what we can now call the Lebesgue integral $\int f$. In both cases, we found that $\int_a^b f = \int f$. We shall now show that this must always happen for a Riemann integrable function. In particular, a continuous function is Lebesgue measurable.

4.3.4. Theorem. *Let* $f : [a, b] \to \mathbb{R}$ *be Riemann integrable. Then it is Lebesgue measurable and its Lebesgue integral agrees with its Riemann integral:*

$$\int\limits_{[a,b]} f = \int f = \int\limits_a^b f.$$

Proof: It is sufficient to consider only nonnegative f. For each $n \in \mathbb{N}$, there exists a partition P_n of $[a, b]$ such that the lower sum $L(f, P_n)$ and upper sum $U(f, P_n)$ satisfy

$$\int\limits_a^b f - \tfrac{1}{n} < L(f, P_n) \le U(f, P_n) < \int\limits_a^b f + \tfrac{1}{n}.$$

Let s_n [resp. t_n] be the function on $[a, b]$ such that, on each left closed subinterval of P_n it agrees with the inf [resp. sup] of f on that subinterval and is 0 at b. By refining each partition P_n to include all the points of P_{n-1} (except when $n = 1$, of course), we can ensure (as noted in the opening paragraph of Sect. 1.1) that

$$0 \le s_n \le s_{n+1} \le f \le t_{n+1} \le t_n. \tag{4.15}$$

Then $s_n \le f \le t_n$ and also, $L(f, P_n) = \int_a^b s_n$ and $U(f, P_n) = \int_a^b t_n$. Hence, by Corollary 4.3.2, we have

$$\int\limits_a^b f - \frac{1}{n} < \int s_n \le \int t_n < \int\limits_a^b f + \frac{1}{n}, \tag{4.16}$$

where the two integrals in the middle are Lebesgue integrals. By (4.15), there exist measurable functions s and t on $[a, b]$ such that $\lim_{n \to \infty} s_n(x) = s(x)$ and $\lim_{n \to \infty} t_n(x) = t(x)$ for every $x \in [a, b]$ and they satisfy

$$s(x) \le f(x) \le t(x) \quad \text{for every } x \in [a, b]. \tag{4.17}$$

By the Dominated Convergence Theorem 3.2.6 (take g there to be the function t_1 here), we have $\lim_{n \to \infty} \int s_n = \int s$ and $\lim_{n \to \infty} \int t_n = \int t$. It follows from (4.16) that

$$\int\limits_a^b f \le \int s \le \int t \le \int\limits_a^b f.$$

Thus

$$\int s = \int\limits_a^b f = \int t. \tag{4.18}$$

Now, the nonnegative function $t - s$ is measurable and satisfies $\int (t - s) + \int s = \int t$, and hence $\int (t - s) = 0$. By Problem 2.3.P1, it follows that the set $X(t > s) = X(t \ne s)$ has Lebesgue measure 0. By (4.17), $X(f \ne s) \subseteq X(t \ne s)$ and therefore has Lebesgue outer measure 0. Using Exercise 4.2.10, we infer that f is Lebesgue measurable. Hence it follows from (4.17) and (4.18) that $\int f = \int_b^a f$. $\qquad\square$

4.3.5. Example. If one were to consider the possibility of extending the above theorem to the case of an unbounded interval, one would have to replace the Riemann integral by an improper integral. We exhibit below a function that has an improper integral over $[0, \infty)$ but no Lebesgue integral over $[0, \infty)$. Proposition 4.4.2 is relevant here.

Suppose $f : [0, \infty) \to \mathbb{R}$ is defined to be equal to $\frac{(-1)^{k-1}}{k}$ on $[k - 1, k)$ for every $k \in \mathbb{N}$. In other words, $f(x) = \frac{(-1)^{\lfloor x \rfloor}}{1 + \lfloor x \rfloor}$ for every $x \in [0, \infty)$, where $\lfloor x \rfloor$ means the integer part of x. It is useful to bear in mind that this function is positive on an interval $[k-1, k)$ if k is odd and negative if k is even; this provides a handle on what f^+ and f^- are.

Regarding the improper integral $\int_0^\infty f$, i.e. $\lim_{B\to\infty} \int_0^B f$, consider any $B > 0$ and let $N = \lfloor B \rfloor$, the integer part of B, so that $0 \le B - N < 1$ and $N \to \infty$ as $B \to \infty$. The restriction of f to $[0, B]$ is a step function and

$$\int_0^B f = \sum_{k=1}^{N} \frac{(-1)^{k-1}}{k} + (B-N)\frac{(-1)^N}{1+N}.$$

As $0 \le B - N < 1$, this lies in an interval having endpoints $\sum_{k=1}^{N} \frac{(-1)^{k-1}}{k}$ and $\sum_{k=1}^{N+1} \frac{(-1)^{k-1}}{k}$ and these endpoints are partial sums of a series that is well known to converge. Since $N \to \infty$ as $B \to \infty$, it follows that $\int_0^\infty f$ converges to the sum of the series.

Regarding Lebesgue integrability of f over $[0, \infty)$, we shall argue that $\int f^+ = \int f^- = \infty$, which means the function has no Lebesgue integral. The function f^+ is the limit of the increasing sequence of functions obtained by multiplying it to the characteristic function of $[0, 2n-1]$. Therefore by the Monotone Convergence Theorem 2.3.2, its integral is the limit of the sequence of integrals of the products. Careful inspection of the function f^+ shows that such a product is a simple function with integral $\sum_{k=1}^{n} \frac{1}{2k-1}$. This is the partial sum of a series that is well known to diverge to ∞, and it follows that $\int f^+ = \infty$. Analogously, the function f^- is the limit of the increasing sequence of functions obtained by multiplying it by the characteristic function of $[0, 2n]$, and the integral of such a product is $\sum_{k=1}^{n} \frac{1}{2k}$. This too is the partial sum of a series that is well known to diverge to ∞, and it follows that $\int f^- = \infty$.

4.3.6. Exercise. The sequence $\{f_n\}_{n\ge 1}$ of functions defined on $X = [0, 1]$ by $f_n(x) = 1 - nx^{n-1}$ converges for each $x \in [0, 1)$ to $f(x)$, where f is the function that equals 1 everywhere on $[0, 1]$. Show that (a) for each $x \in [0, 1)$, there exists an $N \in \mathbb{N}$ such that $n+1 \ge n \ge N$ implies $f_n(x) \le f_{n+1}(x)$ but (b) $\int_{[0,1]} f \ne \lim_{n\to\infty} \int_{[0,1]} f_n$.

Solution: To prove (a), it is enough to show that, for each $x \in [0, 1)$, there exists an $N \in \mathbb{N}$ such that $n + 1 \ge n \ge N$ implies $nx^{n-1} \ge (n+1)x^n$. If x is 0, then $N = 1$ satisfies the requirement. Let $0 < x < 1$. Then for any $n \in \mathbb{N}$, we have $(n+1)x^n \ne 0$ and $nx^{n-1}/(n+1)x^n = n/(n+1)x$, which can be easily proved to be 1 or greater as long as $n \ge x/(1-x)$. To arrive at (b), we note that by Theorem 4.3.4, $\int_{[0,1]} f_n$ agrees with its Riemann integral, which is found to be 0 for every n, while $\int_{[0,1]} f = 1$.

4.3.7. Proposition. *There does not exist a measure α which is defined on the family of all subsets of $\mathbb{R}$ and which also satisfies*

(i) $\ell(I) \le \alpha(I) < \infty$ *for every bounded interval I,*
(ii) $\alpha(A) = \alpha\{a + x : a \in A\}$ *for every $A \subseteq \mathbb{R}$ and every $x \in \mathbb{R}$.*

Proof: Suppose such a measure α exists.

Define an equivalence relation ~ in $\mathbb{R}$ by setting

$$x \sim y \Leftrightarrow x - y \in \mathbb{Q}.$$

Clearly, the relation '~' partitions $\mathbb{R}$ into disjoint equivalence classes E_α. Any two elements of the same class differ by a rational number while those of two different classes differ by an irrational number. Moreover $\cup_\alpha E_\alpha = \mathbb{R}$. For any real number x, let $\lfloor x \rfloor$ denote its integer part. Then

$$x \sim (x - \lfloor x \rfloor) \quad \text{and} \quad (x - \lfloor x \rfloor) \in [0, 1).$$

We construct a set A using the Axiom of Choice to consist of one element from each set $E_\alpha \cap [0, 1)$. Observe that $\cup_\alpha E_\alpha \cap [0, 1) = [0, 1)$. Let the sequence $\{x_k\}_{k \geq 1}$ be an enumeration of the rationals in $(-1, 1)$ and set

$$A_n = A + x_n = \{a + x_n : a \in A\}.$$

Any two among sets $\{A_n\}_{n \geq 1}$ must be disjoint. In fact, if there is a point $x \in A_m \cap A_n$, then $x = x' + x_m = x'' + x_n$, where x' and x'' are points in A. This implies $x' - x'' = x_n - x_m$ is a rational number, which forces $x' = x''$ and hence $x_m = x_n$; this implies $m = n$. Moreover, the following inclusions hold, as we shall show:

$$(0, 1) \subseteq \bigcup_{n=1}^{\infty} A_n \subseteq (-1, 2).$$

To justify the first inclusion, consider any $x \in (0, 1)$. As observed above, $\cup_\alpha E_\alpha = \mathbb{R}$, and therefore $x \in E_\alpha$ for some α; let x' be the unique element of $E_\alpha \cap [0,1)$ that belongs to A. Then $x - x'$ is a rational number; moreover, x and x' both belong to $[0,1)$ and consequently, $x - x' \in (-1, 1)$. This means $x - x' = x_n$ for some n and hence $x \in A_n$. The second inclusion follows from the fact that $A \subseteq [0,1)$ and $x_n \in (-1,1)$.

As α is assumed to be defined on all subsets, it follows from the inclusions established in the preceding paragraph and (i) of the hypothesis that

$$1 = \ell((0, 1)) \leq \alpha((0, 1)) \leq \alpha\left(\bigcup_{n=1}^{\infty} A_n\right) \leq \alpha((-1, 2)) < \infty.$$

Together with (ii) of the hypothesis and the disjointness of the sets A_n, the first part of the above inequality leads to

$$\alpha\left(\bigcup_{n=1}^{\infty} A_n\right) = \sum_{n=1}^{\infty} \alpha(A_n) = \sum_{n=1}^{\infty} \alpha(A) = \infty,$$

which contradicts the second part of the above inequality. The contradiction establishes that there cannot exist a measure α which is defined on all subsets of $\mathbb{R}$ and which also satisfies (i) and (ii). $\qquad\Box$

4.3.8. Corollary. *There exists a subset of $\mathbb{R}$ that is not Lebesgue measurable.*

Proof: Otherwise Lebesgue measure would be the kind of measure that cannot exist according to Proposition 4.3.7. $\qquad\Box$

According to Problem 4.1.P4, the Cantor set C defined there has Lebesgue outer measure 0. It now follows in view of Remark 4.2.3(c) that C is Lebesgue measurable with Lebesgue measure 0. We shall next show that the set is uncountable, whereupon it will have been found to be an example of an uncountable set of Lebesgue measure 0. In Sect. 8.2, we shall show that there is a bijection between the interval [0, 1] and the Cantor set.

Suppose the elements of C can be arranged in a sequence $x_1, x_2,\dots$. Note the consequence that any number that is distinct from x_n for every n cannot be in C. Since the two closed intervals $\left[0, \frac{1}{3}\right]$ and $\left[\frac{2}{3}, 1\right]$ comprising C_1 are disjoint, one of them, which we shall name as I_1, excludes (i.e. does not contain) the number x_1. Similarly, one of the two subintervals of I_1 among the 2^2 that comprise C_2 excludes x_2; being a subinterval of I_1, it also excludes x_1. Name it as I_2. In general, suppose I_n is one among the 2^n intervals comprising C_n that excludes $x_1, x_2,\dots,x_n$. Then one of its two subintervals that are among the 2^{n+1} comprising C_{n+1} excludes $x_1, x_2,\dots,x_{n+1}$. Name it as I_{n+1}. We thus have a nested sequence of closed intervals I_n such that $I_n \subseteq C_n$ for every n. By the Nested Interval Theorem (see Theorem 3.3.6 of Shirali and Vasudeva [9]), there exists some x in the intersection $\cap_{n=1}^{\infty} I_n \subseteq \cap_{n=1}^{\infty} C_n = C$. Since for every n, the interval I_n excludes all the numbers $x_1, x_2,\dots, x_n$, the number x must be distinct from x_n for every n. As noted at the beginning, it follows that x cannot be in C. This is a contradiction, proving that C is uncountable.

Problem Set 4.3

4.3.P1. Suppose that A is a subset of $\mathbb{R}$ with the property that, for every $\varepsilon > 0$, there exist Lebesgue measurable sets B and C such that $B \subseteq A \subseteq C$ and $m(C \cap B^c) < \varepsilon$. (Here m denotes Lebesgue measure as usual.) Show that A is Lebesgue measurable.

4.3.P2. Let $A \subseteq [a, b]$ and m denote Lebesgue measure. Show that there exists a Lebesgue measurable set $B \subseteq [a, b]$ such that $m(B) = m^*(B) = m^*(A)$ and $A \subseteq B$. The same is true with $\mathbb{R}$ in place of $[a, b]$.

4.3.P3. Show that the Lebesgue outer measure m^* is inner continuous. In other words, if $\{S_n\}_{n\geq 1}$ is a sequence of sets such that $S_1 \subseteq S_2 \subseteq \cdots$, then $m^*\left(\cup_{n=1}^{\infty} S_n\right) = \lim_{n\to\infty} m^*(S_n)$.

4.3.P4. Let $E \subseteq [a, b]$. Show that E is a Lebesgue measurable subset of $[a, b]$ if and only if it is a Lebesgue measurable subset of $\mathbb{R}$.

4.3.P5. For any $S \subseteq \mathbb{R}$ and any $x \in \mathbb{R}$, let $S+x$ denote the set $\{s + x : s \in S\}$. Prove the following:

(a) If E is a Lebesgue measurable subset of $\mathbb{R}$, then $E+x$ is also a Lebesgue measurable subset of $\mathbb{R}$ and $m(E+x) = m(E)$.
(b) If E is a Lebesgue measurable subset of $[a, b]$ and $E+x \subseteq [a, b]$, then $E+x$ is also a Lebesgue measurable subset of $[a, b]$.

4.3.P6. Let $\{r_n\}_{n \geq 1}$ be an enumeration of the rationals in $[0, 1]$. For each $i \in \mathbb{N}$, let g_i be a continuous nonnegative function on $[0,1]$ such that

$$g_i(r_i) \geq i \quad \text{and} \quad \int_0^1 g_i(x)dx = 1/2^i.$$

Define $f_n = \sum_{i=1}^{n} g_i$. Then each f_n is nonnegative and continuous. Besides, the sequence $\{f_n\}_{n \geq 1}$ is increasing and therefore has a limit function f that is nonnegative extended real-valued and measurable. Show that the function f is unbounded on every subinterval of $[0, 1]$ and satisfies $\int_{[0,1]} f = 1$.

4.3.P7. For any interval $I \subseteq \mathbb{R}$ of positive length and any $M > 0$, show that there exists a sequence $\{E_j\}_{j \geq 1}$ of disjoint sets such that $\cup_{j=1}^{\infty} E_j = I$ and $\sum_{j=1}^{\infty} m^*(E_j) > M$.

4.3.P8. For an increasing nonnegative real-valued function ϕ on $\mathbb{R}$ satisfying $\phi(0) = 0$ and $\phi(u + v) \leq \phi(u) + \phi(v)$ whenever $u, v \in \mathbb{R}$, show that

$$\phi(1)^{1/2} \leq \left(\int_0^1 \phi(1 - \sqrt{t})dt \right)^{1/2} + \left(\int_0^1 \phi(\sqrt{1 - t})dt \right)^{1/2},$$

where the integrals are understood in the sense of Riemann.

4.4 Induced Measure and an Application

Let $(X, \mathcal{F}, \mu)$ be a measure space and $\emptyset \neq E \in \mathcal{F}$. Then the family of subsets $\mathcal{F}_E = \{F : F \in \mathcal{F} \text{ and } F \subseteq E\} \subseteq \mathcal{F}$ is clearly the same as the family $\{G \cap E : G \in \mathcal{F}\}$. It is easy to verify that this is a σ-algebra of subsets of E and that the restriction of μ to the σ-algebra is a measure. We shall denote the restriction by μ_E; also, we shall denote the restriction to E of any function ϕ on X by $\phi|_E$. If it happens that the function has been described only by its expression $\phi(x)$ without any symbol such as ϕ having been introduced for it, then the restriction may be denoted by $\phi(x)|_E$.

The σ-algebra $\mathcal{F}_E$ is called the σ-algebra induced by $\mathcal{F}$ on E and the measure μ_E is called the **measure induced by μ on E**. It is clear what is meant by **measure space induced on** E.

It is an immediate consequence of Problem 4.3.P4 that the measure induced on an interval $[a, b]$ by Lebesgue measure on $\mathbb{R}$ is nothing but Lebesgue measure on $[a, b]$. This renders it possible to apply Proposition 4.4.1 below in the proof of the subsequent Proposition 4.4.2, which makes the latter an application of the concept of induced measure.

4.4.1. Proposition. *Let $(X, \mathcal{F}, \mu)$ be a measure space and $\emptyset \neq E \in \mathcal{F}$. For any function $\phi: X \to \mathbb{R}^*$, the restriction $\phi|_E$ is $\mathcal{F}_E$-measurable if and only if the product function $\phi \chi_E: X \to \mathbb{R}^*$ is $\mathcal{F}$-measurable, in which case the equality*

$$\int_X \phi \chi_E d\mu = \int_E \phi|_E d\mu_E \qquad (4.19)$$

holds for nonnegative extended real-valued ϕ. For any real-valued function $\phi: X \to \mathbb{R}$, the restriction $\phi|_E$ is integrable if and only if the product function $\phi\chi_E: X \to \mathbb{R}$ is integrable, in which case (4.19) holds.

Proof: Given a function $\phi: X \to \mathbb{R}^*$, assume that $\phi|_E$ is $\mathcal{F}_E$-measurable. Then the $\mathcal{F}$-measurability of $\phi\chi_E$ follows from the observation that, for $\alpha \geq 0$, we have

$$X(\phi\chi_E > \alpha) = \{x \in E: \phi(x) > \alpha\} = \{x \in E: \phi|_E(x) > \alpha\}$$
$$= E(\phi|_E > \alpha) \in \mathcal{F}_E \subseteq \mathcal{F}$$

and for $\alpha < 0$, we have

$$X(\phi\chi_E > \alpha) = E^c \cup \{x \in E: \phi|_E(x) > \alpha\} = E^c \cup E(\phi|_E > \alpha) \in \mathcal{F}.$$

Now assume conversely that $\phi\chi_E$ is $\mathcal{F}$-measurable. Then the $\mathcal{F}_E$-measurability of $\phi|_E$ follows from the observation that, for $\alpha \geq 0$, we have

$$E(\phi|_E > \alpha) = X(\phi\chi_E > \alpha) \in \mathcal{F} \quad \text{and} \quad E(\phi|_E > \alpha) \subseteq E$$

and for $\alpha < 0$, we have

$$E(\phi|_E > \alpha) = E \cap \{x \in X: \phi(x) > \alpha\} = E \cap X(\phi\chi_E > \alpha) \in \mathcal{F}_E.$$

We proceed to establish the equality (4.19) for a function $\phi: X \to \mathbb{R}^{*+}$ when $\phi|_E$ is $\mathcal{F}_E$-measurable, or equivalently, when $\phi\chi_E$ is $\mathcal{F}$-measurable.

As is easily verified, for any $F \in \mathcal{F}$, we have $(\chi_{F\cap E})|_E = (\chi_F\chi_E)|_E = \chi_F|_E$ and all three are equal to the characteristic function of $F \cap E$ defined on the subset E. Therefore $\int_E \chi_F|_E d\mu_E = \mu_E(F \cap E) = \mu(F \cap E) = \int_X (\chi_F\chi_E)d\mu$. This shows that (4.19) holds when ϕ is the characteristic function χ_F of an arbitrary $F \in \mathcal{F}$. It follows that it holds for any simple function ϕ on X and hence for any nonnegative $\mathcal{F}$-measurable function ϕ on X.

Now suppose instead of the $\mathcal{F}$-measurability of ϕ, we assume only the $\mathcal{F}$-measurability of $\phi\chi_E$. Applying what has been proved in the preceding paragraph to $\phi\chi_E$, we get

$$\int_X (\phi\chi_E)\chi_E d\mu = \int_E (\phi\chi_E)|_E d\mu_E.$$

Upon combining this with the straightforward equalities $\phi\chi_E = (\phi\chi_E)\chi_E$ and $(\phi\chi_E)|_E = \phi|_E$, we find that (4.19) holds when $\phi\chi_E$ is $\mathcal{F}$-measurable.

The last part about a real-valued function $\phi\colon X \to \mathbb{R}$ now follows from the above by using the obvious equalities $\phi^+|_E = \phi|_{E^+}, (\phi\chi_E)^+ = \phi^+\chi_E, \phi^-|_E = \phi|_{\bar E}$ and $(\phi\chi_E)^- = \phi^-\chi_E$. $\qquad\square$

Remark. So far we have used the symbol $\int_I f$, where I is an interval, to mean the integral of f with respect to Lebesgue measure m on I. It means the same thing as $\int_I f\,dm$. Since Lebesgue measure on I is the same as the measure m_I induced on I by Lebesgue measure m on $\mathbb{R}$, the symbol $\int_I f$ also means the same thing as $\int_I f\,dm_I$. It is worth bearing this in mind for the next proposition, which is an application of induced measures.

4.4.2. Proposition. *Suppose $f\colon [\beta, \infty) \to \mathbb{R}$ is Riemann integrable over $[\beta, B]$ for every $B > \beta$. If the Lebesgue integral $\int_{[\beta,\infty)} f$ over $[\beta, \infty)$ exists, then the improper integral $\int_\beta^\infty f$ also exists and agrees with the Lebesgue integral.*

Proof: Consider an arbitrary sequence $\{B_n\}_{n\geq 1}$ in $[\beta, \infty)$ tending to ∞. For each $n \in \mathbb{N}$, let $f_n = f\chi_{[\beta, B_n]}$ and $g_n = f|_{[\beta, B_n]}$. Then each f_n is Lebesgue integrable and each g_n is Riemann integrable by hypothesis, with

$$\int_\beta^{B_n} g_n = \int_\beta^{B_n} f. \qquad (4.20)$$

Moreover, $f_n \to f$ everywhere and $|f_n| \leq |f|$ everywhere. By the Dominated Convergence Theorem 3.2.6, we find that

$$\lim_{n\to\infty} \int_{[\beta,\infty)} f_n \text{ exists and equals } \int_{[\beta,\infty)} f. \qquad (4.21)$$

By Proposition 4.4.1 and Theorem 4.3.4 (in that order), for each $n \in \mathbb{N}$, we have

$$\int_{[\beta,\infty)} f_n = \int_{[\beta, B_n]} g_n = \int_\beta^{B_n} g_n$$

$$= \int_\beta^{B_n} f$$

by (4.20).

It follows from here in the light of (4.21) that $\lim_{n\to\infty} \int_\beta^{B_n} f$ exists and equals $\int_{[\beta,\infty)} f$. Since $\{B_n\}_{n\geq 1}$ is an arbitrary sequence tending to ∞, the contention of the proposition stands proved. $\qquad\square$

The integral $\int_E \phi|_E d\mu_E$ is usually abbreviated as $\int_E \phi$. Recall that $\int_X \phi$ is usually written as $\int \phi$. The equality (A) of Proposition 4.4.1 can then be stated as $\int \phi \chi_E = \int_E \phi$. If $E = \emptyset$, we shall understand $\int_E \phi$ to be 0 because $\int \phi \chi_E = 0$.

4.4.3. Proposition. *Let $(X, \mathcal{F}, \mu)$ be a measure space and $H \in \mathcal{F}$ satisfy $\mu(H^c) = 0$. Then for any measurable $f: X \to \mathbb{R}^{**}$, we have $\int f = \int_H f$. The same holds for any integrable $f: X \to \mathbb{R}$.*

Proof: If $H = \emptyset$, it follows that $\mu(X) = 0$ and all integrals are 0; so there is nothing to prove. If $H^c = \emptyset$, again there is nothing to prove. So, assume both H and $H^c = G$ to be nonempty. Then there are induced measures on both H and G. However the induced measure on G is the zero measure and therefore all integrals on it are 0; in particular, $\int_G f = 0$. Now,

$$\int f = \int f\chi_H + \int f(1 - \chi_H) = \int f\chi_H + \int f\chi_G$$
$$= \int_H f + \int_G f \text{ by Proposition 4.4.1}$$
$$= \int_H f.$$

□

We close this section with a proposition that justifies such computations as $\int_{[a,b]} f + \int_{[b,c]} f = \int_{[a,c]} f$, where $a \le b \le c$. However, no reference to it will be given when it is used.

4.4.4. Proposition. *Let $(X, \mathcal{F}, \mu)$ be a measure space and the subsets $E, F \in \mathcal{F}$ satisfy $\mu(E \cap F) = 0$ and $E \cup F = X$. Then for any measurable $f: X \to \mathbb{R}^{**}$, we have*

$$\int_E f + \int_F f = \int_X f. \tag{4.22}$$

For any real-valued function $f: X \to \mathbb{R}$, the restrictions to E and to F are both integrable if and only if f is integrable, in which case (4.22) holds.

Proof: Since $E \cup F = X$, we have

$$\chi_X + \chi_{E \cap F} = \chi_E + \chi_F.$$

Consider a measurable function $f: X \to \mathbb{R}^{**}$. By the above equality, additivity of the integral for nonnegative functions (Theorem 3.2.2) and Proposition 4.4.1,

$$\int_X f + \int_{E \cap F} f = \int_E f + \int_F f.$$

But $\int_{E\cap F} f = 0$ because $\mu(E \cap F) = 0$ and hence (4.22) holds.

Now consider an integrable function $f: X \to \mathbb{R}$. Since $|f\chi_E| \le |f|$, $|f\chi_F| \le |f|$ and $|f\chi_{E\cap F}| \le |f|$, all the three functions $f\chi_E$, $f\chi_F$ and $f\chi_{E\cap F}$ are integrable and the above equality regarding characteristic functions, taken in conjunction with additivity of the integral for integrable functions (Theorem 3.2.5) and Proposition 4.4.1, leads to (4.22) as above.

Finally, suppose the restrictions to E and F of the real-valued function $f: X \to \mathbb{R}$ are both integrable. By Proposition 4.4.1, the products $f\chi_E$ and $f\chi_F$ are both integrable and the inequality $|f\chi_{E\cap F}| \le |f\chi_E|$ implies that $f\chi_{E\cap F}$ is also integrable. The above equality regarding characteristic functions can be written in the form

$$\chi_X = \chi_E + \chi_F - \chi_{E\cap F}.$$

When taken in conjunction with the additivity and homogeneity of the integral for integrable functions (Theorem 3.2.5) and Proposition 4.4.1, the equality leads to

$$\int_X f = \int_E f + \int_F f - \int_{E\cap F} f.$$

□

Problem Set 4.4

4.4.P1. (a) Let $g: [1, \infty) \to \mathbb{R}$ be given by $g(x) = \sin x$ and I_k be the interval $[2k\pi, (2k+1)\pi]$, where $k \in \mathbb{N}$. Show that $\int_{I_k} g(x)dx = 2$.

(b) Let $f: [1, \infty) \to \mathbb{R}$ be given by $f(x) = \frac{1}{x} \sin x$ and I_k be the interval $[2k\pi, (2k+1)\pi]$, where $k \in \mathbb{N}$. Show that $\int_{I_k} f(x)dx \ge \frac{2}{(2k+1)\pi}$.

4.4.P2. Let $(X, \mathcal{F}, \mu)$ be a measure space and $\{f_k\}_{k\ge 1}$ be a sequence of integrable functions on X such that $\sum_{k=1}^{\infty} \int |f_k| < \infty$. Show that there exists an integrable function f on X such that the series $\sum_{k=1}^{\infty} f_k$ converges to f on a measurable subset $H \subseteq X$ satisfying $\mu(H^c) = 0$, and that $\int f = \sum_{k=1}^{\infty} \int f_k$.

4.4.P3. Let $\mathcal{F}$ be a σ-algebra of subsets of X and A, $B \in \mathcal{F}$ be nonempty subsets satisfying $A \cup B = X$. Show that a function $\phi: X \to \mathbb{R}^*$ is $\mathcal{F}$-measurable if and only if the restrictions $\phi|_A$ and $\phi|_B$ are $\mathcal{F}_A$-measurable and $\mathcal{F}_B$-measurable respectively.

Chapter 5
The Counting Measure

5.1 Interchanging the Order of Summation

Since a sequence $a = \{a_j\}_{j\geq1}$ is nothing but a function with domain $\mathbb{N}$, it makes sense to speak of the integral $\int a$ with reference to some measure on the σ-algebra of all subsets of $\mathbb{N}$, in particular the counting measure. It was noted in Problem 3.1.P5 that when each a_j is nonnegative, the integral $\int a$ is just the sum $\sum_{j=1}^{\infty} a_j$.

For this paragraph, drop the requirement that every a_j be nonnegative but require every a_j to be finite. Then $\int a^+ = \sum_{j=1}^{\infty} a_j^+$, the sum of the series formed by the non-negative terms of $\sum_{j=1}^{\infty} a_j$; correspondingly for $\int a^-$. By Definition 2.2.5, $\int a$ is then $\sum_{j=1}^{\infty} a_j^+ - \sum_{j=1}^{\infty} a_j^-$ provided $\sum_{j=1}^{\infty} a_j^-$ is convergent; thus $\int a = \sum_{j=1}^{\infty} (a_j^+ - a_j^-) = \sum_{j=1}^{\infty} a_j$. For a to be integrable with reference to the counting measure, what we need according to Definition 2.2.5 is that $\sum_{j=1}^{\infty} a_j^+$ and $\sum_{j=1}^{\infty} a_j^-$ should both be convergent. It is well known from the elementary theory of series that convergence of both is equivalent to the absolute convergence of $\sum_{j=1}^{\infty} a_j$. In any event, the equivalence can be easily derived from the fact that any $x \in \mathbb{R}$ satisfies $x^+ \leq |x|, x^- \leq |x| = x^+ + x^-$.

It was established in Sect. 1.2 that the equality $\sum_{n=1}^{\infty} \sum_{k=1}^{\infty} a_{n,k} = \sum_{k=1}^{\infty} \sum_{n=1}^{\infty} a_{n,k}$, called interchanging the order of summation, is valid for any function $a : \mathbb{N} \times \mathbb{N} \to \mathbb{R}^{*+}$. We know from Problem 1.2.P1 that this inequality can fail if a is not required to be nonnegative even if it is restricted to being real. However, for $a : \mathbb{N} \times \mathbb{N} \to \mathbb{R}$, an additional hypothesis does permit the interchange (see Problem 5.1.P4).

To reinterpret these repeated sums in terms of integrals with respect to counting measures, it will be convenient to denote $a_{n,k}$ by $a(n, k)$ and introduce some further notation. In taking the sum $\sum_{k=1}^{\infty} a_{n,k}$, we are holding n fixed and regarding $a(n, k)$ as a function of k only. The function will be denoted by a_n. Similarly, in taking the other sum $\sum_{n=1}^{\infty} a_{n,k}$, we are holding k fixed and regarding $a(n, k)$ as a function of n only. It will be denoted by a^k.

© Springer Nature Switzerland AG 2018
S. Shirali, *A Concise Introduction to Measure Theory*,
https://doi.org/10.1007/978-3-030-03241-8_5

5.1.1. Definition. *Given a function* $a\colon \mathbb{N} \times \mathbb{N} \to \mathbb{R} \cup \{\infty\}$ *and* $n \in \mathbb{N}$, *the function with domain* $\mathbb{N}$ *such that each* $k \in \mathbb{N}$ *is mapped into* $a(n, k)$ *will be denoted by* a_n:

$$a_n(k) = a(n, k) \quad \text{for every } k \in \mathbb{N}.$$

Likewise, for any $k \in \mathbb{N}$, *the function with domain* $\mathbb{N}$ *such that each* $n \in \mathbb{N}$ *is mapped into* $a(n, k)$ *will be denoted by* a^k:

$$a^k(n) = a(n, k) \quad \text{for every } n \in \mathbb{N}.$$

5.1.2. Remark. In terms of the notation just introduced, $\sum_{k=1}^{\infty} a_{n,k} = \int a_n = \int a(n, k) d\nu(k)$ and $\sum_{n=1}^{\infty} a_{n,k} = \int a^k = \int a(n, k) d\mu(n)$ when a is nonnegative-valued, μ and ν both being the counting measure on $\mathbb{N}$. Moreover, interchanging the order of summation says that

$$\int \left(\int a_n \right) = \int \left(\int a^k \right),$$

or in standard and more readable notation,

$$\int \left(\int a(n, k) d\nu(k) \right) d\mu(n) = \int \left(\int a(n, k) d\mu(n) \right) d\nu(k)$$

(for nonnegative-valued a). This is interchanging the order of integration in repeated integrals, the integration being with respect to the counting measure on $\mathbb{N}$ every time. It is natural to ask whether such interchanging is valid for any function $f : X \times Y \to \mathbb{R} \cup \{\infty\}$, at least when $\int f(x, y) d\nu(y)$ and $\int f(x, y) d\mu(x)$ describe integrable (or nonnegative measurable) functions of x and y respectively. In standard and more traditional notation, we are asking whether

$$\int d\nu(y) \int f(x, y) d\mu(x) = \int d\mu(x) \int f(x, y) d\nu(y),$$

where μ and ν are measures on some σ-algebras of subsets of X and Y respectively, and it is assumed that all integrals involved make sense.

We shall first give an example to show that interchanging the order of integration need not always be valid even if f takes only the values 0 and 1.

5.1.3. Example. Let $X = Y = [0, 1]$, $\mathcal{F} = \mathcal{G} =$ the σ-algebra of Lebesgue measurable subsets of $[0, 1]$; suppose μ is the Lebesgue measure on $\mathcal{F}$ and ν is the counting measure on $\mathcal{G}$. Let $V = \{(x, y) \in [0, 1] \times [0, 1] \colon x = y\}$. Then as argued below,

$$\int d\mu(x) \int \chi_V(x, y) d\nu(y) = 1 \quad \text{but} \quad \int d\nu(y) \int \chi_V(x, y) d\mu(x) = 0.$$

Indeed, given $x \in X$, we have $\chi_V(x, y) = 1$ for $y = x$ and 0 for $y \neq x$, which implies $\int \chi_V(x, y) d\nu(y) = 1$, whereas given $y \in Y$, we have $\chi_V(x, y) = 1$ for $x = y$ and 0 for $x \neq y$, which implies $\int \chi_V(x, y) d\mu(x) = 0$.

Readers who may be conversant with the Tonelli Theorem 6.5.2 and Fubini Theorem 6.5.5 for what are called σ-*finite* measures may note that the upcoming theorem does not assume the measure μ to be σ-finite and is therefore not a special case of those theorems.

5.1.4. Theorem. *Let μ be a measure on a set X. Consider $\mathbb{N}$ with the counting measure (all subsets measurable) and let $f : X \times \mathbb{N} \to \mathbb{R}^{*+}$ be a function such that, for each $k \in \mathbb{N}$, the function f^k on X described by $f^k(x) = f(x, k)$ is measurable. Then interchanging the order of integration is valid, which is to say,*

$$\int \left(\sum_{k=1}^{\infty} f(x, k) \right) d\mu(x) = \sum_{k=1}^{\infty} \left(\int f(x, k) d\mu(x) \right).$$

Proof: For brevity, we shall write $d\mu(x)$ as simply dx. Define a sequence of functions ϕ_n on X by $\phi_n(x) = \sum_{k=1}^{n} f(x, k)$. Then each ϕ_n is nonnegative-valued and measurable. Besides, $\phi_n(x) \leq \phi_{n+1}(x)$ for every $x \in X$ and every $n \in \mathbb{N}$. By the Monotone Convergence Theorem 2.3.2,

$$\int \left(\sum_{k=1}^{\infty} f(x, k) \right) dx = \int \left(\lim_{n \to \infty} \phi_n(x) \right) dx = \lim_{n \to \infty} \left(\int \phi_n(x) dx \right)$$

$$= \lim_{n \to \infty} \sum_{k=1}^{n} \int f(x, k) dx = \sum_{k=1}^{\infty} \left(\int f(x, k) dx \right).$$

$\square$

5.1.5. Remark. Let us return to the context of the sum $\sum_{j=1}^{\infty} a_j$ being the integral $\int a$ with respect to the counting measure on $\mathbb{N}$, where a takes values in $\mathbb{R}^{*+}$. It is clear that for any finite subset $A \subseteq \mathbb{N}$, we have $\sum_{j=1}^{\infty} a_j \geq \sum_{j \in A} a_j$. Therefore $\sum_{j=1}^{\infty} a_j \geq \sup\{\sum_{j \in A} a_j : A \subseteq \mathbb{N} \text{ finite}\}$. It is quite easy to see that actually equality holds:

$$\sum_{j=1}^{\infty} a_j = \sup\{\sum_{j \in A} a_j : A \subseteq \mathbb{N} \text{ finite}\}$$

Indeed, the right side here is greater than or equal to the supremum taken over the special finite subsets of the form $\{1, 2, \ldots, n\}$, i.e. $\sup\{\sum_{j=1}^{n} a_j : n \in \mathbb{N}\}$, which is the same as $\sum_{j=1}^{\infty} a_j$. It makes no difference whether we allow $A = \emptyset$, because $\sum_{j \in \emptyset} a_j = 0$ by convention.

5.1.6. Exercise. Let S denote a class of finite subsets of $\mathbb{N}$ such that every finite subset is contained in some $F \in S$. (For example, S may consist of all subsets of the form $\{1, 2, \ldots, n\}$ or of the form $\{1, 2, \ldots, n\} \cup \{n + 3\}$.) For any $\mathbb{R}^{*+}$-valued sequence a, show that $\sum_{j=1}^{\infty} a_j = \sup\{\sum_{j \in F} a_j : F \in S\}$.

Solution: Since each $F \in S$ is finite, $\sup\{\sum_{j \in A} a_j : A \subseteq \mathbb{N} \text{ finite}\} \geq \sup\{\sum_{j \in F} a_j : F \in S\}$. Consider any finite $A \subseteq \mathbb{N}$. By hypothesis, there exists some $F \in S$ such that $A \subseteq F$. Since $a_j \geq 0$, it follows that $\sum_{j \in A} a_j \leq \sum_{j \in F} a_j \leq \sup\{\sum_{j \in F} a_j : F \in S\}$. Consequently, $\sup\{\sum_{j \in A} a_j : A \subseteq \mathbb{N} \text{ finite}\} \leq \sup\{\sum_{j \in F} a_j : F \in S\}$. Thus the two suprema are equal. But one of them is known to be $\sum_{j=1}^{\infty} a_j$ (Remark 5.1.5).

The advantage of looking at the sum as the supremum is that the latter makes sense even when $\mathbb{N}$ is replaced by an arbitrary set.

5.1.7. Definition. *For any function* $a : X \to \mathbb{R}^{*+}$, *the* **sum of all** $a(x)$ *over* $x \in X$ *is*

$$\sum_{x \in X} a(x) = \sup\{\sum_{x \in A} a(x) : A \subseteq X \text{ finite}\}.$$

It should be noted that this is not a measure-theoretic definition. Situations in which the need to interchange the order of summations in repeated sums of this kind arises naturally will not be discussed here but the interested reader is referred to Dunford and Schwartz [2, p. 1010]. However, the matter of interchanging the order of summation will be taken up.

5.1.8. Remark. In view of Remark 5.1.5, when $X = \mathbb{N}$, the sum $\sum_{x \in X} a(x)$, or $\sum_{j \in \mathbb{N}} a_j$, is the same as $\sum_{j=1}^{\infty} a_j$. Recall once again from Problem 3.1.P5 that this sum is the same as the integral $\int a$ or $\int a(x) d\mu(x)$, where μ is the counting measure on $X = \mathbb{N}$. Therefore interchange of the order of summation in repeated sums that was proved in Sect. 1.2 can be reformulated in terms of integrals with respect to counting measures. In other words, what was expressed as $\sum_{i \in \mathbb{N}} \sum_{j \in \mathbb{N}} a(i, j) = \sum_{j \in \mathbb{N}} \sum_{i \in \mathbb{N}} a(i, j)$, or equivalently,

$$\sum_{x \in A} \sum_{y \in B} a(x, y) = \sum_{y \in B} \sum_{x \in A} a(x, y)$$

for countable sets A, B and $a : A \times B \to \mathbb{R}^{*+}$, can also be expressed as interchanging the order of integration:

$$\int (\int a(x, y) d\nu(y)) d\mu(x) = \int (\int a(x, y) d\mu(x)) d\nu(y),$$

μ and ν being the counting measure s on A and B respectively.

We are going to justify interchanging the order of summation when the function is defined on the Cartesian product of sets that are not assumed countable, but first a preparatory result.

5.1.9. Proposition. (a) *If a subset $A \subseteq X$ has the property that $a(x)=0$ whenever $x \notin A$, i.e. $a(x)=0$ unless $x \in A$, then $\sum_{x\in A} a(x) = \sum_{x\in X} a(x)$.*

(b) *If $\sum_{x\in X} a(x) < \infty$, then there is a countable subset $A \subseteq X$ with the property that $a(x)=0$ whenever $x \notin A$. Any subset B of X that contains A has the same property.*

Proof: (a) The assertion is clearly true when X is finite. Any finite subset F of A is also a finite subset of X and therefore $\sum_{x\in A} a(x) \leq \sum_{x\in X} a(x)$. To prove the reverse inequality, consider an arbitrary finite subset F of X. The intersection $B = F \cap A$ is a finite subset of A and therefore $\sum_{x\in A} a(x) \geq \sum_{x\in B} a(x)$. Now, $a(x)=0$ unless $x \in A$. Therefore, for $x \in F$, we have $a(x)=0$ unless $x \in F \cap A = B$. Consequently, $\sum_{x\in B} a(x) = \sum_{x\in F} a(x)$, considering that F is finite. It follows that $\sum_{x\in A} a(x) \geq \sum_{x\in F} a(x)$. As this holds for every finite subset F of X, we infer that $\sum_{x\in A} a(x) \geq \sum_{x\in X} a(x)$, which is the desired reverse inequality.

(b) Left to the reader as Problem 5.1.P2. □

With the above preparatory result in hand, we come to the theorem that justifies interchanging the order of summation when the function is defined on the Cartesian product of sets that are not assumed countable. There are authors who regard the matter as being too obvious to be worthy of a proof (Dunford and Schwartz [2, p. 1010]), presumably because the first assertion in part (b) of the above Proposition 5.1.9 "reduces" the matter to the countable case. But the reduction is not quite as straightforward as may seem at first sight, something that will be clear from the argumentation below. The reduction idea will be oppugned further in Remark 6.5.4(a).

5.1.10. Theorem. *For any function $a : X \times Y \to \mathbb{R}^{*+}$,*

$$\sum_{x\in X}\sum_{y\in Y} a(x, y) = \sum_{y\in Y}\sum_{x\in X} a(x, y).$$

Proof: First suppose $S = \sum_{x\in X}\sum_{y\in Y} a(x, y) < \infty$. By Proposition 5.1.9(b), there exists a countable $A \subseteq X$ such that

$$\sum_{y\in Y} a(x, y) = 0 \text{ for every } x \notin A.$$

Hence

$$a(x, y) = 0 \text{ for every } x \notin A \text{ and every } y \in Y.$$

Together with Proposition 5.1.9(a), these statements imply

$$S = \sum_{x\in A}\sum_{y\in Y} a(x, y) \tag{5.1}$$

and

$$\sum_{x \in A} a(x, y) = \sum_{x \in X} a(x, y) \text{ for every } y \in Y. \tag{5.2}$$

Using the finiteness of S again, we have $\sum_{y \in Y} a(x, y) < \infty$ for every $x \in X$. Therefore Proposition 5.1.9(b) yields for each $x \in X$ a countable subset $B_x \subseteq Y$ such that

$$a(x, y) = 0 \text{ for every } y \notin B_x. \tag{5.3}$$

Let $B = \cup_{x \in A} B_x$. Then $B \subseteq Y$ is countable, and from (5.3) we have

$$a(x, y) = 0 \text{ for every } y \notin B \text{ and } x \in A.$$

Hence by (5.2),

$$\sum_{x \in X} a(x, y) = 0 \text{ for every } y \notin B.$$

Together with Proposition 5.1.9(a), these statements imply

$$\sum_{y \in Y} a(x, y) = \sum_{y \in B} a(x, y) \text{ for every } x \in A \tag{5.4}$$

and

$$\sum_{y \in Y} \sum_{x \in X} a(x, y) = \sum_{y \in B} \sum_{x \in X} a(x, y). \tag{5.5}$$

Since A and B are countable sets, we know that

$$\sum_{x \in A} \sum_{y \in B} a(x, y) = \sum_{y \in B} \sum_{x \in A} a(x, y). \tag{5.6}$$

Now,

$$
\begin{aligned}
S &= \sum_{x \in A} \sum_{y \in Y} a(x, y) \ \text{ by } (5.1) \\
&= \sum_{x \in A} \sum_{y \in B} a(x, y) \ \text{ by } (5.4) \\
&= \sum_{y \in B} \sum_{x \in A} a(x, y) \ \text{ by } (5.6) \\
&= \sum_{y \in B} \sum_{x \in X} a(x, y) \ \text{ by } (5.2) \\
&= \sum_{y \in Y} \sum_{x \in X} a(x, y) \ \text{ by } (5.5).
\end{aligned}
$$

This proves the equality in question when $\sum_{x\in X}\sum_{y\in Y}a(x,y)<\infty$. A similar argument goes through when $\sum_{y\in Y}\sum_{x\in X}a(x,y)<\infty$. In the remaining case, the two are equal because both are ∞. ∎

It was noted in Remark 5.1.8 that interchanging the order of summation for a nonnegative function a with domain $A\times B$ with both A and B countable can be interpreted as interchanging the order of integration with respect to the counting measures on A and B. We intend to show in Sect. 5.2 (see Remark 5.2.2) that interchanging the order of summation for a nonnegative function a with domain $X\times Y$ with general X and Y—asserted by Theorem 5.1.10—can also be interpreted as interchanging the order of integration with respect to the counting measures on X and Y.

It will later be shown in Theorem 5.2.3 that the sums in Theorem 5.1.10 are both equal to $\sum_{(x,y)\in X\times Y}a(x,y)$.

Problem Set 5.1

5.1.P1. Let X be any nonempty set. Define $\mathcal{F}$ to consist of those subsets of X that are either countable or have a countable complement. (It is easy to argue that $\mathcal{F}$ is a σ-algebra, sometimes called the "co-countable" σ-algebra). Now let $\phi:X\to\mathbb{R}^{*+}$ be any function and set $\mu(A)=\sum_{x\in A}\phi(x)$ for any $A\in\mathcal{F}$, where the sum is understood in the sense of Definition 5.1.7. Show that μ is a measure on $\mathcal{F}$. (Note: If $\phi>0$ everywhere, then the only subset of measure 0 is the empty set.)

5.1.P2. Prove Proposition 5.1.9(b).

5.1.P3. Let μ be a measure on a set X. Consider $\mathbb{N}$ with the counting measure (all subsets measurable) and let $f:X\times\mathbb{N}\to\mathbb{R}$ be a real-valued function such that, for each $k\in\mathbb{N}$, the function f^k on X described by $f^k(x)=f(x,k)$ is measurable. Suppose also that there is an integrable function $\phi:X\to\mathbb{R}^+$ such that $|\sum_{k=1}^{n}f(x,k)|\leq\phi(x)$ for each $x\in X$ and each $n\in\mathbb{N}$. If the series $\sum_{k=1}^{\infty}f(x,k)$ converges, show that interchanging the order of integration is valid, which is to say,

$$\int\left(\sum_{k=1}^{\infty}f(x,k)\right)d\mu(x)=\sum_{k=1}^{\infty}\left(\int f(x,k)d\mu(x)\right).$$

5.1.P4. Suppose the real-valued function $a:\mathbb{N}\times\mathbb{N}\to\mathbb{R}$ has the property that there exists a sequence $\{b_m\}_{m\geq 1}$ such that $\sum_{m=1}^{\infty}b_m<\infty$ and $|\sum_{k=1}^{n}a_{m,k}|\leq b_m$ for each n and each m. Show that, if the series $\sum_{k=1}^{\infty}a_{m,k}$ converges for each m, then $\sum_{m=1}^{\infty}\sum_{k=1}^{\infty}a_{m,k}=\sum_{k=1}^{\infty}\sum_{m=1}^{\infty}a_{m,k}$. (Note: (1) The hypothesis about $\{b_m\}_{m\geq 1}$ is stronger than convergence of $\sum_{m=1}^{\infty}|\sum_{k=1}^{\infty}a_{m,k}|$. The latter is not sufficient for the purpose at hand, in view of Problem 1.2.P1. (2) A proof of the result of the present problem by using metric spaces is found in Feldman [3, p. 2]. What has been stated there has a stronger but more palatable version of the hypothesis about $\{b_m\}$, in that it is assumed that $\sum_{m=1}^{\infty}\sum_{k=1}^{\infty}|a_{m,k}|<\infty$, so that $\sum_{k=1}^{\infty}a_{m,k}$ not only converges for each m but does so absolutely.)

5.2   Integration with the Counting Measure

With a view to the earlier stated purpose of interpreting Theorem 5.1.10 as interchanging the order of integration in repeated integrals, we show that the sum defined
by Definition 5.1.7 is actually the integral with respect to the counting measure.

5.2.1. Theorem. *For any function* $a : X \to \mathbb{R}^{*+}$,

$$\sum_{x \in X} a(x) = \int a,$$

where the integral is taken with respect to the counting measure on X.

Proof: For any function $a : X \to \mathbb{R}^{*+}$ that takes the value ∞, this is obvious because
the sum and integral are both ∞. So we need consider only real-valued a.

Suppose there is a simple function s such that $0 \le s \le a$ and s is nonzero on an
infinite set. Then $\int a \ge \int s = \infty$. Moreover, s takes some positive value σ on an
infinite subset $B \subseteq X$. Therefore, for any $M > 0$, there exists a finite subset $F \subseteq B$ with
more than M/σ elements. It follows that $\sum_{x \in X} a(x) \ge \sum_{x \in F} a(x) \ge \sum_{x \in F} s(x) >$
$\sigma \cdot (M/\sigma) = M$. Since this holds for every $M > 0$, we have $\sum_{x \in X} a(x) = \infty$. Thus
$\int a = \sum_{x \in X} a(x)$.

Now consider the case when there is no simple function s such that $0 \le s \le a$ and
s is nonzero on an infinite set. In other words, any simple function s such that $0 \le s$
$\le a$ is nonzero only on a finite set F. An application of Proposition 3.1.7 yields $\int s =$
$\sum_{x \in F} s(x)$. But $\sum_{x \in F} s(x) \le \sum_{x \in F} a(x) \le \sum_{x \in X} a(x)$ and hence $\int s \le \sum_{x \in X} a(x)$.
By Theorem 3.2.1 and the Monotone Convergence Theorem 2.3.2, it follows that

$$\int a \le \sum_{x \in X} a(x).$$

To derive the reverse inequality, consider any finite subset $A \subseteq X$. The product $a\chi_A$ is a
nonnegative simple function not exceeding a and hence $\int a \ge \int a\chi_A$. As Proposition
3.1.7 implies $\int a\chi_A = \sum_{x \in A} a(x)$, the desired reverse inequality now follows right
away.   □

5.2.2. Remark. In the light of Theorem 5.2.1, the interchange of the order of summation in Theorem 5.1.10 can be restated as an interchange of the order of integration:

$$\int d\mu(x) \int f(x, y) d\nu(y) = \int d\nu(y) \int f(x, y) d\mu(x),$$

where μ, ν are both counting measures and the function f is $\mathbb{R}^{*+}$-valued. For other
measures, we have seen in Example 5.1.3 that interchanging the order of integration
need not hold. We can nevertheless ask under what conditions it does hold.

Next, we show that in the context of Theorem 5.1.10, the two sums are equal to $\sum_{(x,y)\in X\times Y} a(x, y)$.

5.2.3. Theorem. *For any function* $a : X \times Y \to \mathbb{R}^{*+}$,

$$\sum_{(x,y)\in X\times Y} a(x, y) = \sum_{x\in X}\sum_{y\in Y} a(x, y).$$

Proof: The argument that

$$\sum_{(x,y)\in X\times Y} a(x, y) \le \sum_{x\in X}\sum_{y\in Y} a(x, y)$$

is straightforward and is left as Problem 5.2.P1. We shall show that strict inequality cannot hold. This is trivially true if the left side is ∞. So we assume

$$S = \sum_{(x,y)\in X\times Y} a(x, y) < \infty.$$

The finiteness immediately implies that of $\sum_{y\in Y} a(x, y)$ for each x. To get a contradiction, suppose $\sum_{x\in X}\sum_{y\in Y} a(x, y) > S$. Then there exists a finite set $A \subseteq X$ such that $\sum_{x\in A}\sum_{y\in Y} a(x, y) > S$. The finiteness of A and the finiteness of $\sum_{y\in Y} a(x, y)$ for each x together imply that $\sum_{x\in A}\sum_{y\in Y} a(x, y) < \infty$. So, if we set $\alpha = \sum_{x\in A}\sum_{y\in Y} a(x, y) - S$, we have $0 < \alpha < \infty$. Suppose A contains N elements. For each $x \in X$, finiteness of $\sum_{y\in Y} a(x, y)$ yields a finite subset $B_x \subseteq Y$ such that $\sum_{y\in B_x} a(x, y) > \sum_{y\in Y} a(x, y) - \frac{\alpha}{N}$. So, $\sum_{x\in A}\sum_{y\in B_x} a(x, y) > \sum_{x\in A}\sum_{y\in Y} a(x, y) - \alpha = S$. If we select $F \subseteq X \times Y$ to be $\{(x, y) \in X\times Y : x \in A, y \in B_x\}$, then F is finite and the foregoing inequality leads to $\sum_{(x,y)\in F} a(x, y) > S$. However, this implies $\sum_{(x,y)\in X\times Y} a(x, y) > S$, contrary to the definition of S. $\square$

The left side in the equality asserted by Theorem 5.2.3 can be represented as the integral of a with respect to the counting measure on the Cartesian product $X \times Y$. With this as a backdrop, we are led to the question of whether there is a measure on $X \times Y$ such that the integral with respect to that measure, a "double integral", agrees with the repeated integrals even when the measures on X and Y are not counting measures. These matters will be taken up in the next chapter. The reader is cautioned that the integral with respect to the counting measure on $X \times Y$ is not the same as what is called the sum of a "double series" (see Shirali and Vasudeva [10, pp. 70–71] or Habil [4, pp. 4, 20–21]); the latter takes into account only a limited class of finite subsets of $X \times Y$.

5.2.4. Remark. It is an easy consequence of Theorem 5.2.1 that for any real-valued function $f : X \to \mathbb{R}$ such that $\sum_{x\in X} f^-(x) < \infty$, we have $\sum_{x\in X} f^+(x) - \sum_{x\in X} f^-(x) = \int f$, where the integral is taken with respect to the counting measure on X.

We proceed to discuss how $\int f$, when it is finite and f is real-valued, can be described as a sum over the set X without resort to f^+ and f^-. This kind of sum arises naturally in some contexts where f takes vectors or complex numbers as values and there is no such thing as f^+ and f^-.

Let $f : X \to \mathbb{R}^+$ be a nonnegative real-valued function on a set X. If $\sigma = \sum_{x \in X} f(x) < \infty$, then it is the unique real number with the property that for any $\varepsilon > 0$, there exists a finite $F \subseteq X$ such that

$$\left| \sum_{x \in G} f(x) - \sigma \right| < \varepsilon \quad \text{whenever } F \subseteq G \subseteq X \text{ and } G \text{ is finite.} \tag{5.7}$$

Conversely, if such a real number σ exists, then $\infty > \sum_{x \in X} f(x) = \sigma$. This reformulation of what $\sum_{x \in X} f(x)$ is remains meaningful even if f is permitted to take negative values. With this in mind, we enunciate the following definition.

5.2.5. Definition. *If* $f : X \to \mathbb{R}$ *is a real-valued function, an* **unconditional sum** $\sum_{x \in X}^{\text{un}} f(x)$ *is a real number* σ *having the property* (5.7), *if there is any.*

5.2.6. Remarks. (a)   The resemblance to the definition of a limit of a sequence suggests how to prove by an $\frac{\varepsilon}{2}$-argument, using the fact that the union of two finite sets is a finite set, that there cannot be two different real numbers σ satisfying (5.7). In other words, the unconditional sum, if it exists, is unique. Also, if $\sum_{x \in X}^{\text{un}} f(x)$ and $\sum_{x \in X}^{\text{un}} g(x)$ both exist, then $\sum_{x \in X}^{\text{un}} (f + g)(x)$ exists and equals $\sum_{x \in X}^{\text{un}} f(x) + \sum_{x \in X}^{\text{un}} g(x)$.
(b)   In view of the observation preceding the above definition, when the function f is nonnegative real-valued, $\sum_{x \in X}^{\text{un}} f(x)$ exists if and only if $\sum_{x \in X} f(x)$ is finite, in which case $\sum_{x \in X}^{\text{un}} f(x) = \sum_{x \in X} f(x)$.
(c)   In particular, interchanging the order of summation in Theorem 5.1.10 continues to hold when the function a is nonnegative real-valued and the sums are understood as unconditional sums. Similarly for Theorem 5.2.3.
(d)   If $f : X \to \mathbb{R}$ is any real-valued function and either $\sum_{x \in X} f^+(x)$, $\sum_{x \in X} f^-(x)$ are both finite or $\sum_{x \in X}^{\text{un}} f^+(x)$, $\sum_{x \in X}^{\text{un}} f^-(x)$ are both finite, then $\sum_{x \in X} f^+(x) = \sum_{x \in X}^{\text{un}} f^+(x)$, $\sum_{x \in X} f^-(x) = \sum_{x \in X}^{\text{un}} f^-(x)$ by (b) above. When this is so, once again the resemblance between Definition 5.2.5 and the definition of a limit of a sequence suggests how to prove the following statement by an $\frac{\varepsilon}{2}$-argument, using the fact that the union of two finite sets is a finite set:

$$\sum_{x \in X}^{\text{un}} f(x) = \sum_{x \in X}^{\text{un}} f^+(x) - \sum_{x \in X}^{\text{un}} f^-(x) = \sum_{x \in X} f^+(x) - \sum_{x \in X} f^-(x).$$

In particular, the unconditional sum $\sum_{x \in X}^{\text{un}} f(x)$ exists.

The next result is a sort of a converse.

5.2.7. Theorem. *Suppose* $f : X \to \mathbb{R}$ *is a real-valued function on a set* X *and its unconditional sum* $\sum_{x \in X}^{\text{un}} f(x)$ *exists. Then* $\sum_{x \in X} f^+(x)$, $\sum_{x \in X} f^-(x)$ *are both finite and*

$$\sum_{x \in X} f^+(x) - \sum_{x \in X} f^-(x) = \sum_{x \in X}^{\text{un}} f(x),$$

the unconditional sum. Moreover,

$$\sum_{x \in X}^{\text{un}} f(x) = \sum_{x \in X}^{\text{un}} f^+(x) - \sum_{x \in X}^{\text{un}} f^-(x) = \int f,$$

the integral with respect to the counting measure. In particular, $\int f$ exists.

Proof: Taking $\varepsilon = 1$ in (5.7), we find that X has a finite subset F for which $|\sum_{x \in G} f(x)| < |\sum_{x \in X}^{\text{un}} f(x)| + 1$ whenever $F \subseteq G \subseteq X$ and G is finite. Since F can have only finitely many subsets, some real $\alpha > 0$ satisfies

$$|\sum_{x \in G} f(x)| < \alpha \quad \text{whenever } G \subseteq F \text{ or } F \subseteq G \subseteq X \text{ and } G \text{ is finite.} \qquad (5.8)$$

Now suppose $\sum_{x \in X} f^+(x) = \infty$. Then there exists a finite subset $G' \subseteq X$ for which $\sum_{x \in G'} f^+(x) > 2\alpha$. Let H be the subset of G' on which $f(x) > 0$, so that $f^+ = f$ on H and $f^+ = 0$ on $G' \backslash H$. Then we have $\sum_{x \in H} f(x) = \sum_{x \in H} f^+(x) = \sum_{x \in G'} f^+(x)$ and hence

$$\sum_{x \in H} f(x) > 2\alpha.$$

Let $G_1 = F \backslash H$ and $G_2 = H \cup F$. By (5.8), it follows that

$$|\sum_{x \in G_1} f(x)| < \alpha$$

because G_1 is a subset of F, and also that

$$|\sum_{x \in G_2} f(x)| < \alpha$$

because G_2 is a finite subset containing F. But H and G_1 are disjoint with union G_2, all of them being finite. Therefore

$$\sum_{x \in G_2} f(x) = \sum_{x \in H} f(x) + \sum_{x \in G_1} f(x).$$

This is plainly in contradiction with the preceding three inequalities. Therefore $\sum_{x \in X} f^+(x)$ is finite. A similar argument shows that $\sum_{x \in X} f^-(x)$ is also finite.

Once we know that $\sum_{x \in X} f^+(x)$, $\sum_{x \in X} f^-(x)$ are both finite, it follows from Remark 5.2.6(d) that $\sum_{x \in X} f^+(x) - \sum_{x \in X} f^-(x) = \sum_{x \in X}^{\text{un}} f(x) = \sum_{x \in X}^{\text{un}} f^+(x) - \sum_{x \in X}^{\text{un}} f^-(x)$.

The last part about the integral now follows by virtue of Theorem 5.2.1.   □

5.2.8. Theorem. *Suppose $f : X \to \mathbb{R}$ is a real-valued function on a set X and the integral $\int f$ with respect to the counting measure is finite. Then $\sum_{x \in X} f^+(x)$, $\sum_{x \in X} f^-(x)$ are both finite and $\sum_{x \in X} f^+(x) - \sum_{x \in X} f^-(x) = \int f$. Moreover, the unconditional sum $\sum_{x \in X}^{\text{un}} f(x)$ exists and*

$$\sum_{x \in X}^{\text{un}} f(x) = \sum_{x \in X}^{\text{un}} f^+(x) - \sum_{x \in X}^{\text{un}} f^-(x) = \int f.$$

Furthermore,

$$\int |f| = \sum_{x \in X} |f(x)| = \sum_{x \in X} f^+(x) + \sum_{x \in X} f^-(x) = \sum_{x \in X}^{\text{un}} f^+(x) + \sum_{x \in X}^{\text{un}} f^-(x) = \sum_{x \in X}^{\text{un}} |f(x)|.$$

Proof: Since $\int f$ is finite, $\int f = \int f^+ - \int f^-$ and both the integrals on the right are finite. But by Theorem 5.2.1, $\int f^+ = \sum_{x \in X} f^+(x)$ and $\int f^- = \sum_{x \in X} f^-(x)$. Therefore $\sum_{x \in X} f^+(x)$, $\sum_{x \in X} f^-(x)$ are both finite and $\sum_{x \in X} f^+(x) - \sum_{x \in X} f^-(x) = \int f$. The rest is a consequence of Theorem 5.2.1, the finiteness and Remark 5.2.6(d), (a).   □

Problem Set 5.2

5.2.P1. Complete the proof of Theorem 5.2.3 by showing for any function $a : X \times Y \to \mathbb{R}^{*+}$ that

$$\sum_{(x,y) \in X \times Y} a(x, y) \leq \sum_{x \in X} \sum_{y \in Y} a(x, y).$$

5.2.P2. Let $f : X \times Y \to \mathbb{R}$ be a real-valued function on the Cartesian product $X \times Y$ and suppose that the unconditional sum $\sum_{(x,y) \in X \times Y}^{\text{un}} f(x, y)$ exists (necessarily finite by definition). Show that (i) $\sum_{y \in Y}^{\text{un}} f(x, y)$ exists for each $x \in X$, (ii) $\sum_{x \in X}^{\text{un}} f(x, y)$ exists for each $y \in Y$, and (iii) $\sum_{x \in X}^{\text{un}} \sum_{y \in Y}^{\text{un}} f(x, y) = \sum_{y \in Y}^{\text{un}} \sum_{x \in X}^{\text{un}} f(x, y)$.

5.2.P3. (Cf. Problem 5.1.P4) Suppose the real-valued function $f : X \times Y \to \mathbb{R}$ has the property that there exists a map $\phi : X \to \mathbb{R}^+$ such that $\sum_{x \in X} \phi(x) < \infty$ and $|\sum_{y \in F} f(x, y)| \leq \phi(x)$ for each finite subset $F \subseteq Y$ and each $x \in X$. Show that (i) $\sum_{x \in X}^{\text{un}} f(x, y)$ exists for each $y \in Y$, $\sum_{y \in Y}^{\text{un}} f(x, y)$ exists for each $x \in X$, (ii) the repeated sum $\sum_{x \in X}^{\text{un}} \sum_{y \in Y}^{\text{un}} f(x, y)$ exists, and (iii) if the other repeated sum $\sum_{y \in Y}^{\text{un}} \sum_{x \in X}^{\text{un}} f(x, y)$ also exists, then the two repeated sums are equal: $\sum_{x \in X}^{\text{un}} \sum_{y \in Y}^{\text{un}} f(x, y) = \sum_{y \in Y}^{\text{un}} \sum_{x \in X}^{\text{un}} f(x, y)$.

Chapter 6
Product Measures

6.1 Algebras and Monotone Classes of Sets

After Theorem 5.2.3, the question was raised whether there is a measure on $X \times Y$ such that the integral with respect to that measure, a "double integral", agrees with the repeated integrals even when the measures on X and Y are not counting measures. In this section, we begin to lay out the background for obtaining a σ-algebra in the Cartesian product $X \times Y$, which would be suitable for such a measure. The σ-algebra will only be set up in Sect. 6.2 and the measure in Sect. 6.4.

6.1.1. Definition. *A family $\mathcal{A}$ of subsets of a nonempty set X is called an* **algebra** *if*

(a) $\emptyset \in \mathcal{A}$ (contains the empty set),
(b) $A \cup B \in \mathcal{A}$ whenever $A \in \mathcal{A}$, $B \in \mathcal{A}$ (closed under unions),
(c) $A^c \in \mathcal{A}$ whenever $A \in \mathcal{A}$ (closed under complements).

Since $X = \emptyset^c$, conditions (a) and (c) in the definition together imply that $X \in \mathcal{A}$. Also, conditions (b) and (c) together imply closure under intersections, considering that

$$A \cap B = (A^c \cup B^c)^c.$$

It is immediate that an algebra is closed under finite unions and intersections as well:

$$\text{if } A_k \in \mathcal{A} \text{ for } 1 \leq k \leq n, \text{ then } \bigcup_{k=1}^{n} A_k \in \mathcal{A} \text{ and } \bigcap_{k=1}^{n} A_k \in \mathcal{A}.$$

Moreover, if A and B belong to $\mathcal{A}$, then so does the set-theoretic difference $A\backslash B$, which is defined to be $A \cap B^c$. These facts and their simple consequences will be used without further ado.

© Springer Nature Switzerland AG 2018
S. Shirali, *A Concise Introduction to Measure Theory*,
https://doi.org/10.1007/978-3-030-03241-8_6

Since the intersection of any family of σ-algebras is again a σ-algebra (this is easy to prove), it is clear that given any family S of subsets of X, there always exists a smallest σ-algebra containing S. Here "smallest" means contained in every σ-algebra containing S. The intersection of the family of all σ-algebras containing S serves the purpose. It is usually called the **σ-algebra generated by** S. When S is an algebra, there is an alternative description of the σ-algebra generated by it.

6.1.2. Definition. *A* **monotone class** *is a family* $\mathfrak{M}$ *of subsets of a nonempty set* X *such that*:

(a) $A_1 \subseteq A_2 \subseteq \cdots$ and each $A_j \in \mathfrak{M}$ implies $\bigcup_{j=1}^{\infty} A_j \in \mathfrak{M}$,
(b) $B_1 \supseteq B_2 \supseteq \cdots$ and each $B_j \in \mathfrak{M}$ implies $\bigcap_{j=1}^{\infty} B_j \in \mathfrak{M}$.

Any σ-algebra is a monotone class. When $A \subseteq X$, where X is any nonempty set, the family $\mathfrak{M} = \{A\}$ is a monotone class which is not a σ-algebra. As with σ-algebras, there is always a **monotone class generated by** a family of subsets of X. The alternative description promised above is as follows.

6.1.3. Proposition. *The σ-algebra generated by an algebra is also the smallest monotone class generated by the algebra.*

Proof: Let A be an algebra of subsets of X and $\mathfrak{M}_0$ be the monotone class generated by A, which is to say, the smallest monotone class containing A. Since any σ-algebra is a monotone class, the σ-algebra generated by A certainly contains $\mathfrak{M}_0$. The converse will follow if we show that $\mathfrak{M}_0$ is a σ-algebra.

First we argue that $\mathfrak{M}_0$ is closed under complementation. Let $\mathfrak{M}_0' = \{A \in \mathfrak{M}_0 : A^c \in \mathfrak{M}_0\}$, the family of those subsets in $\mathfrak{M}_0$ whose complements also belong to $\mathfrak{M}_0$. Since A is an algebra and $\mathfrak{M}_0 \supseteq A$, we have $\mathfrak{M}_0' \supseteq A$. Now, it is obvious that $\mathfrak{M}_0'$ is also a monotone class, because $(\bigcup_{j=1}^{\infty} A_j)^c = \bigcap_{j=1}^{\infty} A_j^c$ and $(\bigcap_{j=1}^{\infty} A_j)^c = \bigcup_{j=1}^{\infty} A_j^c$. Therefore $\mathfrak{M}_0' \supseteq \mathfrak{M}_0$, which shows that $\mathfrak{M}_0$ is closed under complementation.

We proceed to argue that $\mathfrak{M}_0$ is closed under unions.

As a preliminary step, we show that $A \cup F \in \mathfrak{M}_0$ for every $A \in A$ and every $F \in \mathfrak{M}_0$. Consider any $A \in A$ and set $U(A) = \{F \in \mathfrak{M}_0 : A \cup F \in \mathfrak{M}_0\}$. We shall first see why $U(A)$ is a monotone class containing A. Since A is an algebra and $\mathfrak{M}_0 \supseteq A$, we have $U(A) \supseteq A$. Now consider sets $E_1 \subseteq E_2 \subseteq \cdots$ with each $E_j \in U(A)$. Since $U(A) \subseteq \mathfrak{M}_0$ by definition and $\mathfrak{M}_0$ is a monotone class, we have $\bigcup_{j=1}^{\infty} E_j \in \mathfrak{M}_0$. The definition of $U(A)$ yields $A \cup E_j \in \mathfrak{M}_0$ for each j; also, $A \cup E_1 \subseteq A \cup E_2 \subseteq \cdots$. As $\mathfrak{M}_0$ is a monotone class, it follows that $\bigcup_{j=1}^{\infty}(A \cup E_j) \in \mathfrak{M}_0$, i.e. $A \cup (\bigcup_{j=1}^{\infty} E_j) \in \mathfrak{M}_0$. Therefore $\bigcup_{j=1}^{\infty} E_j \in U(A)$. The other property required for $U(A)$ to be a monotone class can be established analogously. Thus $U(A)$ is a monotone class containing A, with the result that $U(A) \supseteq \mathfrak{M}_0$. But the reverse inclusion holds by definition of $U(A)$. Therefore $U(A) = \mathfrak{M}_0$. As this has been shown to hold for every $A \in A$, it follows that $A \cup F \in \mathfrak{M}_0$ for every $A \in A$ and every $F \in \mathfrak{M}_0$.

Now consider any $F \in \mathfrak{M}_0$, and set $V(F) = \{G \in \mathfrak{M}_0 : F \cup G \in \mathfrak{M}_0\}$. We shall demonstrate that $V(F)$ is a monotone class containing A. From the previous paragraph, we know that $V(F) \supseteq A$ and the same argument as before enables us to infer that $V(F)$ is a monotone class. Since $V(F) \subseteq \mathfrak{M}_0$ by definition, it follows that $V(F) = \mathfrak{M}_0$. As this has been shown to hold for every $F \in \mathfrak{M}_0$, we have $F \cup G \in \mathfrak{M}_0$ for all $F, G \in \mathfrak{M}_0$. This shows that $\mathfrak{M}_0$ is closed under unions.

We have so far shown that $\mathfrak{M}_0$ is an algebra and it remains to prove that $\mathfrak{M}_0$ is a σ-algebra. Let $F_n \in \mathfrak{M}_0$ for every n. For each n, consider $E_n = \cup_{k=1}^n F_k$. Then $E_1 \subseteq E_2 \subseteq \cdots$ and $\cup_{n=1}^\infty E_n = \cup_{n=1}^\infty F_n$. Since $\mathfrak{M}_0$ has been shown to be an algebra, each E_n belongs to it, which implies $\cup_{n=1}^\infty E_n \in \mathfrak{M}_0$ because $\mathfrak{M}_0$ is a monotone class. $\square$

Problem Set 6.1

6.1.P1. Suppose the set X consists of three elements a, b, c and $\mathcal{B}$ is the family of subsets $\{a\}$ and $\{b, c\}$. List all the subsets in (i) the smallest algebra of sets that contains $\mathcal{B}$ and (ii) the monotone class generated by $\mathcal{B}$.

6.1.P2. Let X consist of four elements a, b, c, d and S consist of the subsets of X described in unorthodox notation as $\emptyset$, a, bcd, bc, ad, $abcd$. Determine (i) whether S is an algebra, (ii) whether S is a monotone class, (iii) what the monotone class generated by S is, and (iv) what the σ-algebra generated by S is.

6.1.P3. Let X be an uncountable set and $\mathcal{B}$ be the family of all subsets that are either finite or have finite complement. Show that $\mathcal{B}$ is an algebra of subsets. Is it a σ-algebra?

6.1.P4. Let $\mathcal{A}$ be an algebra of subsets of a nonempty set X. If there are finitely many sets in $\mathcal{A}$, show that the number of sets in it is 2^n, where n is the number of nonempty sets B in the algebra (called 'atoms') having the property that the only set $B_1 \in \mathcal{A}$ satisfying $\emptyset \subset B_1 \subseteq B$ must be B.

6.1.P5. Let $\mathcal{F}$ be a σ-algebra of subsets of X generated by a subfamily $\mathcal{H}$ and let $A \in \mathcal{F}$. Show that the induced σ-algebra $\mathcal{F}_A$ is the same as the σ-algebra generated by the family $\mathcal{H}_A = \{H \cap A : H \in \mathcal{H}\}$ of intersections of the sets of $\mathcal{H}$ with A.

6.2 Defining a Product σ-Algebra

In what follows, $\mathcal{F}$ denotes a σ-algebra of subsets of X and $\mathcal{G}$ denotes a σ-algebra of subsets of Y. We shall define the anticipated σ-algebra in $X \times Y$ and also give an alternative description of it in terms of monotone classes (Corollary 6.2.9).

6.2.1. Definition. A **rectangle** is a subset $A \times B \subseteq X \times Y$, where $A \in \mathcal{F}$, $B \in \mathcal{G}$.

It is trivial that an intersection of two rectangles is a rectangle.

6.2.2. Exercise. Show that the representation of a rectangle in the form $A \times B$ is unique if and only if it is nonempty.

Solution: The representation of $\emptyset$ in the form $A \times B$ is not unique because $\emptyset = X \times \emptyset = \emptyset \times Y$. However, if $A \times B = P \times Q$ is nonempty, then we can show that $A = P$ and $B = Q$ by arguing thus: Since $A \times B$ is nonempty, B is also nonempty; so choose $y \in B$. Then $x \in A \Rightarrow (x, y) \in A \times B = P \times Q \Rightarrow (x, y) \in P \times Q \Rightarrow x \in P$, thereby validating the inclusion $A \subseteq P$. One may validate the inclusions $B \subseteq Q$, $P \subseteq A$ and $Q \subseteq B$ analogously.

6.2.3. Definition. *The* **product σ-algebra** *of $\mathcal{F}$ and $\mathcal{G}$ is the σ-algebra generated by rectangles and is denoted by $\mathcal{F} \times \mathcal{G}$* [the same symbol as for the Cartesian product although it is not the Cartesian product].

Our next objective is to extend the set function defined on the family of all rectangles $A \times B$ as $\mu(A)\nu(B)$ to a measure on the product σ-algebra $\mathcal{F} \times \mathcal{G}$. We begin by defining a class of sets that provides a stepping stone.

6.2.4. Definition. *A finite union of disjoint rectangles is called an* **elementary set**. *The family of elementary sets will be denoted by $\mathcal{E}$.*

Clearly, any σ-algebra containing all rectangles must also contain all elementary sets and therefore $\mathcal{F} \times \mathcal{G}$ is the same as the σ-algebra generated by elementary sets.

6.2.5. Proposition. *A union of finitely many disjoint elementary sets is an elementary set.*

Proof: Consider finitely many disjoint elementary sets $E_1, E_2, \ldots, E_n$. The rectangles comprising any one of the sets E_i are disjoint not only from each other but also from the rectangles comprising any other $E_j, j \neq i$. Thus the finitely many rectangles occurring in all the n unions forming the respective elementary sets E_i are all disjoint from each other. This means their union F is an elementary set. But F is the same as the union of $E_1, E_2, \ldots, E_n$. □

6.2.6. Proposition. *If $P_1, P_2, \ldots, P_m$ are disjoint rectangles and Q is a rectangle, then $(P_1 \backslash Q) \cup (P_2 \backslash Q) \cup \cdots \cup (P_m \backslash Q)$ is an elementary set.*

Proof: Relegated to Problem 6.2.P2. □

6.2.7. Proposition. *$\mathcal{E}$ is closed under intersections and differences. That is,*

$$E \in \mathcal{E}, F \in \mathcal{E} \Rightarrow E \cap F \in \mathcal{E}, F \backslash E \in \mathcal{E}. \tag{6.1}$$

Proof: Let

$$E = P_1 \cup P_2 \cup \cdots \cup P_m \text{ and } F = Q_1 \cup Q_2 \cup \cdots \cup Q_n,$$

where $P_1, P_2, \ldots, P_m$ are disjoint rectangles and $Q_1, Q_2, \ldots, Q_n$ are also disjoint rectangles. Then

$$E \cap F = \bigcup_{i=1}^{m} \bigcup_{j=1}^{n} (P_i \cap Q_j)$$

and $P_i \cap Q_j$ $(1 \le i \le m, 1 \le j \le n)$ are disjoint rectangles. So, $E \cap F \in \mathcal{E}$.

To complete the proof of (6.1), it remains to show that $E \backslash F \in \mathcal{E}$. We do so by induction on n. For $n = 1$, it is immediate from Proposition 6.2.6 that $E \backslash F \in \mathcal{E}$. So, assume as induction hypothesis that $E \backslash F \in \mathcal{E}$ whenever F is a finite disjoint union of k rectangles. When F is a finite disjoint union F of $k + 1$ rectangles $Q_j (1 \le j \le k+1)$, we have

$$
\begin{aligned}
E \backslash F &= E \backslash (Q_1 \cup Q_2 \cup \cdots \cup Q_{k+1}) = E \cap (Q_1^c \cap Q_2^c \cap \cdots \cap Q_k^c \cap Q_{k+1}^c) \\
&= \left(E \cap (Q_1^c \cap Q_2^c \cap \cdots \cap Q_k^c) \right) \cap Q_{k+1}^c = (E \backslash (Q_1 \cup Q_2 \cup \cdots \cup Q_k)) \cap Q_{k+1}^c.
\end{aligned}
$$

By the induction hypothesis, the set $E \backslash (Q_1 \cup Q_2 \cup \cdots \cup Q_k)$ on the right side here is an elementary set. Therefore the already established case when $n = 1$ leads to the conclusion that the entire right side is an elementary set and hence so is the left side $E \backslash F$. This completes the induction argument. $\qquad \square$

6.2.8. Corollary. $\mathcal{E}$ *is closed under taking finite unions and under complementation. That is,*

$$
E \in \mathcal{E}, F \in \mathcal{E} \Rightarrow E \cup F \in \mathcal{E}, E^c \in \mathcal{E}.
$$

Thus $\mathcal{E}$ is an algebra.

Proof: Since $E \cup F = E \cup (F \backslash E)$, the desired conclusion is a straightforward deduction from Proposition 6.2.5 and Proposition 6.2.7. $\qquad \square$

6.2.9. Corollary. $\mathcal{F} \times \mathcal{G}$ *is the monotone class generated by $\mathcal{E}$.*

Proof: The result follows from Corollary 6.2.8 and Proposition 6.1.3. $\qquad \square$

Problem Set 6.2

6.2.P1. Let $\mathcal{F}$ and $\mathcal{G}$ be σ-algebras of subsets of X and Y respectively. Suppose $X = \cup_{n=1}^{\infty} X_n$ and $Y = \cup_{k=1}^{\infty} Y_k$, where $\{X_n\}$ and $\{Y_k\}$ are disjoint sequences of sets. Then $\mathcal{F}_n = \{F \cap X_n : F \in \mathcal{F}\}$ and $\mathcal{G}_k = \{G \cap Y_k : G \in \mathcal{G}\}$ are σ-algebras of subsets of X_n and Y_k respectively. (If X_n and Y_k are all measurable, $\mathcal{F}_n$ and $\mathcal{G}_k$ are the induced σ-algebras in the sense of Sect. 4.4.) Show that every $E \in \mathcal{F} \times \mathcal{G}$ satisfies

$$
E \cap (X_n \times Y_k) \in \mathcal{F}_n \times \mathcal{G}_k \text{ for every } n \text{ and } k.
$$

6.2.P2. Prove Proposition 6.2.6.

6.2.P3. Let $\mathcal{F}$ and $\mathcal{G}$ be the σ-algebras consisting of all subsets of X and Y respectively, where X and Y are both countable. Show that $\mathcal{F} \times \mathcal{G}$ consists of all subsets of $X \times Y$.

6.2.P4. Let $\mathcal{F}$ and $\mathcal{G}$ be σ-algebras of subsets of X and Y respectively. Suppose $A \in \mathcal{F}$ and $B \in \mathcal{G}$. Then the σ-algebra $(\mathcal{F} \times \mathcal{G})_{A \times B}$ induced by the product σ-algebra $\mathcal{F} \times \mathcal{G}$ on $A \times B$ is the same as the product $\mathcal{F}_A \times \mathcal{G}_B$ of the induced σ-algebras $\mathcal{F}_A$ and $\mathcal{G}_B$. In particular, if $f : X \times Y \to \mathbb{R}^{**}$ is $(\mathcal{F} \times \mathcal{G})$-measurable then the restriction of f to $A \times B$ is $(\mathcal{F}_A \times \mathcal{G}_B)$-measurable.

6.3   Sections of a Subset and of a Function

The reader may recall the notation a_n, a^k introduced in Definition 5.1.1 and f^k in Theorem 5.1.4. The superscript or subscript there represented an element of $\mathbb{N}$. We now introduce the same concept again but with the superscript or subscript representing an element of a more general set. Since measurability, which was not an issue in Sect. 5.1, is now going to be of major concern, it is necessary to have a corresponding concept for sets as well. However, the concept itself does not depend on the availability of a σ-algebra and therefore the same is true of some of the results in this section.

6.3.1. Definition. *For any $x \in X$, the **x-section** of a subset $E \subseteq X \times Y$ is the subset $E_x = \{y \in Y : (x, y) \in E\}$ of Y. Similarly, for any $y \in Y$, the **y-section** is the subset $E^y = \{X \in X : (x, y) \in E\}$ of X.*

6.3.2. Examples.   (a)   Let $A = S \times T$, where $S \subseteq X$ and $T \subseteq Y$. Then

$$A_x = \left\{ \begin{array}{l} T \text{ if } x \in S \\ \emptyset \text{ if } x \notin S \end{array} \right\} \text{ and } A^y = \left\{ \begin{array}{l} S \text{ if } y \in T \\ \emptyset \text{ if } y \notin T \end{array} \right\}$$

(b)   Suppose $X = Y = \mathbb{R}$ and $A = \{(x, y) : x^2 + 4y^2 \le 4\} \subseteq \mathbb{R} \times \mathbb{R}$. Then

$$A_x = \left\{ \begin{array}{ll} [-\frac{1}{2}(4 - x^2)^{1/2}, \frac{1}{2}(4 - x^2)^{1/2}] \text{ if } |x| \le 2 \\ \emptyset & \text{otherwise} \end{array} \right.$$

and

$$A^y = \left\{ \begin{array}{ll} [-2(1 - y^2)^{1/2}, 2(1 - y^2)^{1/2}] \text{ if } |y| \le 1 \\ \emptyset & \text{otherwise.} \end{array} \right.$$

(c)   Let $A = \{(x, y) : 0 \le y \le x^2 \text{ and } x \ge 0\} \subseteq \mathbb{R} \times \mathbb{R}$. Then

$$A_x = \left\{ \begin{array}{ll} [0, x^2] \text{ if } x \ge 0 \\ \emptyset & \text{otherwise} \end{array} \right.$$

and

$$A^y = \begin{cases} (y^{1/2}, \infty) & \text{if } y \geq 0 \\ \emptyset & \text{otherwise.} \end{cases}$$

6.3.3. Proposition. $(E_x)^c = (E^c)_x$ and $(E^y)^c = (E^c)^y$. *Moreover, for any sequence of sets* $E_1, E_2, \ldots$, *we have* $\cup_{j=1}^{\infty}(E_j)_x = (\cup_{j=1}^{\infty}E_j)_x$ *and* $\cap_{j=1}^{\infty}(E_j)_x = (\cap_{j=1}^{\infty}E_j)_x$. *Similarly for y-sections and for finite sequences. In particular, when E and F are disjoint, E_x and F_x are disjoint and E^y and F^y are disjoint. If $A \subseteq X$ and $B \subseteq Y$, then*

$$\chi_{(A \times B)_x}(y) = \chi_A(x)\chi_B(y) = \chi_{(A \times B)^y}(x).$$

Proof: The assertions that $y \in (E_x)^c$ and that $y \in (E^c)_x$ are both equivalent to $(x, y) \notin E$. The argument that $(E^y)^c = (E^c)^y$ is analogous.

The assertions that $y \in \cup_{j=1}^{\infty}(E_j)_x$ and that $y \in (\cup_{j=1}^{\infty}E_j)_x$ are both equivalent to $(x, y) \in E_j$ for some j. This means $\cup_{j=1}^{\infty}(E_j)_x = (\cup_{j=1}^{\infty}E_j)_x$.

The assertions that $y \in \cap_{j=1}^{\infty}(E_j)_x$ and that $y \in (\cap_{j=1}^{\infty}E_j)_x$ are both equivalent to $(x, y) \in E_j$ for every j. This means $\cap_{j=1}^{\infty}(E_j)_x = (\cap_{j=1}^{\infty}E_j)_x$.

The argument for y-sections is similar. The equalities for finite sequences follow because $\emptyset_x$ and $\emptyset^y$ are both empty. In particular, $(E_x \cap F_x) = (E \cap F)_x$ and $(E^y \cap F^y) = (E \cap F)^y$, which justifies the assertion regarding disjointness.

The argument for the last part is that

$$\chi_{(A \times B)_x}(y) = 1 \Leftrightarrow y \in (A \times B)_x \Leftrightarrow (x, y) \in A \times B$$
$$\Leftrightarrow x \in A \text{ as well as } y \in B$$
$$\Leftrightarrow \chi_A(x) = 1 \text{ as well as } \chi_B(y) = 1$$
$$\Leftrightarrow \chi_A(x)\chi_B(y) = 1,$$

while at the same time,

$$(x, y) \in A \times B \Leftrightarrow x \in (A \times B)^y \Leftrightarrow \chi_{(A \times B)^y}(x) = 1.$$

$\square$

6.3.4. Lemma. *Suppose $E = A \times B$, where $A \subseteq X$ and $B \subseteq Y$. Then $E_x = B$ whenever $x \in A$ and $E_x = \emptyset$ whenever $x \notin A$; also, $E^y = A$ whenever $y \in B$ and $E^y = \emptyset$ whenever $y \notin B$. If $A \in \mathcal{F}$ and $B \in \mathcal{G}$, then $\nu(E_x) = \nu(B)\chi_A(x)$ and $\mu(E^y) = \mu(A)\chi_B(y)$.*

Proof: (The first part has been seen in Example 6.3.2(a) but we nevertheless give a detailed argument here.) By definition, $y \in E_x$ if and only if $(x, y) \in E$. Consider any $x \in A$. Since $E = A \times B$, therefore $(x, y) \in E$ if and only if $y \in B$. Thus $y \in E_x$ if and only if $y \in B$. This means $E_x = B$. Now consider any $x \notin A$. Then $(x, y) \notin E = A \times B$ for every $y \in Y$, which means $E_x = \emptyset$. The proof for E^y is analogous. The last part is now trivial. $\square$

6.3.5. Proposition. *Let $E \in \mathcal{F} \times \mathcal{G}$. Then $E_x \in \mathcal{G}$ for every $x \in X$, and $E^y \in \mathcal{F}$ for every $y \in Y$.*

Proof: Let $\mathcal{H}$ be the family of all $E \in \mathcal{F} \times \mathcal{G}$ for which $E_x \in \mathcal{G}$ for every $x \in X$ and $E^y \in \mathcal{F}$ for every $y \in Y$. It is a consequence of Proposition 6.3.3 that $E \in \mathcal{H} \Rightarrow E^c \in \mathcal{H}$ and that for any sequence of sets $E_1, E_2, \ldots$ in $\mathcal{H}$ we have $\cup_{j=1}^{\infty} E_j \in \mathcal{H}$. Thus $\mathcal{H}$ is a σ-algebra. Moreover, in view of Lemma 6.3.4, it contains all rectangles. By definition of the σ-algebra $\mathcal{F} \times \mathcal{G}$, it follows that $\mathcal{F} \times \mathcal{G} \subseteq \mathcal{H}$. On the other hand, $\mathcal{H} \subseteq \mathcal{F} \times \mathcal{G}$ by the definition of $\mathcal{H}$. Therefore $\mathcal{H} = \mathcal{F} \times \mathcal{G}$.   □

6.3.6. Definition. *Given $x \in X$, the **x-section of** an extended real-valued function f on $X \times Y$ is the extended real-valued function f_x on the domain Y such that $f_x(y) = f(x, y)$ for all $y \in Y$. Similarly for the **y-section** f^y with a given $y \in Y$.*

As the y-section and the yth power are denoted by the same symbol, one has to judge from the context which one is meant.

When f_x is integrable or has integral ∞, we shall denote $\int_Y f_x$ by the more elaborate symbol $\int_Y f(x, y) d\nu(y)$, which resembles the notation used in Calculus. Similarly, we shall denote the integral $\int_X f^y$ by $\int_X f(x, y) d\mu(x)$.

6.3.7. Exercise. If $E \subseteq X \times Y$, show that

$$\chi_{E_x}(y) = (\chi_E)_x(y) = (\chi_E)^y(x) = \chi_{E^y}(x) = \chi_E(x, y) \quad \text{for all } x \in X \text{ and } y \in Y.$$

If moreover $E \in \mathcal{F} \times \mathcal{G}$, then show that

$$\int_Y \chi_E(x, y) \, d\nu(y) = \int_Y (\chi_E)_x = \int_Y \chi_{E_x} = \nu(E_x) \text{ for all } x \in X$$

and

$$\int_X \chi_E(x, y) d\mu(x) = \int_X (\chi_E)^y = \int_X \chi_{E^y} = \mu(E^y) \quad \text{for all } y \in Y.$$

Solution: From the definition of section, we have

$$\chi_{E_x}(y) = 1 \text{ if } y \in E_x \quad \text{and} \quad 0 \text{ if } y \notin E_x$$
$$= 1 \text{ if}(x, y) \in E \quad \text{and} \quad 0 \text{ if}(x, y) \notin E,$$

and

$$(\chi_E)_x(y) = (\chi_E)^y(x) = \chi_E(x, y) = 1 \text{ if}(x, y) \in E \quad \text{and} \quad 0 \text{ if}(x, y) \notin E.$$

Similarly,

$$\chi_{E^y}(x) = 1 \text{ if}(x, y) \in E \quad \text{and} \quad 0 \text{ if}(x, y) \notin E.$$

This proves the first part.

For the second part, the equality of the integrals follows from the first part. We need to show only that $\int_Y \chi_{E_x} = \nu(E_x)$ and $\int_X \chi_{E^y} = \mu(E^y)$. This follows from the fact that the integral of the characteristic function of a measurable set is always equal to measure of that set.

6.3.8. Remark. It is trivial that $(\phi + \psi)_x = \phi_x + \psi_x$ and $(\alpha\phi)_x = \alpha(\phi_x)$, where α is a real number; correspondingly for y-sections.

It is a simple consequence of Exercise 6.3.7 and Lemma 6.3.4 that, if E is a rectangle $A \times B$, then

$$(\chi_E)_x = \chi_{E_x} = \chi_B \text{ when } x \in A \text{ and } (\chi_E)^y = \chi_{E^y} = \chi_A \text{ when } y \in B.$$

6.3.9. Proposition. *If the extended real-valued function f on $X \times Y$ is measurable, then its sections f_x and f^y are also measurable.*

Proof: Let α be any real number; we must show that $\{y \in Y : f_x(y) > \alpha\} \in \mathcal{G}$. This set is nothing but the section E_x, where $E = \{(x, y) \in X \times Y : f(x, y) > \alpha\}$. Since f is measurable, $E \in \mathcal{F} \times \mathcal{G}$. It follows by Proposition 6.3.5 that $E_x \in \mathcal{G}$. Thus f_x is a measurable function. A similar argument shows that f^y is also measurable. □

Problem Set 6.3
6.3.P1. Let X, Y and V be as in Example 5.1.3. For each $x \in X$, describe the set V_x and for each $y \in Y$, describe the set V^y. If $f = \chi_V$, describe the function f_x for each $x \in X$.

6.4 Defining a Product Measure

We now proceed to set forth a measure on $X \times Y$ that is intended to serve the purpose described in the opening paragraph of this chapter, but we do so under a certain restriction called "σ-finiteness". The question of whether it actually serves the intended purpose will be taken up in Sect. 6.5.

The definition of the measure in question presupposes a certain result involving sections of sets, which will be taken up shortly.

6.4.1. Definition. *A measure space $(X, \mathcal{F}, \mu)$ is called σ-finite if there exists a sequence $\{X_n\}_{n \geq 1}$ of subsets such that $\bigcup_{n=1}^{\infty} X_n = X$ and each X_n is $\mathcal{F}$-measurable with $\mu(X_n) < \infty$.*

One also speaks of the measure μ itself as being σ-finite.

The measure space $(\mathbb{R}, \mathcal{M}, m)$ is σ-finite [here m means Lebesgue measure] because the sets $X_n = [-n, n]$, where $n \in \mathbb{N}$, satisfy $\bigcup_{n=1}^{\infty} X_n = \mathbb{R}$, $X_n \in \mathcal{M}$ and $m(X_n) < \infty$ for each n. Clearly, the same is true if $\mathbb{R}$ is replaced by an interval. Let μ be the counting measure on the σ-algebra $\mathcal{P}(\mathbb{N})$ of all subsets of $\mathbb{N}$. Then the sets $X_n = \{n\}$, consisting of a single point each, satisfy the requirements that $\bigcup_{n=1}^{\infty} X_n = \mathbb{N}$

and $\mu(X_n) = 1 < \infty$ for each n. So, $(\mathbb{N}, \mathcal{P}(\mathbb{N}), \mu)$ is a σ-finite measure space. Of course, σ-finiteness holds for the counting measure on any countable set.

In the σ-finiteness condition, we are free to take the X_n to be disjoint because each X_n with $n > 1$ can be replaced by $X_n \setminus \cup_{k=1}^{n-1} X_k$.

6.4.2. Proposition. *Let $(X, \mathcal{F}, \mu)$ and $(Y, \mathcal{G}, \nu)$ be σ-finite measure spaces and $E \in \mathcal{F} \times \mathcal{G}$. Then*

(a) *the function that maps each $x \in X$ into $\nu(E_x)$ is measurable,*
(b) *the function that maps each $y \in Y$ into $\mu(E^y)$ is measurable,*
(c) $\int_X \nu(E_x) d\mu(x) = \int_Y \mu(E^y) d\nu(y).$

Proof: We first consider finite measures μ and ν.

Let Ω be the family of all subsets $E \in \mathcal{F} \times \mathcal{G}$ such that assertions (a), (b) and (c) hold. We shall show that $\Omega = \mathcal{F} \times \mathcal{G}$. It is enough to show that Ω contains every elementary set and that it is a monotone class contained in $\mathcal{F} \times \mathcal{G}$, because Corollary 6.2.9 will then imply $\Omega = \mathcal{F} \times \mathcal{G}$.

Consider a rectangle $E = A \times B$, where $A \in \mathcal{F}$ and $B \in \mathcal{G}$. By Lemma 6.3.4, $\nu(E_x) = \nu(B)\chi_A(x)$ and therefore $\nu(E_x)$ is a measurable function on X satisfying

$$\int_X \nu(E_x) d\mu(x) = \nu(B)\mu(A). \tag{6.2}$$

Similarly, $\mu(E^y)$ is a measurable function on Y and its integral over Y equals the right side of (6.2). Therefore a rectangle always belongs to Ω.

Now consider an elementary set $E = \cup_{n=1}^N (A_n \times B_n)$, a finite union of disjoint rectangles. By Proposition 6.3.3, we have

$$\chi_{E_x}(y) = \sum_{n=1}^N \chi_{(A_n \times B_n)_x}(y) = \sum_{n=1}^N \chi_{A_n}(x)\chi_{B_n}(y),$$

and hence

$$\nu(E_x) = \sum_{n=1}^N \left(\chi_{A_n}(x) \int_Y \chi_{B_n}(y) d\nu(y) \right) = \sum_{n=1}^N \nu(B_n)\chi_{A_n}(x).$$

This shows that $\nu(E_x)$ is a sum of measurable functions and therefore measurable. Moreover,

$$\int_X \nu(E_x) d\mu(x) = \sum_{n=1}^N \nu(B_n)\mu(A_n).$$

Similarly, $\mu(E^y)$ is a measurable function on Y and

$$\int_Y \mu(E^y) d\nu(y) = \sum_{n=1}^{N} \nu(B_n)\mu(A_n).$$

Thus the assertions (a), (b), (c) of the proposition hold for every elementary set, which means Ω contains every elementary set.

We proceed to show that Ω is a monotone class. Let $E_n \in \Omega, n \geq 1$, satisfy $E_n \subseteq E_{n+1}$ for every n, and put $E = \cup_{n=1}^{\infty} E_n$. Then $(E_n)_x \subseteq (E_{n+1})_x$ and $(E_n)^y \subseteq (E_{n+1})^y$ for every $x \in X$ and $y \in Y$. Hence the sequences of nonnegative measurable functions $\{\nu((E_n)_x)\}_{n\geq 1}$ and $\{\mu((E_n)^y)\}_{n\geq 1}$ are increasing, and $\lim_{n\to\infty} \nu((E_n)_x) = \nu(E_x)$, $\lim_{n\to\infty} \mu((E_n)^y) = \mu(E^y)$. This means $\nu(E_x)$ and $\mu(E^y)$ are limits of measurable functions and are therefore measurable, i.e. E satisfies (a) and (b). We have to show that E satisfies (c) as well.

By the Monotone Convergence Theorem 2.3.2, we have

$$\int_X \nu(E_x) d\mu(x) = \lim_{n\to\infty} \int_X \nu((E_n)_x) d\mu(x)$$

and

$$\int_Y \mu(E^y) d\nu(y) = \lim_{n\to\infty} \int_Y \mu((E_n)^y) d\nu(y).$$

Since $E_n \in \Omega$ for each $n \geq 1$, we have

$$\int_X \nu((E_n)_x) d\mu(x) = \int_Y \mu((E_n)^y) d\nu(y).$$

Therefore, it follows from the limits noted above that

$$\int_X \nu(E_x) d\mu(x) = \int_Y \mu(E^y) d\nu(y).$$

Consequently, $E = \cup_{n=1}^{\infty} E_n$ satisfies (c) as well, so that $E \in \Omega$. Similarly, if $E_n \in \Omega$ and $E_n \supseteq E_{n+1}$, $n \geq 1$, we can conclude that $E = \cap_{n=1}^{\infty} E_n \in \Omega$ by using the Dominated Convergence Theorem 3.2.6 and the hypothesis that μ and ν are both finite measures. This completes the proof that Ω is a monotone class, thereby establishing the contention of the proposition for the case of finite measures.

For the infinite but σ-finite case, we have $X = \cup_{n=1}^{\infty} X_n$, $Y = \cup_{k=1}^{\infty} Y_k$, where $\{X_n\}$ and $\{Y_k\}$ are disjoint sequences of nonempty sets of finite measure. Denote by $(X_n, \mathcal{F}_n, \mu_n)$ and $(Y_k, \mathcal{G}_k, \nu_k)$ the measure spaces induced on X_n and Y_k respectively. Consider an arbitrary $E \in \mathcal{F} \times \mathcal{G}$ and take $E_{n,k} = E \cap (X_n \times Y_k)$. Then $E_{n,k} \in \mathcal{F} \times \mathcal{G}$ and hence by Proposition 6.3.5, its sections are measurable. Moreover, considering that the sets $E_{n,k}$ are disjoint and their union is E, we infer on the basis of Proposition

6.3.3 that the x-sections $(E_{n,k})_x$ are disjoint and their union is E_x; correspondingly for y-sections. Therefore

$$\nu(E_x) = \sum_{n=1}^{\infty}\sum_{k=1}^{\infty}(\nu(E_{n,k})_x) \quad\text{and}\quad \mu(E^y) = \sum_{n=1}^{\infty}\sum_{k=1}^{\infty}\mu((E_{n,k})^y). \tag{6.3}$$

By Problem 6.2.P1, we have $E_{n,k} \in \mathcal{F}_n \times \mathcal{G}_k$, and hence by what has been proved in the case of finite measures, it follows that the functions of x and y defined by $\nu_k((E_{n,k})_x)$ and $\mu_n((E_{n,k})^y)$ on X_n and Y_k respectively are measurable, and that

$$\int_{X_n} \nu((E_{n,k})_x)d\mu_n(x) = \int_{Y_k}\mu((E_{n,k})^y)d\nu_k(y). \tag{6.4}$$

Now, the functions defined on the whole of X and the whole of Y by $\nu((E_{n,k})_x)$ and $\mu((E_{n,k})^y)$ respectively agree with the aforementioned functions on X_n and Y_k while being zero elsewhere. In other words, they agree with their own respective products with the characteristic functions of X_n and Y_k and their restrictions to the latter are precisely the aforementioned functions. Upon invoking Proposition 4.4.1, we find that they are measurable and

$$\int_X \nu((E_{n,k})_x)d\mu(x) = \int_{X_n}\nu((E_{n,k})_x)d\mu_n(x)$$

and

$$\int_Y \mu((E_{n,k})^y)d\nu(y) = \int_{Y_k}\mu((E_{n,k})^y)d\nu_k(y).$$

However, the last recorded equality of integrals above, (6.4), says that the right sides here are equal. It therefore follows that the left sides here are equal:

$$\int_X \nu((E_{n,k})_x)d\mu(x) = \int_Y\mu((E_{n,k})^y)d\nu(y). \tag{6.5}$$

Since the terms in the series $\sum_{k=1}^{\infty}\nu((E_{n,k})_x)$ and $\sum_{k=1}^{\infty}\mu((E_{n,k})^y)$ have been shown to be measurable functions on X and Y respectively, the same is true of the functions represented by their sums and hence also of the functions represented by their repeated sums $\sum_{n=1}^{\infty}\sum_{k=1}^{\infty}\nu((E_{n,k})_x)$ and $\sum_{n=1}^{\infty}\sum_{k=1}^{\infty}\mu((E_{n,k})^y)$. However, the repeated sums are precisely the right sides in (6.3). Therefore the functions represented by the left sides $\nu(E_x)$ and $\mu(E^y)$ in (6.3) are measurable, showing that assertions (a) and (b) of the proposition hold.

Finally, in view of (6.3), (6.5) and the Monotone Convergence Theorem,

$$\int_X \nu(E_x)d\mu(x) = \sum_{n=1}^{\infty}\sum_{k=1}^{\infty}\int_{X_n} \nu\big((E_{n,k})_x\big)d\mu(x)$$

$$= \sum_{n=1}^{\infty}\sum_{k=1}^{\infty}\int_{Y_k} \mu\big((E_{n,k})^y\big)d\nu(y) = \int_Y \mu(E^y)d\nu(y),$$

showing that assertion (c) of the proposition also holds. $\square$

Example 5.1.3 shows that the σ-finiteness condition in the above proposition is essential, because the set V there can be shown to be measurable as follows.

For any positive integer n, set $I_j = \left[\frac{j-1}{n}, \frac{j}{n}\right]$ and $V_n = \cup_{j=1}^{n}(I_j \times I_j)$. Clearly, each V_n is a union of measurable rectangles and therefore measurable, and hence so is $\cap_{n=1}^{\infty}V_n$. But this intersection is precisely V.

In the definition about to be presented, the equality of the integrals is assured by Proposition 6.4.2.

6.4.3. Definition. *Let $(X, \mathcal{F}, \mu)$ and $(Y, \mathcal{G}, \nu)$ be σ-finite measure spaces. Then the set function $\mu \times \nu$ defined on $\mathcal{F} \times \mathcal{G}$ by*

$$(\mu \times \nu)(E) = \int_X \nu(E_x)\,d\mu(x) = \int_Y \mu(E^y)d\nu(y)$$

is called the **product measure** *of μ and ν.*

6.4.4. Theorem. *Let $(X, \mathcal{F}, \mu)$ and $(Y, \mathcal{G}, \nu)$ be σ-finite measure spaces. Then the product $\mu \times \nu$ defined on the σ-algebra $\mathcal{F}$ as*

$$(\mu \times \nu)(E) = \int_X \nu(E_x)d\mu(x) = \int_Y \mu(E^y)d\nu(y)$$

is a σ-finite measure.

Proof: Relegated to Problem 6.4.P1. $\square$

6.4.5. Remark. In view of Exercise 6.3.7, $\int_Y \chi_E(x, y)d\nu(y) = \int_Y (\chi_E)_x = \nu(E_x)$ while $\int_X \chi_E(x, y)d\mu(x) = \int_X (\chi_E)^y = \mu(E^y)$. Parts (a) and (b) of Proposition 6.4.2 assert that these two functions of x and y respectively are measurable; part (c) asserts that

$$\int_X \left(\int_Y (\chi_E)_x\right) = \int_Y \left(\int_X (\chi_E)^y\right),$$

or in alternate notation,

$$\int_X d\mu(x) \int_Y \chi_E(x,y) dv(y) = \int_Y dv(y) \int_X \chi_E(x,y) d\mu(x)$$

Since $\mu \times v$ is a measure by Theorem 6.4.4, we have

$$(\mu \times v)(E) = \int_{X \times Y} \chi_E d(\mu \times v) = \int_{X \times Y} \chi_E(x,y) d(\mu \times v)(x,y).$$

Therefore on the basis of Definition 6.4.3,

$$\int_X \left(\int_Y (\chi_E)_x \right) = \int_{X \times Y} \chi_E d(\mu \times v) = \int_Y \left(\int_X (\chi_E)^y \right),$$

which may also be written as

$$\int_X d\mu(x) \int_Y \chi_E(x,y) dv(y) = \int_{X \times Y} \chi_E(x,y) d(\mu \times v)(x,y) = \int_Y dv(y) \int_X \chi_E(x,y) d\mu(x).$$

In other words, the "double integral" of the characteristic function of a measurable subset of the product space equals both repeated integrals of that function. It should be noted that measurability of the functions $\int_Y \chi_E(x,y) dv(y)$ and $\int_X \chi_E(x,y) d\mu(x)$ is not assumed but claimed. In the next section, we shall prove the equality of double and repeated integrals for a larger class of measurable functions on the product space.

6.4.6 Remark. When $(X, \mathcal{F}, \mu)$ is a measure space, a measurable subset of X is said to be σ-finite if the measure space induced on it is σ-finite.

Consider an integrable function $f : X \to \mathbb{R}^*$. By Problem 2.2.P12, for each $n \in \mathbb{N}$, we have $\mu(X(|f| > \frac{1}{n})) \leq n \int |f| < \infty$. Therefore the set $X(|f| > 0) = \cup_{n=1}^\infty X(|f| > \frac{1}{n})$ is σ-finite. This is the analog of Proposition 5.1.9(b) for a general measure that need not be a counting measure.

Problem Set 6.4
6.4.P1. Prove Theorem 6.4.4.

6.4.P2. For any positive integer n, set $I_j = \left[\frac{j-1}{n}, \frac{j}{n} \right]$. For the set V of Example 5.1.3, show that $V = \cap_{n=1}^\infty V_n$, where $V_n = \cup_{j=1}^n (I_j \times I_j)$.

6.5 The Tonelli and Fubini Theorems

We now have the requisite machinery for formally defining what we have informally called a "double integral" so far, and for relating it to repeated integrals. There are two well-known theorems in this regard. Both of them validate interchanging the order of integration under certain conditions provided the measures are σ-finite.

6.5.1. Definition. *Given σ-finite measure spaces* $(X, \mathcal{F}, \mu)$ *and* $(Y, \mathcal{G}, \nu)$, *any integral with respect to the product measure* $\mu \times \nu$ *is called a* **double integral**.

6.5.2. Tonelli's Theorem. *Let* $(X, \mathcal{F}, \mu)$ *and* $(Y, \mathcal{G}, \nu)$ *be σ-finite measure spaces and f be a nonnegative extended real-valued* $(\mathcal{F} \times \mathcal{G})$-*measurable function on* $X \times Y$. *Then* $\int_Y f(x, y)d\nu(y)$ *is an* $\mathcal{F}$-*measurable function on* X, $\int_X f(x, y)d\mu(x)$ *is a* $\mathcal{G}$-*measurable function on* Y, *and the double integral of f agrees with each of its repeated integrals*:

$$\int_X d\mu(x) \int_Y f(x, y)d\nu(y) = \int_{X \times Y} f(x, y)d(\mu \times \nu)(x, y) = \int_Y d\nu(y) \int_X f(x, y)d\mu(x).$$

Proof: It is sufficient to prove the first equality. By Proposition 6.3.9, the integral $\int_Y f(x, y)d\nu(x)$ exists for each $x \in X$. We need to prove that it provides a measurable function on X with integral equal to $\int_{X \times Y} f d(\mu \times \nu)$.

As observed in Remark 6.4.5, this is true when f is the characteristic function of an $(\mathcal{F} \times \mathcal{G})$-measurable set. It follows immediately from Remark 6.3.8 that the same is true also when f is any simple function. Now let f be an arbitrary non-negative extended real-valued $(\mathcal{F} \times \mathcal{G})$-measurable function on $X \times Y$. There exists an increasing sequence of nonnegative simple functions s_k, $k \in \mathbb{N}$, having limit f (see Theorem 3.2.1). Since the required conclusion is known to be valid for simple functions, each $\int_Y s_k(x, y)d\nu(y)$ is an $\mathcal{F}$-measurable function on X and

$$\int_X d\mu(x) \int_Y s_k(x, y)d\nu(y) = \int_{X \times Y} s_k(x, y)d(\mu \times \nu)(x, y). \tag{6.6}$$

By applying the Monotone Convergence Theorem 2.3.2, we obtain

$$\lim_{k \to \infty} \int_Y s_k(x, y)d\nu(y) = \int_Y f(x, y)d\nu(y),$$

from which we know that $\int_Y f(x, y)d\nu(y)$ is an $\mathcal{F}$-measurable function on X. More-over, the $\mathcal{F}$-measurable functions $\int_Y s_k(x, y)d\nu(y)$ form an increasing sequence, and hence by another application of the Monotone Convergence Theorem, we obtain

$$\lim_{k \to \infty} \int_X d\mu(x) \int_Y s_k(x, y)d\nu(y) = \int_X d\mu(x) \int_Y f(x, y)d\nu(y),$$

which is the same as

$$\lim_{k \to \infty} \int_{X \times Y} s_k(x, y)d(\mu \times \nu)(x, y) = \int_X d\mu(x) \int_Y f(x, y)d\nu(y)$$

in the light of (6.6). Finally, one further application of the Monotone Convergence Theorem yields

$$\lim_{k\to\infty} \int_{X\times Y} s_k(x,y)d(\mu\times\nu)(x,y) = \int_{X\times Y} f(x,y)d(\mu\times\nu)(x,y).$$

The required equality now springs forth from the preceding two. □

6.5.3. Remark. Given any measure space $(X,\mathcal{F},\mu)$ and an integrable function f on X, the subset $E \subseteq X$ on which f takes the value ∞, obviously measurable, must have measure 0. This is because $f^+ \geq n\chi_E$ for every positive integer n, which implies $\int_X f^+ \geq n\int_X \chi_E \geq n\mu(E)$ for every positive integer, thereby contradicting the finiteness of $\int_X f$ unless $\mu(E)=0$. It is a consequence that there exists a measurable subset F (e.g. E^c) on which f is finite and $\mu(F^c)=0$. Such an F need not be precisely the complement of E and could be a smaller set instead, because "removing" a nonempty subset of measure 0 from any such F leads to a smaller subset with the property that f is finite at each point of it and its complement has measure 0.

It is often convenient to avoid explicit mention of the set of measure 0 by describing the relevant property, such as being finite, that holds at each point of its complement as holding **almost everywhere**. Phrased in this terminology, the remark in the preceding paragraph asserts that an integrable function is finite almost everywhere.

If the property has been formulated in terms of an explicitly named point $x \in X$, then one can express the same thing by saying that the property holds for **almost all** x. The standard abbreviation for "almost everywhere" is **a.e.** Thus for instance, the assertion that $f = g$ a.e. means the same thing as "$f(x) = g(x)$ a.e." or "$f(x) = g(x)$ for almost all x".

6.5.4. Remarks. (a) Before stating Theorem 5.1.10 about interchanging the order of summation, i.e. integration with the counting measure, over *un*countable sets, we had mentioned the idea of reducing the matter to the countable case by exploiting Proposition 5.1.9(b). Now that we have the analog of Proposition 5.1.9(b) for a general measure (see Remark 6.4.6), the same idea would suggest reducing the matter of interchanging the order of integration with *non*-σ-finite measures to the σ-finite case, which has already been established in Tonelli's Theorem 6.5.2. Accepting such a reduction would lead to the conclusion that interchanging the order of integration with respect to measures that may not be σ-finite is valid. However, we know from Example 5.1.3 that this conclusion is definitely erroneous. This shows that reduction to the countable case is a fallacious line of reasoning. In particular, interchanging the order of summation over uncountable sets is not a special case of Tonelli's Theorem.

(b) Suppose ϕ and ψ are integrable nonnegative extended real-valued functions. It makes no sense to speak of a function $\phi - \psi$, because ψ may take the value ∞. By Remark 6.5.3, the functions ϕ and ψ are finite a.e. That is, there exist measurable sets E and F such that $\mu(E^c)=\mu(F^c)=0$, and the functions ϕ and ψ are finite-valued on E and F respectively. Then they are both finite-valued on the set $E \cap F$ and $\mu((E \cap F)^c)=0$. Thus, it is true almost everywhere that both are finite-valued. There is a function $\phi - \psi$ on the domain $E \cap F$, that is to say, $\phi - \psi$ is defined a.e. If we extend it to all of X by setting it equal to 0 outside $E \cap F$, the resulting function Φ is the same as $\phi\chi_{E\cap F} - \psi\chi_{E\cap F}$, which is a measurable function on X. Now, $\int_X \phi\chi_{E\cap F} = \int_X \phi$ and $\int_X \psi\chi_{E\cap F} = \int_X \psi$ by Problem 2.2.P7. Since ϕ and ψ are integrable, so are $\phi\chi_{E\cap F}$ and $\psi\chi_{E\cap F}$, and therefore also Φ; moreover, $\int_X \Phi = \int_X (\phi\chi_{E\cap F} - \psi\chi_{E\cap F}) = \int_X \phi\chi_{E\cap F} - \int_X \psi\chi_{E\cap F} = \int_X \phi - \int_X \psi$. We thus see that there exists an integrable function Φ that agrees with $\phi - \psi$ almost everywhere and satisfies $\int_X \Phi = \int_X \phi - \int_X \psi$.

(c) Let $\{f_n\}_{n\geq 1}$ be a sequence of measurable nonnegative functions on a measure space X and f a measurable function such that $\lim_{n\to\infty} f_n(x) = f(x)$ for almost all $x \in X$. It follows from Fatou's Lemma 3.2.3, as we shall presently show, that

$$\int f \leq \liminf_{n\to\infty} \int f_n.$$

By definition of almost all, there exists a measurable set $A \subseteq X$ such that $\mu(A^c)=0$ and $\lim_{n\to\infty} f_n(x) = f(x)$ for all $x \in A$. It follows that $\lim_{n\to\infty} f_n(x)\chi_A(x) = f(x)\chi_A(x)$ for all $x \in A$. Therefore by Fatou's Lemma, $\int f\chi_A \leq \liminf_{n\to\infty} \int f_n\chi_A$. But also, $f(x)\chi_A(x)=f(x)$ and $f_n(x)\chi_A(x)=f_n(x)$ for all $x \in A$. Therefore by Problem 2.2.P7, $\int f\chi_A = \int f$ and $\int f_n\chi_A = \int f_n$.

(d) Now suppose Φ and Φ_1 are integrable functions and g is some function defined almost everywhere such that both Φ and Φ_1 agree with g almost everywhere. Then Φ and Φ_1 agree with each other almost everywhere, that is, they agree on some set G with $\mu(G^c)=0$. By Problem 2.2.P7,

$$\int_X \Phi = \int_X \Phi^+ - \int_X \Phi^- = \int_X \Phi^+\chi_G - \int_X \Phi^-\chi_G = \int_X \Phi_1^+\chi_G - \int_X \Phi_1^-\chi_G$$

$$= \int_X \Phi_1^+ - \int_X \Phi_1^- = \int_X \Phi_1.$$

By $\int_X g$ we shall understand $\int_X \phi$, where Φ is *any* integrable function that agrees with g almost everywhere. In part (c) of Fubini's Theorem below, the first and third integrals are understood in this sense, keeping part (b) of the theorem in mind.

6.5.5. Fubini's Theorem. *Let $(X, \mathcal{F}, \mu)$ and $(Y, \mathcal{G}, \nu)$ be σ-finite measure spaces and the real-valued function f on $X \times Y$ be integrable. Then*

(a) *f_x is integrable for almost all x, and f^y is integrable for almost all y;*
(b) *there exist integrable functions on X and Y, which agree almost everywhere with $\int_Y f(x, y)d\nu(y)$ and $\int_X f(x, y)d\mu(x)$ respectively;*

(c) *the double integral of f equals both its repeated integrals*:

$$\int_X d\mu(x) \int_Y f(x,y)dv(y) = \int_{X\times Y} f(x,y)d(\mu\times v)(x,y) = \int_Y dv(y) \int_X f(x,y)d\mu(x).$$

Proof: By Proposition 6.3.9, the extended real-valued function f_x on Y is measurable for each $x\in X$. Correspondingly for f^y.

Since $f^+ \geq 0$ is measurable (see the observation just before Definition 2.2.5), it follows by Tonelli's Theorem 6.5.2 that $\int_Y f^+(x,y)dv(y)$ is a measurable function on X and that its integral over X satisfies $\int_X d\mu(x)\int_Y f^+(x,y)dv(y) = \int_{X\times Y} f^+ d(\mu\times v)(x,y)$. Since f is integrable, it follows that

$$\int_X d\mu(x) \int_Y f^+(x,y)dv(y) < \infty. \tag{6.7}$$

By Remark 6.5.3, this has the consequence that the measurable function on X given by $\int_Y f^+(x,y)dv(y)$ is finite almost everywhere. Since $\int_Y f^+(x,y)dv(y)$ means the same thing as $\int_Y (f^+)_x$, the preceding assertion is precisely that $(f^+)_x$ is integrable for almost all x. By a similar argument,

$$\int_X d\mu(x) \int_Y f^-(x,y)dv(y) < \infty, \tag{6.8}$$

and the measurable function on X given by $\int_Y f^-(x,y)dv(y)$ is finite almost everywhere, which means $(f^-)_x$ is integrable for almost all x. From the obvious equality (see Proposition 2.2.4)

$$f_x = (f^+ - f^-)_x = (f^+)_x - (f^-)_x, \tag{6.9}$$

we deduce that f_x is integrable for almost all x. This proves the first part of (a) and the second part follows by an analogous argument.

By (6.7) and (6.8), $\int_Y f^+(x,y)dv(y)$ and $\int_Y f^-(x,y)dv(y)$ represent integrable functions on X. By Remark 6.5.4(b), there is an integrable function Φ on X such that

$$\Phi(x) = \int_Y f^+(x,y)dv(y) - \int_Y f^-(x,y)dv(y) \tag{6.10}$$

almost everywhere and has integral equal to

$$\int_X \Phi = \int_X d\mu(x) \int_Y f^+(x,y)dv(y) - \int_X d\mu(x) \int_Y f^-(x,y)dv(y). \tag{6.11}$$

From (6.9) we also deduce that, for almost all x, we have

$$\int_Y f_x = \int_Y (f^+)_x - \int_Y (f^-)_x,$$

which can be expressed as

$$\int_Y f(x, y)dv(y) = \int_Y f^+(x, y)dv(y) - \int_Y f^-(x, y)dv(y).$$

Together with (6.10), this leads to

$$\Phi(x) = \int_Y f(x, y)dv(y) \text{ for almost all } x. \tag{6.12}$$

Thus there exists an integrable function on X that agrees with $\int_Y f(x, y)dv(y)$ almost everywhere. By a similar argument, there exists an integrable function on Y that agrees with $\int_X f(x, y)d\mu(x)$ almost everywhere. This proves (b).

It follows from Remark 6.5.4(d), (6.11) and (6.12) that

$$\int_X d\mu(x) \int_Y f(x, y)dv(y) = \int_X d\mu(x) \int_Y f^+(x, y)dv(y) - \int_X d\mu(x) \int_Y f^-(x, y)dv(y). \tag{6.13}$$

By definition of integral (applied to $\mu \times v$ on $X \times Y$), we have

$$\int_{X\times Y} f(x, y)d(\mu \times v)(x, y) = \int_{X\times Y} f^+(x, y)d(\mu \times v)(x, y) - \int_{X\times Y} f^-(x, y)d(\mu \times v)(x, y).$$

By Tonelli's Theorem, the right side here agrees with that of (6.13). Therefore the left side here agrees with that of (6.13), which means the first equality of part (c) is valid. The second equality is also valid for analogous reasons. $\square$

Problem Set 6.5

6.5.P1. Let $X = Y = [1, \infty)$ with Lebesgue measure m. Suppose $f: X \times Y \to \mathbb{R}$ satisfies

$$f(0, 0) = 0 \text{ and } f(x, y) = \frac{x^2 - y^2}{(x^2 + y^2)^2}.$$

Show that f cannot be integrable even if it is measurable.

6.5.P2. If X and Y are both countable, show that the product measure of the counting measures on them is the counting measure on $X \times Y$.

6.5.P3. Show that repeated integrals with respect to counting measures can be unequal even if the function on the product space is measurable.

Hint: See Problem 1.2.P1.

6.5.P4. Give an example of two functions on a measure space that are equal almost everywhere, one is measurable but the other is not.

6.5.P5. Suppose $\{f_n\}_{n \geq 1}$ is a sequence of measurable functions such that $\lim\limits_{n \to \infty} f_n(x)$ exists for all x in some measurable set of convergence A. Show that the function defined to agree with the limit on A and to be 0 elsewhere is measurable. Show also that two different measurable functions obtained in this manner by using two different sets of convergence, both having complement of measure 0, agree a.e.

6.5.P6. Let $(X, \mathcal{F}, \mu)$ and $(Y, \mathcal{G}, \nu)$ be measure spaces and $f : X \to \mathbb{R}$, $\phi : X \to \mathbb{R}$ be measurable functions such that $f = \phi$ a.e. Suppose $g : Y \to X$ has the property that $F \in \mathcal{F} \Rightarrow g^{-1}(F) \in \mathcal{G}$, so that the compositions $f \circ g : Y \to \mathbb{R}$ and $\phi \circ g : Y \to \mathbb{R}$ are both measurable. Is it necessary that $f \circ g = \phi \circ g$ a.e.?

6.5.P7. Let $f:[0, 1] \times [0, 1] \to \mathbb{R}$ be the function given by

$$f(x, y) = (x - \frac{1}{2})^{-3} \text{ if } 0 < y < |x - \frac{1}{2}| \text{ and } f(x, y) = 0 \text{ otherwise.}$$

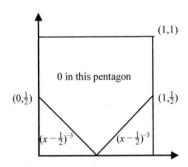

Show that the repeated integral $\int_{[0,1]} dx \int_{[0,1]} f(x, y) dy$ does not even exist but the repeated integral $\int_{[0,1]} dy \int_{[0,1]} f(x, y) dx$ exists.

6.6   Borel Sets

Although a product σ-algebra $\mathcal{F} \times \mathcal{G}$ is easy to define, it can be intractable in the concrete situation when $\mathcal{F}$ and $\mathcal{G}$ are the σ-algebra of Lebesgue measurable sets in $\mathbb{R}$ or an interval. This can make it difficult to verify measurability of a function on the product space $\mathbb{R} \times \mathbb{R}$ for applying the Fubini or Tonelli Theorem. In order to alleviate this difficulty, we introduce a smaller but more tractable σ-algebra. The benefits will be harvested in Chap. 7.

So far, we have used a symbol such as "(a, b)" to mean either the open interval from a to b or the ordered pair with a as the first element and b the second, as there was no possibility of confusion. However, in the present section, confusion can arise and we shall denote the ordered pair by "$\langle a, b \rangle$". But we shall continue to denote the value of a function f at $\langle a, b \rangle$ by the symbol "$f(a, b)$" as usual.

6.6.1. Definition. *The σ-algebra of Lebesgue measurable subsets of $\mathbb{R}$ will be denoted by $\mathfrak{M}$. The σ-algebra generated by all open intervals is called the* **Borel σ-algebra** *and will be denoted by $\mathfrak{B}$. A set belonging to $\mathfrak{B}$ is called a* **Borel set**.

Since every interval is a Lebesgue measurable set, it is immediate that $\mathfrak{B} \subseteq \mathfrak{M}$ and hence that $\mathfrak{B} \times \mathfrak{B} \subseteq \mathfrak{M} \times \mathfrak{M}$. Since a Borel set B is a Lebesgue measurable set, it has a Lebesgue measure $m(B)$, which is sometimes referred to as its Borel measure. In other words, the restriction to $\mathfrak{B}$ of the Lebesgue measure m is sometimes called the "Borel measure".

It is left to the reader as Problem 6.6.P1 to show that every interval is a Borel set.

6.6.2. Proposition. *Let $I \subseteq \mathbb{R}$ be an open interval. Then*

$$\{\langle x, y \rangle \in \mathbb{R} \times \mathbb{R} : x + y \in I\} \in \mathfrak{B} \times \mathfrak{B}.$$

Proof: Denote the set in question by A and consider an arbitrary point $\langle \xi, \eta \rangle \in A$. Then $\xi + \eta \in I$. Therefore there exists a $\delta > 0$ such that

$$(\xi + \eta - 2\delta, \xi + \eta + 2\delta) \subseteq I.$$

It follows that

$$
\begin{aligned}
\langle x, y \rangle \in (\xi - \delta, \xi + \delta) \times (\eta - \delta, \eta + \delta) & \Rightarrow -\delta < x - \xi < \delta \text{ and } -\delta < y - \eta < \delta \\
& \Rightarrow -2\delta < (x + y) - (\xi + \eta) < 2\delta \\
& \Rightarrow (\xi + \eta) - 2\delta < x + y < (\xi + \eta) + 2\delta \\
& \Rightarrow x + y \in (\xi + \eta - 2\delta, \xi + \eta + 2\delta) \\
& \Rightarrow x + y \in I \\
& \Rightarrow \langle x, y \rangle \in A.
\end{aligned}
$$

Thus, $(\xi - \delta, \xi + \delta) \times (\eta - \delta, \eta + \delta) \subseteq A$. Now let p, q, r, s be rational numbers such that $\xi - \delta < p < \xi < q < \xi + \delta$ and $\eta - \delta < r < \eta < s < \eta + \delta$. We then have

$$\langle \xi, \eta \rangle \in (p, q) \times (r, s) \subseteq (\xi - \delta, \xi + \delta) \times (\eta - \delta, \eta + \delta) \subseteq A.$$

This shows that each $\langle \xi, \eta \rangle \in A$ belongs to a Cartesian product $(p, q) \times (r, s) \subseteq A$ of intervals with rational endpoints p, q, r, s. The union of all these Cartesian products is therefore precisely A. Now all such Cartesian products $(p, q) \times (r, s)$ belong to $\mathfrak{B} \times \mathfrak{B}$ and there can be only countably many of them, as their endpoints p, q, r, s are rational. Thus, A is a countable union of sets belonging to $\mathfrak{B} \times \mathfrak{B}$ and hence must belong to $\mathfrak{B} \times \mathfrak{B}$. $\square$

6.6.3. Exercise. Let $\mathcal{D}$ be a σ-algebra of subsets of $\mathbb{R} \times \mathbb{R}$. Show that the family $\mathcal{E}$ of subsets E of $\mathbb{R}$ such that

$$\{\langle x, y \rangle \in \mathbb{R} \times \mathbb{R} : x + y \in E\} \in \mathcal{D}$$

is a σ-algebra.

Solution: For any $E \subseteq \mathbb{R}$, denote the associated set $\{\langle x, y \rangle \in \mathbb{R} \times \mathbb{R} : x + y \in E\}$ by E^*. We have to show that the family $\mathcal{E}$ of subsets E of $\mathbb{R}$ such that $E^* \in \mathcal{D}$ is a σ-algebra.

First we argue that

$$\emptyset^* = \emptyset.$$

When $E = \emptyset$, there cannot be any $\langle x, y \rangle \in \mathbb{R} \times \mathbb{R}$ for which $x + y \in E$, which means $\emptyset^* = E^* = \{\langle x, y \rangle \in \mathbb{R} \times \mathbb{R} : x + y \in E\} = \emptyset$.

Next, let $\{E_k\}_{k \geq 1}$ be a sequence of subsets of $\mathbb{R}$. Then

$$\left(\bigcup_{k=1}^{\infty} E_k \right)^* = \bigcup_{k=1}^{\infty} E_k^*,$$

because for any $\langle x, y \rangle \in \mathbb{R} \times \mathbb{R}$, we have

$$\langle x, y \rangle \in \left(\bigcup_{k=1}^{\infty} E_k \right)^* \Leftrightarrow x + y \in \bigcup_{k=1}^{\infty} E_k \Leftrightarrow x + y \in E_k \text{ for some } k \in \mathbb{N}$$

$$\Leftrightarrow \langle x, y \rangle \in E_k^* \text{ for some } k \in \mathbb{N} \Leftrightarrow \langle x, y \rangle \in \bigcup_{k=1}^{\infty} E_k^*.$$

It is equally easy to see that

$$(E^*)^c = (E^c)^*.$$

In fact,

$$\langle x, y \rangle (E^*)^c \Leftrightarrow \langle x, y \rangle \notin E^* \Leftrightarrow x + y \notin E$$
$$\Leftrightarrow x + y \in E^c \Leftrightarrow \langle x, y \rangle \in (E^c)^*.$$

It follows from the equalities established in the foregoing three paragraphs that $\mathcal{E}$ is a σ-algebra because $\mathcal{D}$ is.

6.6.4. Proposition. *Let* $B \subseteq \mathbb{R}$ *be any Borel set. Then*

$$\{\langle x, y \rangle \in \mathbb{R} \times \mathbb{R} : x + y \in B\} \in \mathfrak{B} \times \mathfrak{B}.$$

Proof: Let $\mathcal{E}$ be the family of subsets E of $\mathbb{R}$ such that

$$\{\langle x, y \rangle \in \mathbb{R} \times \mathbb{R} : x + y \in E\} \in \mathfrak{B} \times \mathfrak{B}.$$

By the result of Exercise 6.6.3, the family $\mathcal{E}$ is a σ-algebra, and by Proposition 6.6.2, it contains every open interval. By Definition 6.6.1 of the σ-algebra $\mathfrak{B}$, we have $\mathcal{E} \supseteq \mathfrak{B}$. This is precisely the conclusion we wanted to arrive at. □

Recall the terminology introduced in Remark 6.5.3 that in the context of a measure space $(X, \mathcal{F}, \mu)$, a property involving a point of X is said to hold *almost everywhere*, abbreviated as "a.e.", if it holds at each point of X outside some set of measure 0.

Suppose there is a finite or infinite sequence of properties, all of which hold a.e., but each with its own set of measure 0. Using the fact that a union of a finite or infinite sequence of sets of measure 0 is again a set of measure 0, we can deduce that all the properties hold simultaneously outside a single set of measure 0, and thus hold a.e. Such deductions will often be carried out with no express mention of the sets of measure 0, as is customary.

In the lemma below, the hypothesis that the measurable set under consideration has finite measure can be removed but we shall not need to.

6.6.5. Lemma. *Let $E \subseteq \mathbb{R}$ be a Lebesgue measurable set of finite measure. Then for any $\eta > 0$, there exists a sequence $\{I_n\}_{n \geq 1}$ of open intervals such that $E \subseteq \cup_{n=1}^{\infty} I_n$ and $m\big((\cup_{n=1}^{\infty} I_n) \backslash E\big) \leq \eta$.*

Proof: Left to the reader as Problem 6.6.P2. □

6.6.6. Proposition. *Any Lebesgue measurable set $E \subseteq \mathbb{R}$ is contained in a Borel set F such that $m(F \backslash E) = 0$; in particular, $m(F) = m(E)$.*

Proof: First suppose that E has finite measure. By Lemma 6.6.5, for every $p \in \mathbb{N}$, there exists a Borel set $B_p \supseteq E$ such that $m(B_p \backslash E) \leq \frac{1}{p}$. The intersection $F = \cap_{p=1}^{\infty} B_p$ is then a Borel set such that $F \supseteq E$. Moreover, $F \backslash E \subseteq B_p \backslash E$ for every $p \in \mathbb{N}$ and therefore $m(F \backslash E) \leq m(B_p \backslash E) \leq \frac{1}{p}$ for every $p \in \mathbb{N}$, so that $m(F \backslash E) = 0$.

Now consider a Lebesgue measurable set E of infinite measure. For each $n \in \mathbb{N}$, let $E_n = E \cap [-n, n]$, so that $\cup_{n=1}^{\infty} E_n = E$. Furthermore, each E_n has finite measure and therefore the result of the preceding paragraph yields a Borel set F_n such that

$$F_n \supseteq E_n \text{ and } m(F_n \backslash E_n) = 0.$$

It follows that

$$\bigcup_{n=1}^{\infty} F_n \supseteq \bigcup_{n=1}^{\infty} E_n = E \tag{6.14}$$

and $m(\cup_{n=1}^{\infty}(F_n \backslash E_n)) = 0$. But $\cup_{n=1}^{\infty} F_n \backslash \cup_{n=1}^{\infty} E_n \subseteq \cup_{n=1}^{\infty}(F_n \backslash E_n)$, as is easy to check. Therefore $m(\cup_{n=1}^{\infty} F_n \backslash \cup_{n=1}^{\infty} E_n) = 0$. By (6.14), this means $m((\cup_{n=1}^{\infty} F_n) \backslash E) = 0$. Therefore the Borel set $F = \cup_{n=1}^{\infty} F_n$ satisfies $m(F \backslash E) = 0$; moreover, it satisfies $F \supseteq E$ by virtue of (6.14). □

We can now assert that a property holds outside some Lebsgue measurable set of measure 0 if and only if it holds outside some Borel set of measure 0. Indeed, the "if" part is obvious because $\mathfrak{B} \subseteq \mathfrak{M}$, and the converse is a consequence of Proposition 6.6.6 applied to a Lebesgue measurable set of measure 0.

Any function that is $\mathcal{F}$-measurable in the sense of Definition 2.1.4 with $\mathcal{F} = \mathfrak{B}$ is said to be **Borel measurable** (or $\mathfrak{B}$-**measurable** or just **Borel**).

6.6.7. Proposition. *Given a real-valued Lebesgue measurable function f defined on $\mathbb{R}$, there exists a real-valued Borel function g on $\mathbb{R}$ such that $f = g$ a.e.*

Proof: We begin by proving this when $f = \chi_E$, the characteristic function of a Lebesgue measurable set E. By Proposition 6.6.6, there exists a Borel set F such that $E \subseteq F$ and $m(F \backslash E) = 0$. Select $g = \chi_F$. We claim that $f = g$ a.e., that is $\chi_E = \chi_F$ a.e. More precisely, we claim that $\chi_E = \chi_F$ outside $F \backslash E$. To demonstrate why this is so, we first split $\mathbb{R}$ as

$$\mathbb{R} = F^c \cup F = F^c \cup \big(F \cap (E^c \cup E)\big) = F^c \cup \big((F \cap E^c) \cup (F \cap E)\big) = F^c \cup (F \backslash E) \cup E.$$

Next, we note that it follows from the inclusion $E \subseteq F$ that χ_E and χ_F are both 0 on F^c and are both 1 on E. Upon combining this with the above splitting of $\mathbb{R}$, we find that $f(x) = g(x)$ as long as $x \notin (F \backslash E)$. But $m(F \backslash E) = 0$ and therefore $f = g$ a.e. So far, we have proved that the result in question holds for the characteristic function of a Lebesgue measurable set. It is a straightforward consequence that it holds for all simple Lebesgue measurable functions.

Now consider any nonnegative Lebesgue measurable function f. By Theorem 3.2.1, there exists a sequence $\{s_n\}_{n \geq 1}$ of Lebesgue measurable simple functions converging pointwise to f. According to what has been established in the preceding paragraph, each s_n has a Borel function g_n and a Borel set G_n of measure 0 such that $s_n(x) = g_n(x)$ whenever $x \notin G_n$. The Borel set $G = \cup_{n=1}^{\infty} G_n$ then has the property that $m(G) = 0$ and also $g_n(x) \to f(x)$ for each $x \notin G$. So, $g_n(x)\chi_{G^c}(x) \to f(x)\chi_{G^c}(x)$ for each $x \in \mathbb{R}$. Since each $g_n \chi_{G^c}$ is a Borel function, so is their limit $f \chi_{G^c}$. However $f(x) = f(x)\chi_{G^c}(x)$ for each $x \notin G$. Taking into account that $m(G) = 0$, we have $f = f\chi_{G^c}$ a.e. Thus the function $g = f\chi_{G^c}$ is a Borel function such that $f = g$ a.e. Since f is real-valued, the product $f\chi_{G^c}$ is also real-valued.

For a Lebesgue measurable function f that need not be nonnegative-valued, we obtain the result by considering the related functions f^+ and f^-. $\qquad\square$

We note in passing that Proposition 6.6.7 holds even if f is extended real-valued but with the proviso that g may also be extended real-valued.

Problem Set 6.6

6.6.P1. (Cf. Problem 6.5.P4) Suppose f and g are functions on $[a, b]$ or $\mathbb{R}$ such that $f = g$ a.e. If f is Lebesgue measurable, show that g is also Lebesgue measurable.

6.6.P2. Prove Lemma 6.6.5.

6.6.P3. Let $\mathcal{F}$ be a σ-algebra of subsets of a nonempty set X and $F:X \to \mathbb{R}$ be an $\mathcal{F}$-measurable map. Show that any Borel set A satisfies $F^{-1}(A) \in \mathcal{F}$.

Chapter 7
Differentiation

7.1 Integrability and Step Functions

The two fundamental theorems of calculus are about the derivative of the Riemann integral and Riemann integral of the derivative. There are corresponding results about the Lebesgue integral addressing the same matters, but they require a considerable amount of spadework.

The foundational result of this section is Theorem 7.1.5 and its proof requires the following fact: The characteristic function on $[a, b]$ of a subset which is a finite union of disjoint subintervals is a step function. It is fairly easy to convince oneself of this via a figure, and furthermore that the same is true even if the subintervals are not assumed disjoint because the union can be rewritten as the finite union of disjoint subintervals. The reader who is prepared to accept this assertion as being too obvious to need a proof is welcome to proceed directly to Theorem 7.1.5. The more skeptical reader may want to ratiocinate over the material leading up to Proposition 7.1.4, wherein the assertion is proved but on the assumption that the subintervals involved are of positive length, which is adequate for our purpose. However, the material can profitably be postponed until after the rest of the section has been read.

7.1.1. Remark. If the intersection of an open interval I with a closed interval J of positive length is nonempty, we shall show that it is an interval of positive length. That the intersection is an interval is trivial. Suppose $u \in I \cap J$. Since I is open, some interval of the form $(u - \delta, u + \delta)$, where $\delta > 0$, must be contained in I; and since J is closed with positive length, some interval of the form either $[u, u + \delta')$ or $(u - \delta', u]$, where $\delta' > 0$, must be contained in J. It now follows easily that an interval of one of the latter two forms must be contained in $I \cap J$.

7.1.2. Exercise. It is known from elementary Analysis that a subset of $\mathbb{R}$ is an interval if and only if any number lying between two numbers belonging to the subset also belongs to the subset. Using this, prove that the union of two nondisjoint intervals is an interval.

© Springer Nature Switzerland AG 2018
S. Shirali, *A Concise Introduction to Measure Theory*,
https://doi.org/10.1007/978-3-030-03241-8_7

Solution: Consider intervals I and J with $u \in I \cap J$ and $x, y \in I \cup J$ satisfying $x < y$. We shall show that any $z \in \mathbb{R}$ such that $x < z < y$ lies in $I \cup J$. If $z = u$, there is nothing to prove. If x, y both belong to I or both belong to J, then again there is nothing to prove. So, suppose $x \in I$, $y \in J$. If $z < u$, we have $x < z < u$ with $x, u \in I$, which implies $z \in I \subseteq I \cup J$, and if $z > u$, we have $u < z < y$ with $u, y \in J$, which implies $z \in J \subseteq I \cup J$. The remaining possibility that $x \in J$, $y \in I$ can be argued analogously.

7.1.3. Proposition. (a) *A finite union of intervals is also a finite union of disjoint intervals.*

(b) *A finite union of intervals of positive length is a finite union of disjoint intervals of positive length.*

Proof: (a) When the union is empty or when there is only one interval, the matter is trivial. First consider a union of two intervals. If they are not disjoint, then Exercise 7.1.2 ensures that their union is an interval, which makes it a union of disjoint intervals. Now assume as induction hypothesis that any union of $n+1$ intervals ($n \geq 1$) is a union of disjoint intervals, and consider a union of $n+2$ intervals $I_1, ..., I_{n+2}$.

If I_{n+2} is disjoint from $\cup_{k=1}^{n+1} I_k$, then the induction hypothesis yields the desired conclusion immediately. If on the contrary I_{n+2} is not disjoint from $\cup_{k=1}^{n+1} I_k$, then we may take $I_{n+1} \cap I_{n+2}$ to be nonempty, whereupon it follows by using Exercise 7.1.2 that $I_{n+1} \cup I_{n+2}$ is an interval J, leading to the consequence that $\cup_{K=1}^{n+2} I_k = \left(\cup_{K=1}^{n} I_k\right) \cup (I_{n+1} \cup I_{n+2}) = \left(\cup_{K=1}^{n} I_k\right) \cup J$, which is a union of $n+1$ intervals, and thus a union of disjoint intervals by the induction hypothesis.

(b) This follows upon merely reinterpreting the word "interval" in the reasoning above to mean "interval of positive length". □

In what follows, an interval named as $[a, b]$ will be assumed to be of positive length, i.e. $a < b$, but not in Sect. 7.6.

7.1.4. Proposition. *The characteristic function on $[a, b]$ of a finite union of subintervals thereof, each having positive length, is a step function.*

Proof: On the basis of Proposition 7.1.3(b), we may take the subintervals to be disjoint and with positive length. We shall refer to them as *given* subintervals, because there will be occasion to mention other subintervals too. Let the left endpoints of the given subintervals be $a_1, ..., a_n$ in increasing order and the corresponding right endpoints be $b_1, ..., b_n$.

If $n = 1$, we have $a \leq a_1 < b_1 \leq b$; this gives rise to the four cases

$$a = a_1 < b_1 = b, \quad a = a_1 < b_1 < b, \quad a < a_1 < b_1 = b, \quad a < a_1 < b_1 < b.$$

It is easy to verify in each case that the characteristic function of any subinterval with endpoints a_1, b_1 is a step function.

Now assume as induction hypothesis that the result in question holds whenever there are n given subintervals, and suppose we have $n + 1$ given subintervals at hand with endpoints a_j, b_j, $1 \leq j \leq n + 1$. We claim that $b_1 \leq a_2$. If not, then $a_2 < \min\{b_1, b_2\}$ and any $u \in (a_2, \min\{b_1, b_2\})$ satisfies $a_1 < u < b_1$ as well as $a_2 < u < b_2$, contradicting disjointness. Consider the characteristic function of the union of the $n + 1$ given subintervals at hand. Its restriction to the interval $[a, b_1]$ agrees with the characteristic function of the given subinterval with endpoints a_1, b_1, except perhaps at b_1 if $b_1 = a_2$, and is therefore seen to be a step function on $[a, b_1]$ on the grounds of what has been proved in the paragraph above. As for the restriction to $[b_1, b]$, since it agrees with the characteristic function of the union of the remaining n given subintervals, except perhaps at b_1, it must also be a step function by the induction hypothesis. Thus the restrictions to $[a, b_1]$ and $[b_1, b]$ are both step functions. It follows that the characteristic function under consideration is a step function, as was to be proved.   □

For reasons that need not delay us here, the content of the next theorem and its corollary will be summarized by saying that simple functions are "dense".

Note that, whenever p, q and M are real numbers such that $p \leq M$, we can show that $|p - \min\{q, M\}| \leq |p - q|$ in the following manner: If $q \leq M$, then $\min\{q, M\} = q$, so that $|p - \min\{q, M\}| = |p - q|$, whereas if $q > M$, then $p \leq M = \min\{q, M\} < q$, so that $|p - \min\{q, M\}| < |p - q|$.

7.1.5. Theorem. *Given an integrable function $f \colon [a, b] \to \mathbb{R}$ and any $\varepsilon > 0$, there exists a step function $s \colon [a, b] \to \mathbb{R}$ such that $\int_{[a,b]} |f - s| < \varepsilon$. Moreover, if $|f| \leq M$ for some $M \geq 0$, then s can be chosen to satisfy $|s| \leq M$.*

Proof: First we prove this when $f = \chi_E$, the characteristic function of a measurable set $E \subseteq [a, b]$.

Set $\eta = \frac{\varepsilon}{2}$. By Lemma 6.6.5, there exists a sequence $\{I_n\}_{n \geq 1}$ of open intervals such that $E \subseteq \bigcup_{n=1}^{\infty} I_n$ and

$$m\left(\left(\bigcup_{n=1}^{\infty} I_n\right) \backslash E\right) \leq \eta. \tag{7.1}$$

Consider the intersections $J_n = I_n \cap [a, b]$. They are all subintervals of $[a, b]$ and the nonempty ones have positive length (see Remark 7.1.1 above). Moreover, they satisfy $\bigcup_{n=1}^{\infty} J_n \supseteq E$, which ensures that $\bigcup_{n=1}^{\infty} (J_n \cap E) = E$. By Proposition 2.3.1, we have $m\left(\bigcup_{n=1}^{\infty} (J_n \cap E)\right) = \lim_{N \to \infty} m\left(\bigcup_{n=1}^{N} (J_n \cap E)\right)$. Hence there exists an $N \in \mathbb{N}$ such that $0 \leq m(E) - m\left(\bigcup_{n=1}^{N} (J_n \cap E)\right) < \eta$. This implies

$$m\left(E \backslash \bigcup_{n=1}^{N} (J_n \cap E)\right) < \eta. \tag{7.2}$$

Denoting the finite union $\bigcup_{n=1}^{N} J_n \subseteq [a, b]$ by G, we have $E \backslash G = E \backslash \bigcup_{n=1}^{N} (J_n \cap E)$ and hence $m(E \backslash G) < \eta$ by (7.2). On the other hand, $G \backslash E = \left(\bigcup_{n=1}^{N} J_n\right) \backslash E \subseteq$

$\left(\cup_{n=1}^{\infty} I_n\right) \backslash E$ and therefore $m(G \backslash E) \leq m\left(\left(\cup_{n=1}^{\infty} I_n\right) \backslash E\right) \leq \eta$ by (7.1). Consequently, the symmetric difference $E \Delta G = (E \backslash G) \cup (G \backslash E)$ satisfies $m(E \Delta G) < 2\eta$. The easily proven equality $\chi_{E \Delta G} = |\chi_E - \chi_G|$ now leads to $\int_{[a,b]} |\chi_E - \chi_G| < 2\eta = \varepsilon$. But G is a finite union of subintervals of $[a, b]$ of positive length, and hence Proposition 7.1.4 shows that the function $\chi_G : [a, b] \to \mathbb{R}$ is a step function.

The assertion of the theorem has thus been proved when the function f is the characteristic function of a measurable set $E \subseteq [a, b]$.

This extends to all simple functions because they are finite linear combinations of characteristic functions of measurable sets.

Next, consider an integrable nonnegative f. By Theorem 3.2.1 and the Monotone Convergence Theorem 2.3.2, there exists a simple function $\sigma : [a, b] \to \mathbb{R}$ such that $0 \leq \sigma \leq f$ and $0 \leq \int_{[a,b]} f - \int_{[a,b]} \sigma = \int_{[a,b]} |f - \sigma| < \frac{\varepsilon}{2}$. Since the assertion of the theorem has been shown to hold for simple functions, the simple function σ yields a step function $s : [a, b] \to \mathbb{R}$ such that $\int_{[a,b]} |\sigma - s| < \frac{\varepsilon}{2}$. This step function s then satisfies $\int_{[a,b]} |f - s| < \varepsilon$. Thus the assertion in question is found to hold for any integrable nonnegative f. Extending this to an arbitrary integrable function is now a matter of employing the usual kind of $\frac{\varepsilon}{2}$-argument after writing $f = f^+ - f^-$.

For the last part, first suppose f is nonnegative and satisfies $f \leq M$ for some $M \geq 0$. Then $\min\{s, M\}$ is a step function which serves our purpose because $|f - \min\{s, M\}| \leq |f - s|$. For an arbitrary integrable f, the required step function is obtained by considering f_+ and f^- separately.   □

Since a step function, by definition, has a closed bounded interval as domain, the above theorem does not carry over directly to the case when the domain is $\mathbb{R}$. However, if we agree to extend a step function to be zero outside its domain, it does carry over, as we shall see in Corollary 7.1.7.

7.1.6. Notation. *Given a function* $g: [a, b] \to \mathbb{R}$, *its extension to all of* $\mathbb{R}$ *obtained by setting it equal to* 0 *outside* $[a, b]$ *will be denoted by* g^*.

It is clear that $g^*|_{[a,b]} = g$ and $g^* \chi_{[a,b]} = g^*$. Recall from Sect. 4.4 that the measure induced on $[a, b]$ by Lebesgue measure on $\mathbb{R}$ is nothing but Lebesgue measure on $[a, b]$. It follows upon using Proposition 4.4.1 that $g : [a, b] \to \mathbb{R}$ is measurable if and only if $g^* : \mathbb{R} \to \mathbb{R}$ is. Moreover, when this is so, g is integrable if and only if g^* is, in which case $\int_{[a,b]} g = \int_{\mathbb{R}} g^*$.

It is trivial to see that $|g_1 - g_2|^* = |g_1^* - g_2^*|$.

7.1.7. Corollary. *Given an integrable function* $f: \mathbb{R} \to \mathbb{R}$ *and any* $\varepsilon > 0$, *there exists an interval* $[a, b]$ *and a step function* $s: [a, b] \to \mathbb{R}$ *such that* $\int_{\mathbb{R}} |f - s^*| < \varepsilon$. *Moreover, if* $|f| \leq M$ *for some* $M \geq 0$, *then* s^* *can be chosen to satisfy* $|s^*| \leq M$.

Proof: The sequence of functions $\{f_n\}_{n \geq 1}$, where $f_n = f \chi_{[-n,n]}$, converges pointwise to f on $\mathbb{R}$ and it follows from the Dominated Convergence Theorem 3.2.6 that $\int_{\mathbb{R}} |f_n - f| \to 0$ as $n \to \infty$. Therefore there exists a $k \in \mathbb{N}$ such that

$$\int_{\mathbb{R}} |f_k - f| < \frac{\varepsilon}{2}.$$

Now f_k is integrable and hence, so is its restriction g to $[-k, k]$; moreover, since f_k vanishes outside $[-k, k]$, we have $g^* = f_k$. By Theorem 7.1.5, there exists a step function $s: [-k, k] \to \mathbb{R}$ such that $\int_{[-k,k]} |g - s| < \frac{\varepsilon}{2}$. However, $\int_{[-k,k]} |g - s| = \int_{\mathbb{R}} |g^* - s^*| = \int_{\mathbb{R}} |f_k - s^*|$, so that

$$\int_{\mathbb{R}} |f_k - s^*| < \frac{\varepsilon}{2},$$

which leads to

$$\int_{\mathbb{R}} |f - s^*| \leq \int_{\mathbb{R}} |f_k - f| + \int_{\mathbb{R}} |f_k - s^*| < \varepsilon.$$

The interval $[a, b] = [-k, k]$ has thus been found to be of the required kind.

If $|f| \leq M$ for some $M \geq 0$, then the same holds for $f_k = f \chi_{[-k,k]}$, and for its restriction g. Therefore by the last part of Theorem 7.1.5, the step function $s: [-k, k] \to \mathbb{R}$ can be chosen to satisfy the same inequality; the extension s^* then also satisfies the inequality. □

7.1.8. Mean Continuity Theorem. *For an integrable function $f: \mathbb{R} \to \mathbb{R}$,*

$$\lim_{h \to 0} \int_{\mathbb{R}} |f(x+h) - f(x)| dx = 0.$$

Proof: First suppose f is the characteristic function of an interval $[a, b]$. Then $f(x+h)$ is the characteristic function of the interval $[a-h, b-h]$ and consequently, $|f(x+h) - f(x)|$ is the characteristic function of the symmetric difference of the two intervals. Consider h such that $0 < h < b - a$; then $a - h < a < b - h < b$ and the symmetric difference is the union $[a-h, a) \cup (b-h, b]$, the measure of which is $2h$. Therefore

$$\int_{\mathbb{R}} |f(x+h) - f(x)| dx = 2h$$

and the desired limit must hold from the right. A similar argument shows that it also holds from the left. Thus the limit in question holds for the characteristic function of a closed bounded interval. If the interval is not closed but is bounded, then its characteristic function differs from that of the corresponding closed interval at finitely many points whereas the integral involved has the same value for both. So no separate argumentation is required in this case.

It now follows in the obvious manner that the limit holds for any step function. Its validity for an arbitrary integrable function f now follows from the fact that step functions s are dense (Corollary 7.1.7) and that

$$\int_{\mathbb{R}} |f(x+h) - f(x)|dx \le \int_{\mathbb{R}} |f(x+h) - s(x+h)|dx$$

$$+ \int_{\mathbb{R}} |s(x+h) - s(x)|dx + \int_{\mathbb{R}} |s(x) - f(x)|dx$$

and

$$\int_{\mathbb{R}} |f(x+h) - s(x+h)|dx = \int_{\mathbb{R}} |s(x) - f(x)|dx.$$

$\square$

Problem Set 7.1

7.1.P1. A subset $U \subseteq \mathbb{R}$ is said to be *open* if for every $u \in U$, there exists a $\delta > 0$ such that $(u - \delta, u + \delta) \subseteq U$. (Obviously, any open interval is an open subset; an interval that is an open subset is an open interval; any union of open subsets (or open intervals) is an open subset.) Show that any open subset $U \subseteq \mathbb{R}$ is a countable union of disjoint open intervals.

7.1.P2. Let $X = [0, 1]$ and consider the intervals $\left[\frac{j-1}{k}, \frac{j}{k}\right]$, $1 \le j \le k, k \in \mathbb{N}$, arranged in some order (without repetition). Show that the sequence $\{f_n\}_{n \ge 1}$ of characteristic functions on X of the respective intervals does not converge at any point, although their integrals converge to zero.

7.1.P3. (a) Both the figures below show a stick and a second stick with the overlap of their shadows underneath. They also show a third stick whose shadow overlaps with some part of the existing overlap of the shadows of the first two sticks (i.e. all three shadows together have some overlap). If in the figure on the left, the second stick is removed, the combined shadow of the remaining two sticks is the same as that of all three sticks. If in the figure on the right, the first stick is removed, the combined shadow of the remaining two sticks is the same as that of all three sticks. Question: Can you place three sticks in such a way that all three shadows together have some overlap but if any one among the three sticks is removed, the combined shadow of the remaining two sticks is no longer the same as that of all three sticks?

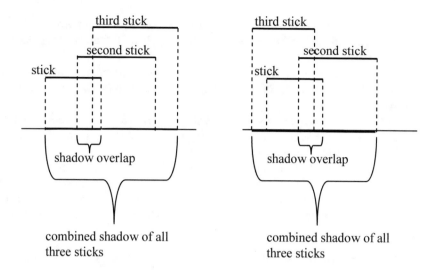

combined shadow of all
three sticks

combined shadow of all
three sticks

(b) Suppose $\mathcal{F}$ is a finite nonempty family of bounded intervals. Show that there
     exists a subfamily $\mathcal{G} \subseteq \mathcal{F}$ such that $\cup \mathcal{G} = \cup \mathcal{F}$ and no interval of $\mathcal{G}$ is contained
     in the union of two other intervals of $\mathcal{G}$.
(c) Can the finiteness hypothesis of part (b) be dropped?

7.2   The Vitali Covering Theorem

The theorem of the title is an important stepping stone in our approach to the remark-
able theorem of Lebesgue (next section) that a function that is increasing or decreasing (monotone) is differentiable almost everywhere. However, there are several other
approaches that do not use the theorem.

A "Vitali covering" of a subset $A \subseteq \mathbb{R}$ is a family of closed intervals such that
every number in A belongs to an arbitrarily short interval of the family, but with
positive length. A trivial example would be A consisting of a single point and the
family of all closed intervals centered at x with lengths $\frac{1}{n}$, where $n \in \mathbb{N}$. A less
trivial example would be $A = [0, 1]$ and the family of all possible closed intervals
with rational centers in A and rational lengths. One variation on this is to drop the
requirement that the centers and lengths be rational; another is to restrict the lengths
to be less than some specified positive number.

The concept can be described formally keeping in mind that the length of an
interval is the same as its Lebesgue measure:

7.2.1. Definition. *A family C of closed intervals is a* **Vitali covering** *of a subset
$A \subseteq \mathbb{R}$ if, for any given $x \in A$ and $\varepsilon > 0$, there exists an $I \in C$ for which $x \in I$ and
$0 < m(I) < \varepsilon$.*

Since closed intervals of length 0 or ∞ do not serve any purpose in a Vitali covering, it will be tacitly assumed that all intervals in a Vitali covering have finite positive length.

The substance of the Vitali Covering Theorem to be proved below, and to be invoked later in the proof of the Lebesgue Differentiability Theorem, can be summarized as follows:

Any Vitali covering of a set of finite (Lebesgue) outer measure contains finitely many disjoint intervals such that removing them all from the set results in a subset of arbitrarily small outer measure.

The proof involves some facts about intervals that the reader will easily believe on the basis of a figure. However, we prefer to spell out the precise reasons, which undoubtedly make for dreary reading. It is possible to proceed directly to Exercise 7.2.5 from here and decide at a later stage whether to suffer through the intervening material.

7.2.2. Remarks. (a) Recall that the interval obtained by deleting the endpoints (if any) of a given interval is called its *interior*. It will be convenient to denote the interior of an interval I by I°. To belong to the interior of an interval is to belong to the interval while being at a positive distance from any endpoint(s) it may have, which is the same as being the center of a suitable closed subinterval of positive length. Surely, an interval must have positive length if its interior is nonempty.

(b) A point which is outside a closed interval I belongs to the interior of a closed bounded interval that is disjoint from I: First suppose $I = [a, \infty)$ and consider $x \notin [a, \infty)$. Then $x < a$ and for $\delta = a - x > 0$, we have $y \in \left[x - \frac{1}{2}\delta, x + \frac{1}{2}\delta\right] \Rightarrow y - x < \delta = a - x \Rightarrow y < a \Rightarrow y \notin [a, \infty)$, so that $[x - \frac{1}{2}\delta, x + \frac{1}{2}\delta] \cap [a, \infty) = \emptyset$. Thus, x belongs to the interior of the closed bounded interval $\left[x - \frac{1}{2}\delta, x + \frac{1}{2}\delta\right]$, which is disjoint from $[a, \infty)$. Similar reasoning works when I is a closed interval of any other type.

(c) The intersection of a finite number of closed intervals of positive length whose interiors all have one given point in common can be shown to be a closed interval which has that point in its interior. In this connection, it is easy to see that the intersection of a finite number of closed intervals is a closed interval, perhaps empty. (However, it need not be of positive length even if all the intersecting intervals have positive length and have a common point, as is illustrated by the example $[0, 1] \cap [1, 2] = [1, 1]$.) Suppose now that their *interiors* have a common point. By (a), each of them contains a closed subinterval of positive length with that point as center. The intersection of the finitely many subintervals so obtained is then a closed interval of positive length contained in the intersection of the original intervals; moreover, it has the common point as its center. Thus the intersection of the original intervals has the common point in its interior.

(d) If $x \in (a, b)$, then any interval to which x belongs and has length less than the positive number $\min\{x - a, b - x\}$ can be shown to be contained in (a, b). To wit, let $[a_0, b_0]$ be such an interval. Then

$$b_0 - a_0 < x - a \quad \text{and} \quad a_0 \le x \le b_0.$$

The second of these implies $x - a_0 \le b_0 - a_0$, which when combined with the first yields $x - a_0 < x - a$. This further yields $a < a_0$. The reason why $b_0 < b$ is similar.

7.2.3. Exercise. Show that a point which lies outside each of a finite number of closed intervals belongs to the interior of a closed bounded interval that is disjoint from each one of the finitely many closed intervals.

Solution: By Remark 7.2.2(b), to each of the finitely many closed intervals corresponds a closed bounded interval of positive length that is disjoint from the former and having the point in its interior. The intersection of the intervals so obtained is then disjoint from each of the given intervals and is bounded. By Remark 7.2.2(c), it is a closed interval whose interior contains the point.

7.2.4. Exercise. Suppose the distance of a point $x \in \mathbb{R}$ from the center of a closed interval does not exceed α times the length ℓ of the interval (here $\alpha > 0$ of course). Show that x lies in a closed interval with the same center but having length $2\alpha\ell$.

Solution: Call the center y, so that the interval is $\left[y - \frac{1}{2}\ell, y + \frac{1}{2}\ell\right]$. Then $|y - x| \le \alpha\ell$. Therefore $x \in \left[y - \alpha\ell, y + \alpha\ell\right]$, which is precisely the closed interval with the same center but length $2\alpha\ell$.

7.2.5. Exercise. Let C be a Vitali covering of a subset $A \subseteq \mathbb{R}$ and U be a union of open intervals (i.e. is an *open subset* as in Problem 7.1.P1) containing A. Show that the family of those intervals belonging to C that are contained in U is a Vitali covering of A.

Solution: Consider any $x \in A$ and $\varepsilon > 0$. We have to conclude that there exists an $I \in C$ for which $x \in I$, $0 < m(I) < \varepsilon$ and also $I \subseteq U$.

Since $x \in U$, a union of open intervals, there exists a $\delta > 0$ such that

$$(x - \delta, x + \delta) \subseteq U. \tag{7.3}$$

Take $\eta = \min\{\varepsilon, \delta\} > 0$. Now C is a Vitali covering of A and hence there exists an $I \in C$ for which $x \in I$ and $0 < m(I) < \eta \le \varepsilon$. All we need to show is that

$$I \subseteq (x - \delta, x + \delta), \tag{7.4}$$

because the required conclusion will then follow from (7.3). To this end, let

$$I = [a, b].$$

Since $b - a = m(I) < \eta \le \delta$, we have

$$b < a + \delta, \quad \text{equivalently,} \quad a > b - \delta. \tag{7.5}$$

Since $x \in I = [a, b]$, we also have $a \leq x \leq b$. Hence by (7.5), $b - \delta < x < a + \delta$. This implies $x - \delta < a$ and $b < x + \delta$. These two inequalities together yield

$$[a, b] \subseteq (x - \delta, x + \delta).$$

Since $I = [a, b]$, this is the same as (7.4), which has already been noted as all that needs to be shown.

7.2.6. Lemma. *Suppose C is a Vitali covering of $A \subseteq \mathbb{R}$ and $I_1, \ldots, I_n$ are finitely many disjoint intervals belonging to C. Then for any $x \in A\backslash\cup_{j=1}^n I_j$, there exists an interval belonging to C that has x as element of it and is disjoint from each of the I_j, $1 \leq j \leq n$.*

Proof: Consider any $x \in A\backslash\cup_{j=1}^n I_j$. It lies outside each of the finitely many closed intervals I_j, $1 \leq j \leq n$, and therefore belongs to an interval (a, b) that is disjoint from each of the I_j, $1 \leq j \leq n$ (Exercise 7.2.3). Now, every interval to which x belongs and has sufficiently small length must be contained in (a, b) (Remark 7.2.2(d)). Since C is a Vitali covering of A, there certainly exists such an interval belonging to C. Being contained in (a, b), it is disjoint from each of the I_j, $1 \leq j \leq n$. Thus, we obtain an interval belonging to C that has x as an element of it and is disjoint from each of the I_j, $1 \leq j \leq n$. $\qquad\square$

7.2.7. Vitali Covering Theorem. *Suppose C is a Vitali covering of $A \subseteq \mathbb{R}$ and $m^*(A) < \infty$. Then for any $\varepsilon > 0$, there exist finitely many disjoint intervals $I_1, \ldots, I_n$ belonging to C such that*

$$m^*(A\backslash\bigcup_{j=1}^n I_j) < \varepsilon.$$

Proof: In the event that either $m^*(A) = 0$ or there exist finitely many disjoint intervals $I_1, \ldots, I_n$ belonging to C such that $A \subseteq \cup_{j=1}^n I_j$, there is nothing to prove. So we may assume that $m^*(A) > 0$ and that whenever $I_1, \ldots, I_n$ are finitely many disjoint intervals belonging to C, the difference $A\backslash\cup_{j=1}^n I_j$ is nonempty.

By definition of Lebesgue outer measure, there exists a countable union G of open intervals such that $A \subseteq G$ and $m(G) < \infty$. By Exercise 7.2.5, we may legitimately assume that every interval belonging to C is contained in G. The consequence that the sum of the measures of any sequence of disjoint intervals belonging to C is finite will be used below.

First consider an arbitrary finite sequence $\{I_j\}_{1 \leq j \leq n}$ of disjoint intervals belonging to C. The difference $A\backslash\cup_{j=1}^n I_j$ is nonempty and therefore there exists some $x \in A\backslash\cup_{j=1}^n I_j$. By Lemma 7.2.6, we obtain an interval belonging to C that has x as an element of it and is disjoint from each of the I, $1 \leq j \leq n$. Let σ_n be the supremum of the lengths of all intervals belonging to C that are disjoint from each of the I_j, $1 \leq j$

$\leq n$, but not necessarily having x as an element. Then $0 < \sigma_n < \infty$ and there exists an interval $I_{n+1} \in \mathcal{C}$ that is disjoint from $\cup_{j=1}^n I_j$ and satisfies $2 \cdot m(I_{n+1}) > \sigma_n$.

A straightforward inductive construction now leads to an infinite sequence $\{I_n\}_{n \geq 1}$ of disjoint intervals belonging to $\mathcal{C}$ and a sequence $\{\sigma_n\}_{n \geq 1}$ of positive real numbers such that

$$I \in \mathcal{C} \text{ and } I \cap I_j = \emptyset \text{ for } 1 \leq j \leq n \implies \sigma_n \geq m(I) \quad \text{for every } n \in \mathbb{N} \tag{7.6}$$

and

$$2 \cdot m(I_{n+1}) > \sigma_n \quad \text{for every } n \in \mathbb{N}. \tag{7.7}$$

We shall show by contradiction that

$$m^*(A \backslash \bigcup_{j=1}^n I_j) < \varepsilon \quad \text{for some } n \in \mathbb{N}. \tag{7.8}$$

Suppose not. Then

$$m^*(A \backslash \bigcup_{j=1}^n I_j) \geq \varepsilon \quad \text{for every } n \in \mathbb{N}. \tag{7.9}$$

As noted at the end of the second paragraph, $\sum_{j=1}^\infty m(I_j) < \infty$. This implies on the one hand that

$$m(I_j) \to 0 \text{ as } j \to \infty \tag{7.10}$$

and on the other hand that some $N \in \mathbb{N}$ has the property that

$$\sum_{j=N+1}^\infty m(I_j) < \frac{\varepsilon}{5}. \tag{7.11}$$

Now, $A \backslash \cup_{j=1}^N I_j \neq \emptyset$. Taking an arbitrary $x \in A \backslash \cup_{j=1}^N I_j$, we deduce from Lemma 7.2.6 that some interval I belonging to $\mathcal{C}$ has x as an element and is disjoint from each of the I_j, $1 \leq j \leq N$. By (7.6), we have $m(I) \leq \sigma_N$ and from (7.7), we have $\sigma_N < 2 \cdot m(I_{N+1})$. If I were to be disjoint from all the intervals I_j, then by (7.6), it would have to satisfy $m(I) \leq \sigma_j$ for all $j \in \mathbb{N}$, and hence by (7.7), also $m(I) < 2 \cdot m(I_{j+1})$ for all $j \in \mathbb{N}$. But this would contradict (7.10) because $m(I) > 0$. Thus there must be some $j > N$ for which $I \cap I_j \neq \emptyset$. Denote the smallest such integer j by k. Then

$$k > N$$

and I is disjoint from each of the I_j, $1 \leq j \leq k - 1$, and therefore by (7.6) and (7.7),

$$m(I) \leq \sigma_{k-1} < 2 \cdot m(I_k). \tag{7.12}$$

At the same time, $I \cap I_k \neq \emptyset$ and therefore the distance of $x \in I$ from the center of I_k is at most $m(I) + \frac{1}{2}m(I_k)$ (see Problem 7.2.P2 below), and hence by (7.12) is less than $2 \cdot m(I_k) + \frac{1}{2}m(I_k) = \frac{5}{2}m(I_k)$. It follows that the closed interval I_k' with the same center as I_k but length $5m(I_k)$ has x as an element (Exercise 7.2.4, wherein $\alpha = \frac{5}{2}$). Taking into account that $x \in A\backslash\bigcup_{j=1}^{N}I_j$ was arbitrary and $k > N$, we obtain $A\backslash\bigcup_{j=1}^{N}I_j \subseteq \bigcup_{j>N}I_j'$. On the basis of (7.9) and (7.11), we further obtain

$$\varepsilon \leq m^* \left(A\backslash\bigcup_{j=1}^{N}I_j \right) \leq \sum_{j=N+1}^{\infty} m(I_j') = 5 \sum_{j=N+1}^{\infty} m(I_j) < \varepsilon.$$

This contradiction proves (7.8) and hence also the theorem. □

Problem Set 7.2

7.2.P1. Let $\mathcal{C}$ be a Vitali covering of a subset $A \subseteq \mathbb{R}$ and $\mathcal{J}$ be a family of closed intervals such that the union of their interiors contains A. Show that the family

$$\mathcal{C}_0 = \{I \cap J : I \in \mathcal{C}, J \in \mathcal{J}\}$$

of closed intervals consisting of the intersections of intervals belonging to $\mathcal{C}$ with intervals belonging to $\mathcal{J}$ is a Vitali covering of A.

7.2.P2. If I and J are intervals of finite positive lengths and $I \cap J \neq \emptyset$, show that the distance of any point in I from the center of J is at most $m(I) + \frac{1}{2}m(J)$.

7.3 Dini Derivatives

So far, we have not spoken of $-\infty$, at least not as a mathematical entity subject to any computational rules. In the present section we shall speak of both ∞ and $-\infty$ as possible limits, the latter as an entity not to compute with but only to compare, as for instance, $-\infty < a < \infty$ for every real number a. It then follows that the supremum of any set to which ∞ belongs is ∞ and the infimum of any set to which $-\infty$ belongs is $-\infty$. Since every element of $\mathbb{R} \cup \{\infty, -\infty\}$ is "vacuously" a lower bound as well as an upper bound of $\emptyset$, it follows that $\inf \emptyset = \infty$ and $\sup \emptyset = -\infty$. The empty set is the only one for which the infimum exceeds the supremum. The implication that

$$A \subseteq B \quad \Rightarrow \quad \inf A \geq \inf B \text{ and } \sup A \leq \sup B$$

continues to hold even if $A = \emptyset$ or $B = \emptyset$.

Suppose the domain of the real-valued function f is some interval $I \subseteq \mathbb{R}$ of positive length and let $c \in I$. Then the so-called "difference quotient at c", $\frac{f(c+h)-f(c)}{h}$, makes sense for all sufficiently small h unless c is an endpoint. We speak of right and left difference quotients depending on whether h is positive or negative. At the left endpoint of I, only right difference quotients make sense and we therefore take any set of left difference quotients there to be $\emptyset$. Similarly, we take any set of right difference quotients at the right endpoint of I to be $\emptyset$.

When $\delta \in \mathbb{R}^{*+}$, the set

$$Q_\delta = \left\{ \frac{f(c+h) - f(c)}{h} : 0 < h < \delta \right\}$$

has a supremum, although it may be $\pm\infty$, as for instance when c is the right endpoint of I, in which case $Q_\delta = \emptyset$. It is easy to check that for the square root function on the interval $\mathbb{R}^+$, the supremum turns out to be ∞ at $c = 0$, whatever the positive number δ may be. This is because the difference quotient equals $h^{-\frac{1}{2}}$. In contrast, the infimum is $\delta^{-\frac{1}{2}}$, with the understanding that this means 0 in the event that $\delta = \infty$.

7.3.1. Exercise. Let f be the function on $\mathbb{R}$ defined by $f(0) = 0$ and $f(x) = |x|^{\frac{1}{2}} \cos\frac{1}{x}$ for $x \neq 0$. Show that, for any $\delta > 0$, the supremum of the set Q_δ of difference quotients at $c = 0$ is ∞. What about the infimum?

Solution: When $h > 0$, the difference quotient works out to be $h^{-\frac{1}{2}} \cos\frac{1}{h}$, which has the value $(2\pi n)^{\frac{1}{2}}$ when $h = (2\pi n)^{-1}$ and $n \in \mathbb{N}$. Given any $K > 0$, one can choose n such that the inequalities $(2\pi n)^{-1} < \delta$ and $(2\pi n)^{\frac{1}{2}} > K$ both hold. So, $\sup Q_\delta = \infty$. On the other hand, the infimum is $-\infty$, because $h^{-\frac{1}{2}} \cos\frac{1}{h}$ has the value $-(\pi(2n+1))^{\frac{1}{2}}$ when $h = (\pi(2n+1))^{-1}$ and $n \in \mathbb{N}$.

In what follows, any number denoted by δ or a variant thereof will be understood to be positive.

7.3.2. Remark. If $\delta'' \leq \delta'$, then $Q_{\delta''} \subseteq Q_{\delta'}$ and therefore $\inf Q_{\delta'} \leq \inf Q_{\delta''}$ and $\sup Q_{\delta''} \leq \sup Q_{\delta'}$. When c is not an endpoint, $Q_{\delta''} \neq \emptyset$ and therefore we also have $\inf Q_{\delta''} \leq \sup Q_{\delta''}$, with the consequence that $\inf Q_{\delta'} \leq \sup Q_{\delta''}$ as well as $\inf Q_{\delta''} \leq \sup Q_{\delta'}$.

We go on to consider infima of sets of the form $\{\sup Q_\delta : \delta < \delta_1\}$, where δ_1 is given. Note that these sets are never empty even though Q_δ can be.

7.3.3. Proposition. *For any δ_1 and $\delta_2 \in \mathbb{R}^{*+}$,*

$$\inf\{\sup Q_\delta : \delta < \delta_1\} = \inf\{\sup Q_\delta : \delta < \delta_2\}.$$

Proof: We need prove the equality only for $\delta_1 < \delta_2$. When this is so,

$$\{\sup Q_\delta : \delta < \delta_1\} \subseteq \{\sup Q_\delta : \delta < \delta_2\}. \tag{7.13}$$

It is immediate from this inclusion that

$$\inf\{\sup Q_\delta : \delta < \delta_1\} \geq \inf\{\sup Q_\delta : \delta < \delta_2\}.$$

To arrive at the reverse inequality, it is sufficient to show that for any q in the bigger of the two sets in (7.13), there exists a p in the smaller set such that $p \leq q$. Accordingly, consider any q in the bigger set. There exists $\delta' < \delta_2$ for which

$$q = \sup Q_{\delta'}.$$

Put $\Delta = \min\{\delta', \frac{1}{2}\delta_1\}$ and

$$p = \sup Q_\Delta.$$

Since $\Delta \leq \delta'$, we have $Q_\Delta \subseteq Q_{\delta'}$ and hence $\sup Q_\Delta \leq \sup Q_{\delta'}$, which is the same as $p \leq q$. Since $\Delta < \delta_1$, we have $\sup Q_\Delta \in \{\sup Q_\delta : \delta < \delta_1\}$, which means exactly that p belongs to the smaller of the two sets in (7.13).   □

The above Proposition 7.3.3 allows for $\delta_2 = \infty$. Therefore, in taking the infimum of the set of suprema, it makes no difference whether we impose any such restriction as $\delta < \delta_1$ or not. Accordingly, in Definition 7.3.4 below, we do not impose a restriction but it can be introduced if needed.

Similar considerations apply with inf and sup interchanged and/or $h > 0$ replaced by $h < 0$. Thus four possibilities can occur. When $h < 0$, we work with

$$Q_{-\delta} = \left\{ \frac{f(c+h)-f(c)}{h} : 0 < -h < \delta \right\}.$$

When c is not an endpoint of the domain of f, the sets Q_δ and $Q_{-\delta}$ are never empty and therefore the infimum of either of them cannot exceed its supremum.

7.3.4. Definition. *Let f be a real-valued function on an interval I of positive length. The four **Dini derivatives** of f at an interior point $c \in I$ are defined as*

$$D^+ f(c) = \inf\{\sup Q_\delta : 0 < \delta\},$$
$$D_+ f(c) = \sup\{\inf Q_\delta : 0 < \delta\},$$
$$D^- f(c) = \inf\{\sup Q_{-\delta} : 0 < \delta\},$$

and

$$D_- f(c) = \sup\{\inf Q_{-\delta} : 0 < \delta\},$$

where

$$Q_\delta = \left\{ \frac{f(c+h)-f(c)}{h} : 0 < h < \delta \right\} \quad and \quad Q_{-\delta} = \left\{ \frac{f(c+h)-f(c)}{h} : 0 < -h < \delta \right\}.$$

By Remark 7.3.2, we always have $D_+ f(c) \leq D^+ f(c)$ and analogously, $D_- f(c) \leq D^- f(c)$.

For an increasing function, all numbers in Q_δ and $Q_{-\delta}$ are nonnegative and therefore so are all the four Dini derivatives at any c.

7.3.5. Examples. (a) Consider the function f of Exercise 7.3.1 again. It was seen there that $\sup Q_\delta = \infty$ and $\inf Q_\delta = -\infty$ for every δ. By Definition 7.3.4, this implies $D^+f(c) = \inf\{\infty\} = \infty$ and $D_+f(c) = \sup\{-\infty\} = -\infty$. It is left to the reader to check that $D^-f(c) = \infty$ and $D_-f(c) = -\infty$.

(b) Let f be the absolute value of the function considered in part (a) and in Exercise 7.3.1. It may be helpful to visualize its graph near the origin. It bounces off the x-axis at the points $((n + \frac{1}{2})\pi)^{-1}$ and reaches the graph of the square root function at points $(n\pi)^{-1}$ between bounces. In other words,

- the sequence $\{h_n\}_{n\geq 1}$ of positive numbers with $h_n = ((n + \frac{1}{2})\pi)^{-1}$, which obviously satisfies $h_n \to 0$ and $h_n < 1$, has the property that $\frac{f(0+h_n)-f(0)}{h_n} = 0$ for every n,

and

- the sequence $\{h_n\}_{n\geq 1}$ of positive numbers with $h_n = (n\pi)^{-1}$, which obviously satisfies $h_n \to 0$ and $h_n < 1$, has the property that $\frac{f(0+h_n)-f(0)}{h_n} = h_n^{-\frac{1}{2}} = (n\pi)^{\frac{1}{2}}$ for every n.

The existence of such sequences shows that when $\delta < 1$, we have $\inf Q_\delta = 0$ and $\sup Q_\delta = \infty$. It follows that

$$D_+f(0) = \sup\{0\} = 0 \quad \text{and} \quad D^+f(0) = \inf\{\infty\} = \infty.$$

Here we have found the values of $D_+f(0)$ and $D^+f(0)$ from certain kinds of sequences $\{h_n\}_{n\geq 1}$. It turns out that sequences of this nature must always exist. We justify this below and record a consequence that will play a role in the proof of the Lebesgue Differentiability Theorem 7.4.1.

7.3.6. Remark. Let c be an interior point and suppose $D^+f(c) < \infty$. By definition of $D^+f(c)$, for any $M > D^+f(c)$, there is some $\delta > 0$ such that $D^+f(c) \leq \sup Q_\delta < M$. By Remark 7.3.2, any smaller positive δ satisfies the same inequality. It follows that there is a sequence $\{h_n\}_{n\geq 1}$ of positive numbers such that

$$h_n \to 0 \quad \text{and} \quad \frac{f(c+h_n)-f(c)}{h_n} \to D^+f(c).$$

If $D^+f(c) = \infty$, the same conclusion can be arrived at even more easily by simply dropping the reference to M.

Consequence: for any $\alpha < D^+f(c)$, there is a sequence $\{h_n\}_{n\geq 1}$ of positive numbers such that

$$h_n \to 0 \quad \text{and} \quad \frac{f(c+h_n)-f(c)}{h_n} > \alpha \text{ for every } n \in \mathbb{N}.$$

Analogously, when $D_-f(c) > -\infty$, for any $M < D_-f(c)$, there is some $\delta > 0$ such that $M < \inf Q_{-\delta} \leq D_-f(c)$. By the obvious analog of Remark 7.3.2, any smaller positive δ satisfies the same inequality. As before, we have the following consequence whether or not $D_-f(c) > -\infty$: for any $\beta > D_-f(c)$, there is a sequence $\{h_n\}_{n\geq 1}$ of negative numbers such that

$$h_n \to 0 \quad \text{and} \quad \frac{f(c+h_n)-f(c)}{h_n} < \beta \text{ for every } n \in \mathbb{N}.$$

Two other such assertions can of course be proved regarding $D_+f(c)$ and $D^-f(c)$.

7.3.7. Exercise. If $g=-f$, show that $D^+g(c) = -D_+f(c)$.

Solution: For any $\delta>0$, it is immediate from the definition that $Q_\delta(g) = -Q_\delta(f)$ and hence that $\sup Q_\delta(g) = -\inf Q_\delta(f)$. Therefore $\inf\{\sup Q_\delta(g) : 0 < \delta\} = -\sup\{\inf Q_\delta : 0 < \delta\}$. By Definition 7.3.4, this means $D^+g(c) = -D_+f(c)$.

Problem Set 7.3

7.3.P1. Prove the following partial converse of the "consequence" in Remark 7.3.6 in the case of $D^+f(c)$: If $\alpha \in \mathbb{R} \cup \{\infty, -\infty\}$ and there exists a sequence $\{h_n\}_{n\geq 1}$ of positive numbers as in the remark (in which case $\alpha \neq \infty$ of course), then $\alpha \leq D^+f(c)$.

7.3.P2. Show that $\alpha < D^+f(c)$ if and only if for some $\beta > \alpha$ there is a sequence $\{h_n\}_{n\geq 1}$ of positive numbers such that

$$h_n \to 0 \quad \text{and} \quad \frac{f(c+h_n)-f(c)}{h_n} > \beta \text{ for every } n \in \mathbb{N}.$$

7.3.P3. Give an example of continuous functions f and g on an interval containing 0 such that $D^+f(0) + D^+g(0) \neq D^+(f+g)(0)$.

7.3.P4. If $g=-f$, show that $D^-g(c) = -D_-f(c)$.

7.3.P5. Suppose f and g are defined on a common interval I and $c \in I^\circ$. Show that $D_+(f+g)(c) \geq D_+f(c) + D_+g(c)$, provided that the sum on the right side is meaningful. What about $D^+(f+g)(c) \leq D^+f(c) + D^+g(c)$?

7.4 The Lebesgue Differentiability Theorem

Recall the notions of left and right derivatives from elementary Analysis (see e.g., Shirali and Vasudeva [9, p. 200]). The reader is called upon to make the obvious modification to permit $\pm\infty$ as their values. A little reflection will then show that $D_+f(c) = D^+f(c)$ if and only if f has a right derivative $f'_+(c)$ at c, in which case $D_+f(c) = D^+f(c) = f'_+(c)$. Similarly, $D_-f(c) = D^-f(c)$ if and only if f has a left derivative $f'_-(c)$ at c, in which case $D_-f(c) = D^-f(c) = f'_-(c)$. Moreover, f has a (two sided) derivative $f'(c)$ at c if and only if $f'_+(c) = f'_-(c)$, in which case $f'_+(c) = f'_-(c) = f'(c)$. In terms of Dini derivatives, this says that f has a (two sided) derivative $f'(c)$ at c if and only if $D_+f(c) = D^+f(c) = D_-f(c) = D^-f(c)$, in which case $D_+f(c) = D^+f(c) = D_-f(c) = D^-f(c) = f'(c)$. Differentiability at c means $f'(c)$ is finite, which is equivalent to all four Dini derivatives at c being equal and finite.

Differentiability here is understood as two sided. So endpoints of the interval are automatically out of the running.

7.4.1. Lebesgue Differentiability Theorem. *If a real-valued function on an interval is monotone, then it is differentiable almost everywhere.*

Proof: It is sufficient to prove the result either for an increasing function or for a decreasing function. So, let f be an increasing function on an interval I. We may assume that I has positive length because otherwise its measure is 0 and there is nothing to prove. It is sufficient to consider the case when I is bounded, because in the unbounded case, I can be represented as the countable union of the disjoint bounded intervals $I \cap (n, n+1]$, $n \in \mathbb{Z}$, to each of which the bounded case applies; the required conclusion then follows for I. So, suppose I is bounded. Since there cannot be two sided differentiability at endpoints (if any), we need consider only interior points, which means we can take I to be open.

Let $A = \{x \in I : D^+f(x) > D_-f(x)\}$. We shall demonstrate that $m^*(A) = 0$. Now A is the union of the countably many sets $A_{p,q} = \{x \in I : D^+f(x) > p > q > D_-f(x)\}$, where $p, q \in \mathbb{Q}$. It is sufficient therefore to show that each $A_{p,q}$ has outer measure 0. The increasing nature of the function f guarantees that all Dini derivatives are nonnegative and hence $A_{p,q} \neq \emptyset$ only if $p > q > 0$. So we need consider only those p, q for which this inequality holds.

With this in view, take any $A_{p,q}$ for which

$$p > q > 0.$$

and consider an arbitrary $\varepsilon > 0$.

Being a subset of the bounded interval I, the set $A_{p,q}$ has finite outer measure and therefore there exists a countable union G of open intervals such that $A_{p,q} \subseteq G$ and $m^*(A_{p,q}) \leq m(G) < m^*(A_{p,q}) + \varepsilon$. By Remark 7.3.6, to each $x \in A_{p,q}$ correspond intervals $[x + h, x] \subseteq G$ with $h < 0$ and arbitrarily small in absolute value such that $\frac{f(x+h)-f(x)}{h} < q$, i.e. such that

$$f(x+h) - f(x) > qh.$$

Since this holds for each $x \in A_{p,q}$, the intervals constitute a Vitali covering of $A_{p,q}$. By the Vitali Covering Theorem 7.2.7, there exist finitely many disjoint intervals $[x_j + h_j, x_j]$, $1 \leq j \leq J$, in the Vitali covering such that

$$m^*(A_{p,q} \backslash \bigcup_{j=1}^{J} [x_j + h_j, x_j]) < \varepsilon. \tag{7.14}$$

The choice of the intervals of the Vitali covering ensures that

$$f(x_j + h_j) - f(x_j) > qh_j \text{ for every } j. \tag{7.15}$$

Now, observe that since the disjoint intervals are contained in G, (remember $h_j < 0$) we have $\sum_{i=1}^{J} h_j \geq -m(G)$. Since $q > 0$, taken in tandem with (7.15), the preceding observation leads to

$$\sum_{j=1}^{J} \left(f(x_j + h_j) - f(x_j) \right) > q \cdot \sum_{j=1}^{J} h_j \geq -q \cdot m(G) > -q \cdot \left(m^*(A_{p,q}) + \varepsilon \right). \tag{7.16}$$

In conjunction with (7.14), the obvious equality

$$m^* \left(A_{p,q} \backslash \bigcup_{j=1}^{J} (x_j + h_j, x_j) \right) = m^* \left(A_{p,q} \backslash \bigcup_{j=1}^{J} [x_j + h_j, x_j] \right),$$

gives rise to

$$m^* \left(A_{p,q} \backslash \bigcup_{j=1}^{J} (x_j + h_j, x_j) \right) < \varepsilon. \tag{7.17}$$

By Remark 7.3.6, to each $y \in A_{p,q} \cap \bigcup_{j=1}^{J} (x_j + h_j, x_j)$ correspond intervals $[y, y + h'] \subseteq \bigcup_{j=1}^{J} (x_j + h_j, x_j)$ with arbitrarily small $h' > 0$ such that $\frac{f(y+h')-f(y)}{h'} > p$, i.e. such that

$$f(y + h') - f(y) > ph'.$$

Since this holds for each $y \in A_{p,q} \cap \bigcup_{j=1}^{J} (x_j + h_j, x_j)$, the intervals constitute a Vitali covering of $A_{p,q} \cap \bigcup_{j=1}^{J} (x_j + h_j, x_j)$. By the Vitali Covering Theorem 7.2.7, there exist finitely many disjoint intervals $[y_k, y_k + h'_k]$, $1 \leq k \leq K$, in the Vitali covering such that

$$m^* \left((A_{p,q} \bigcup_{j=1}^{J} (x_j + h_j, x_j)) \backslash \bigcup_{k=1}^{K} [y_k, y_k + h'_k] \right) < \varepsilon. \tag{7.18}$$

The choice of the intervals of the Vitali covering ensures that

$$f(y_k + h'_k) - f(y_k) > ph'_k \text{ for every } k. \tag{7.19}$$

Since the Vitali covering consists of intervals contained in $\bigcup_{j=1}^{J} (x_j + h_j, x_j)$, we have

$$\bigcup_{k=1}^{K} [y_k, y_k + h_k'] \subseteq \bigcup_{j=1}^{J} (x_j + h_j, x_j). \tag{7.20}$$

We claim that

$$\sum_{k=1}^{K} h_k' > m^*(A_{p,q}) - 2\varepsilon. \tag{7.21}$$

To prove (7.21), let us denote $\cup_{j=1}^{J}(x_j + h_j, x_j)$ by $\mathcal{J}$ and $\cup_{k=1}^{K}[y_k, y_k + h_k']$ by $\mathcal{K}$. Then

$$m^*(\mathcal{K}) = m(\mathcal{K}) = \sum_{k=1}^{K} h_k'. \tag{7.22}$$

Also, $\mathcal{K} \subseteq \mathcal{J}$ in view of (7.20), so that $\mathcal{J}^c = \mathcal{J}^c \cap \mathcal{K}^c$. Together with (7.17), this implies

$$m^*(A_{p,q} \cap \mathcal{J}^c \cap \mathcal{K}^c) = m^*(A_{p,q} \cap \mathcal{J}^c) < \varepsilon.$$

From (7.18), we have

$$m^*((A_{p,q} \cap \mathcal{J}) \cap \mathcal{K}^c) < \varepsilon.$$

Since

$$(A_{p,q} \cap \mathcal{J}^c \cap \mathcal{K}^c) \cup ((A_{p,q} \cap \mathcal{J}) \cap \mathcal{K}^c) = ((A_{p,q} \cap \mathcal{K}^c) \cap \mathcal{J}^c) \cup ((A_{p,q} \cap \mathcal{K}^c) \cap \mathcal{J})$$
$$= A_{p,q} \cap \mathcal{K}^c,$$

the preceding two inequalities yield

$$m^*(A_{p,q} \cap \mathcal{K}^c) < 2\varepsilon.$$

In combination with the fact that $A_{p,q} = (A_{p,q} \cap \mathcal{K}) \cup (A_{p,q} \cap \mathcal{K}^c)$, the above inequality leads us to

$$m^*(A_{p,q}) \leq m^*(A_{p,q} \cap \mathcal{K}) + m^*(A_{p,q} \cap \mathcal{K}^c)$$
$$\leq m^*(\mathcal{K}) + m^*(A_{p,q} \cap \mathcal{K}^c)$$
$$< m^*(\mathcal{K}) + 2\varepsilon$$

and hence $m^*(A_{p,q}) < \sum_{k=1}^{K} h_k' + 2\varepsilon$ by (7.22). This proves our claim (7.21).

It follows from (7.20) that each interval $[y_k, y_k + h_k']$ on the left side therein is contained in the union of finitely many disjoint intervals on the right side. The nature of an interval necessitates that each interval $\left[y_k, y_k + h_k'\right]$ is actually contained in a single interval on the right side. This fact and the increasing nature of the function f together justify the inequality (recall that $h_j < 0$)

$$\sum_{j=1}^{J} \left(f(x_j) - f(x_j + h_j)\right) \geq \sum_{k=1}^{K} \left(f(y_k + h_k') - f(y_k)\right).$$

By (7.19) and (7.21), this implies

$$\sum_{j=1}^{J} \left(f(x_j) - f(x_j + h_j)\right) > p \cdot \sum_{k=1}^{K} h_k' > p \cdot \left(m^*(A_{p,q}) - 2\varepsilon\right).$$

Therefore by (7.16),

$$q \cdot \left(m^*(A_{p,q}) + \varepsilon\right) > p \cdot \left(m^*(A_{p,q}) - \varepsilon\right).$$

Since this has been established for an arbitrary $\varepsilon > 0$, it follows that $q \cdot m^*(A_{p,q}) \geq p \cdot m^*(A_{p,q})$, which is only possible if $m^*(A_{p,q}) = 0$, because $p > q$.

This completes the demonstration that $m^*(A) = 0$, where $A = \{x \in I : D^+f(x) > D_-f(x)\}$.

By similar reasoning, we can show that the set on which any pair of Dini derivatives differ has outer measure 0. Thus the four Dini derivatives agree almost everywhere. It remains to show that the set on which any one of them is ∞ has outer measure 0. We set $B = \{x \in I : D^+f(x) = \infty\}$ and proceed to argue that $m^*(B) = 0$. Any nonempty interval is a countable union of closed subintervals, and it is sufficient therefore to prove the result for the subintervals. In effect, we may replace I by a closed subinterval $[\alpha, \beta]$. Note that $m^*(B) \leq \beta - \alpha < \infty$.

Assuming $m^*(B) > 0$ for the purpose of deriving a contradiction, let

$$\eta = \frac{1}{2} m^*(B) > 0$$

and

$$\infty > M > \frac{2(f(\beta) - f(\alpha))}{m^*(B)} \geq 0. \tag{7.23}$$

By Remark 7.3.6, to each $z \in B$ correspond intervals $\left[z, z + h''\right] \subseteq [\alpha, \beta]$ with arbitrarily small $h'' > 0$ such that $\frac{f(z+h'')-f(z)}{h''} > M$, i.e. such that

$$f(z + h'') - f(z) > Mh''.$$

Since this holds for each $y \in B$, the intervals constitute a Vitali covering of the set. By the Vitali Covering Theorem 7.2.7, there exist finitely many disjoint intervals $[z_r, z_r + h_r'']$, $1 \leq r \leq R$, in the Vitali covering such that

$$m^*(B \setminus \bigcup_{r=1}^{R} [z_r, z_r + h_r'']) < \eta. \tag{7.24}$$

The choice of the intervals of the Vitali covering ensures that

$$f(z_r + h_r'') - f(z_r) > M h_r'' \text{ for every } r. \tag{7.25}$$

We claim that

$$\sum_{r=1}^{R} h_r'' > m^*(B) - \eta. \tag{7.26}$$

To prove (7.26), let us denote $\bigcup_{r=1}^{R} [z_r, z_r + h_r'']$ by $\mathcal{L}$. Then

$$m^*(\mathcal{L}) = m(\mathcal{L}) = \sum_{r=1}^{R} h_r'', \tag{7.27}$$

and by (7.24),

$$m^*(B \cap \mathcal{L}^c) < \eta.$$

In combination with the fact that $B = (B \cap \mathcal{L}) \cup (B \cap \mathcal{L}^c)$, the above inequality leads to

$$\begin{aligned} m^*(B) &\leq m^*(B \cap \mathcal{L}) + m^*(B \cap \mathcal{L}^c) \\ &\leq m^*(\mathcal{L}) + m^*(B \cap \mathcal{L}^c) \\ &< m^*(\mathcal{L}) + \eta, \end{aligned}$$

which implies $m^*(B) < \sum_{r=1}^{R} h_r'' + \eta$ by (7.27). This proves our claim (7.26).

The disjointness of the intervals $[z_r, z_r + h_r'']$ and the increasing nature of f together justify the inequality

$$f(\beta) - f(\alpha) \geq \sum_{r=1}^{R} \left(f(z_r + h_r'') - f(z_r) \right).$$

By (7.25) and (7.26), this implies

$$f(\beta) - f(\alpha) > M \cdot \sum_{r=1}^{R} h_r'' > M \cdot \left(m^*(B) - \eta \right).$$

Recalling that η was defined to be $\frac{1}{2}m^*(B)$, we deduce from the above inequality that $f(\beta) - f(\alpha) > M \cdot \left(\frac{1}{2}m^*(B) \right)$. However, this conflicts with (7.23) and thereby completes the contradiction proof that $m^*(B) = 0$.   □

7.4.2. Proposition. *If a real-valued function f on an interval is a sum of an increasing and a decreasing function, then there is a Borel function on that interval which agrees with the derivative f′ a.e. Any two such functions are equal a.e.*

Proof: It is sufficient to prove the result for an increasing function f. By the Lebesgue Differentiability Theorem 7.4.1, there is a subset A of the domain on which f has a derivative f' and the complement of which has Lebesgue measure 0. By Proposition 6.6.6, we may take the subset A to be a Borel set. Define g to agree with the derivative on A and to be 0 on A^c. We shall show that g is a Borel function.

An increasing function is certainly a Borel function because the set on which $f(x) > t$ is an interval and therefore a Borel set. So, the functions

$$\frac{f\left(x \pm \frac{1}{n}\right) - f(x)}{\pm \frac{1}{n}}, \quad n \in \mathbb{N}$$

are Borel functions. On the set A, the limit of the sequence of these functions exists and equals g. Since g is 0 outside the Borel set A, the deduction is implemented by appealing to Problem 6.5.P5.

That any two such functions must be equal a.e. is trivial.   □

Problem Set 7.4

7.4.P1. For an increasing function f on a bounded open interval I, show by a direct argument, as in the proof of the Lebesgue Differentiability Theorem 7.4.1, that the set $A = \{x \in I : D_+f(x) < D^-f(x)\}$ has Lebesgue outer measure 0.

7.4.P2. Assuming that the result $m^*\{x \in I : D^+f(x) > D_-f(x)\} = 0$ established in the course of proving the Lebesgue Differentiability Theorem 7.4.1 is valid for decreasing functions, deduce the result of Problem 7.4.P1.

7.4.P3. Suppose f and g are defined on a common interval I and $c \in I^\circ$ is such that g has a finite right derivative $g'_+(c)$ at c. Show that $D^+(f + g)(c) = D^+f(c) + g'_+(c)$.

7.5 The Derivative of the Integral

One of the fundamental theorems of calculus says that the derivative of the "indefinite" Riemann integral of a continuous function exists everywhere and agrees with that function. The analog for the Lebesgue integral is that the derivative of the indefinite integral of an integrable function exists almost everywhere and agrees with that function.

We begin by showing that the indefinite integral of an integrable function is differentiable almost everywhere.

7.5.1. Proposition. *Suppose* $f : [a, b] \to \mathbb{R}$ *is integrable and* $F : [a, b] \to \mathbb{R}$ *is the function given by*

$$F(x) = \int_{[a,x]} f \quad \text{for all } x \in [a, b].$$

Then F *is differentiable almost everywhere and there exists a Borel function on* $[a, b]$ *that agrees with the derivative a.e. Moreover, any two such Borel functions are equal a.e.*

Proof: By definition of integrability, the functions $G : [a, b] \to \mathbb{R}$ and $H : [a, b] \to \mathbb{R}$ given by

$$G(x) = \int_{[a,x]} f^+ \text{ and } H(x) = \int_{[a,x]} f^- \quad \text{for all } x \in [a, b]$$

are real-valued and $F = G - H$. Since f^+ and f^- are nonnegative-valued, G and H are increasing functions. It follows by the Lebesgue Differentiability Theorem 7.4.1 that both are differentiable a.e. and hence their difference F is also differentiable a.e.

The existence of a Borel function of the kind claimed in the theorem follows from Proposition 7.4.2, considering that F is the sum of the increasing function G and the decreasing function $-H$. So does the assertion about any two functions of the kind being equal a.e. □

In order to simplify the notation, any Borel function of the kind in the above proposition will be denoted by F'. Any two such functions agree a.e. and the kind of thing we have to say about any one of them is equally true of the others. So, there is no harm in speaking of such a function as though it is unique. We do so in the impending proof.

7.5.2. Theorem (Derivative of the Integral). *Suppose* $f : [a, b] \to \mathbb{R}$ *is integrable and* $F : [a, b] \to \mathbb{R}$ *is the function given by*

$$F(x) = \int_{[a,x]} f \quad \text{for all } x \in [a, b].$$

Then for almost all $x \in [a, b]$, *the derivative* $F'(x)$ *exists and equals* $f(x)$.

Proof: That the derivative exists for almost all $x \in [a, b]$ is known from Proposition 7.5.1. It suffices to show that the function F' (understood to be a Borel function as in Proposition 7.5.1) agrees a.e. with some function, which in turn, agrees a.e. with f.

By Proposition 6.6.7, there exists a real-valued Borel function g on $\mathbb{R}$ that agrees a.e. with the extension f^* (see 7.1.6 for the meaning of f^*). By Problem 3.2.P1, it is sufficient to establish that

$$\int_{[a,b]} |F' - g| = 0. \tag{7.28}$$

Let $\{h_n\}_{n \geq 1}$ be a sequence of negative numbers tending to zero and $\{F_n\}_{n \geq 1}$ be the sequence of functions on $\mathbb{R}$ for which

$$F_n(x) = -h_n^{-1} \int_{[x+h_n,x]} g \quad \text{for all } x \in \mathbb{R}.$$

Then for all $x \in (a, b)$ and sufficiently large n, the number $F_n(x)$ agrees with $h_n^{-1}(F(x+h_n) - F(x))$. Hence by Proposition 7.5.1,

$$\lim_{n \to \infty} F_n = F' \text{ a.e. on } [a, b]. \tag{7.29}$$

We shall first prove that

$$\lim_{n \to \infty} \int_{[a,b]} |F_n - g| = 0. \tag{7.30}$$

For all $x \in \mathbb{R}$ and $n \in \mathbb{N}$, we have

$$F_n(x) - g(x) = -h_n^{-1} \int_{[x+h_n,x]} (g(t) - g(x))dt = -h_n^{-1} \int_{[h_n,0]} (g(x+\tau) - g(x))d\tau.$$

Therefore

$$\int_{\mathbb{R}} |F_n - g| \leq |h_n^{-1}| \int_{\mathbb{R}} dx \int_{[h_n,0]} |g(x+\tau) - g(x)|d\tau \quad \text{for all } n \in \mathbb{N}. \tag{7.31}$$

Now, the real-valued function on $\mathbb{R} \times \mathbb{R}$ given by $g(x+\tau)$ is $(\mathfrak{B} \times \mathfrak{B})$-measurable by Proposition 6.6.4 and hence $(\mathfrak{M} \times \mathfrak{M})$-measurable. It follows by Tonelli's Theorem 6.5.2 that

$$|h_n^{-1}| \int_{\mathbb{R}} dx \int_{[h_n,0]} |g(x+\tau) - g(x)|d\tau = |h_n^{-1}| \int_{[h_n,0]} d\tau \int_{\mathbb{R}} |g(x+\tau) - g(x)|dx.$$

Therefore by (7.31),

$$\int_{\mathbb{R}} |F_n - g| \le |h_n^{-1}| \int_{[h_n,0]} d\tau \int_{\mathbb{R}} |g(x+\tau) - g(x)| dx \quad \text{for all } n \in \mathbb{N}. \tag{7.32}$$

Now, consider any $\varepsilon > 0$. Since $h_n \to 0$, by the Mean Continuity Theorem 7.1.8, there exists an $N \in \mathbb{N}$ such that $n \ge N$ implies

$$\int_{\mathbb{R}} |g(x+\tau) - g(x)| dx < \varepsilon \quad \text{whenever } h_n \le \tau < 0,$$

and hence by (7.32) implies

$$\int_{\mathbb{R}} |F_n - g| \le |h_n^{-1}| \int_{[h_n,0]} \varepsilon d\tau = \varepsilon.$$

Thus, (7.30) has been shown to hold.

Now it emerges from (7.29), Remark 6.5.4(c) (a consequence of Fatou's Lemma) and (7.30) that equality (7.28) must hold. As already noted when (7.28) was formulated above, this is all that needed to be established. $\qquad\square$

7.6 Bounded Variation

By the Lebesgue Differentiability Theorem, a sum of an increasing and a decreasing function is differentiable a.e. Although an increasing or a decreasing function is usually easy to identify, their sum may not be.

In order to get a feel for what may be involved in seeking to identify a function f as a sum of an increasing function g and a decreasing function h, we enter into a heuristic discussion with a concrete example. Of course, g and h cannot be unique.

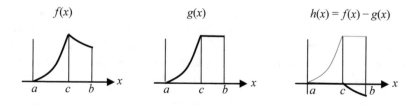

Consider the function $f: [a, b] \to \mathbb{R}$ with a graph as in the figure above, extreme left. To keep matters simple, we take $f(a)$ to be 0. In seeking to represent it as $f = g + h$, with increasing g and decreasing h, it is natural to try first to take g as in the middle of the figure. What g "does" is to mimic f as long as f is increasing but freeze into a constant as soon as f starts decreasing. The function g that does this will serve our purpose provided that $h = f - g$ turns out to be decreasing. The graph on the extreme right in the above figure shows the function $h = f - g$; the graph of g is included there in a lighter and thinner line in order to facilitate understanding why the graph of $f - g$ is indeed as shown. Unsurprisingly, h "does" the exact opposite of what g does, which is to say, it remains frozen as a constant as long as f is increasing and begins to mimic (run parallel to) f as soon as the latter starts decreasing.

$$V(x)$$

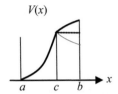

$$a \quad c \quad b \qquad x$$

Now consider the related function V as in the figure just above this line. An informal description of it is that it is obtained by reversing even the minutest decrease in f into an increase of the same magnitude, but retaining even the minutest increase in f exactly as it is. Another way to put it is that $V(x)$ is the accumulated *absolute* variation in f from a to x. In particular, $V(a)=0$ because there is no change in the value of f from a to a. On the interval of increase, which is $[a, c]$, the function V mimics f and happens to agree with f, because $f(a) = 0 = V(a)$. It can now be read off the figure quite easily that g is the mean between V and f on the entire domain.

On the interval of decrease, which is $[c, b]$, the function V is readily described as being given by

$$V(x) = f(c) + |f(x) - f(c)|.$$

On this interval, $g(x) = f(c)$, a constant, and $V(x) = 2f(c) - f(x)$. As a result, a simple computation confirms what had been read off the figure earlier:

$$g(x) = \tfrac{1}{2}(V(x) + f(x)) \quad \text{for } x \in [c, b].$$

This is trivially true on the interval $[a, c]$ of increase, because $g = V = f$ there.

For the above equality to hold even when $f(a) \neq 0$, which does not alter the fact that $V(a)=0$, we have to have $g(a) = \tfrac{1}{2}f(a)$. This only entails that the description of what g "does" be modified to say that g begins at a by being equal to $\tfrac{1}{2}f(a)$ and then mimics f on $[a, c]$.

7.6.1. Example. Let $f : [0, 3] \to \mathbb{R}$ be given by $f(x) = 3 - x$ for $x \in [0, 2]$ and $f(x) = 2x - 4$ for $x \in (2, 3]$. Note that f is left continuous at 2 but not right continuous. We identify the function $V(x)$ on the basis of its informal description as the accumulated absolute variation in f from 0 to x.

If $x \in [0, 2]$, then

$$V(x) = |f(x) - f(0)| = f(0) - f(x) = 3 - f(x) = x.$$

Across $x = 2$, the function jumps up by -1 because $f(2) = 1$ and the right limit $f(2+) = 0$. Therefore, if $x \in (2, 3]$, then

$$V(x) = V(2) + |-1| + (f(x) - f(2+)) = 2x - 1.$$

Note that the function $g : [0, 3] \to R$ given by $g(x) = \frac{1}{2}(V(x) + f(x))$ works out to be $\frac{3}{2}$ on $[0, 2]$ and to be $2x - \frac{5}{2}$ on $(2, 3]$. This is an increasing function. Moreover, $h(x) = f(x) - g(x) = f(x) - \frac{1}{2}(V(x) + f(x)) = \frac{1}{2}(f(x) - V(x))$ works out to be $\frac{3}{2} - x$ on $[0, 2]$ and to be $-\frac{3}{2}$ on $(2, 3]$. This is a decreasing function.

It should be noted how the jump was taken into account in the description of V over $(2, 3]$. In this connection, the reader will have no difficulty agreeing that $V(x)$ can be approximated on $(2, 3]$ by

$$|f(2 - \delta) - f(0)| + |f(2 + \delta) - f(2 - \delta)| + |f(x) - f(2 + \delta)|$$

with "small" positive δ. In fact, drawing a figure will suggest that the supremum of such sums over $\delta > 0$ is nothing but $V(x)$ when $x \in (2, 3]$.

7.6.2. Exercise. Let $f : [0, 4] \to \mathbb{R}$ be given by $f(x) = x^2 + 1$ for $x \in [0, 1)$ and $f(x) = 5 - x$ for $x \in [1, 4]$. If $0 < 1 - \delta < 1 + \delta < x_0 \leq 4$, compute the sum

$$S = |f(1 - \delta) - f(0)| + |f(1 + \delta) - f(1 - \delta)| + |f(x_0) - f(1 + \delta)|$$

and find its supremum over $\delta > 0$.

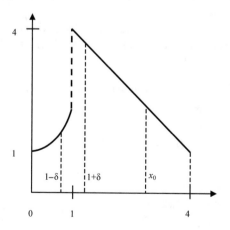

Solution: Since 0 and $1 - \delta$ both belong to the interval $[0, 1)$, we have $f(1 - \delta) - f(0) = (1 - \delta)^2 + 1 - f(0) = (1 - \delta)^2 \geq 0$. Therefore the first term in the sum S is

$$|f(1 - \delta) - f(0)| = 1 - 2\delta + \delta^2. \tag{7.33}$$

Now, $1 - \delta \in [0, 1)$ and $1 + \delta \in [1, 4]$; therefore

$$f(1 + \delta) - f(1 - \delta) = 5 - (1 + \delta) - ((1 - \delta)^2 + 1)$$
$$= 2 + \delta - \delta^2 = 2 + \delta(1 - \delta).$$

Since $0 < \delta < 1$, we have $\delta(1 - \delta) > 0$ and consequently, the second term in the sum S is

$$|f(1 + \delta) - f(1 - \delta)| = 2 + \delta - \delta^2. \tag{7.34}$$

Next, $1 + \delta$ and x_0 both belong to the interval $[1, 4]$. So, we have $f(x_0) - f(1 + \delta) = 5 - x - (5 - (1 + \delta)) = (1 + \delta) - x_0 < 0$ because it is given that $1 + \delta < x_0$. Therefore the third (and final) term in the sum S is

$$|f(x_0) - f(1 + \delta)| = -1 - \delta + x_0. \tag{7.35}$$

It follows from (7.33), (7.34) and (7.35) that

$$S = (1 - 2\delta + \delta^2) + (2 + \delta - \delta^2) + (-1 - \delta + x_0) = x_0 + 2 - 2\delta.$$

Hence the required supremum is $x_0 + 2$.

The reader may note how the "jump" at 1 is "captured" by $1 - \delta$ and $1 + \delta$ in the process of taking the supremum, because arbitrarily small positive values of δ are taken into account thereby.

With the foregoing considerations as backdrop, it would seem reasonable that a mathematically precise way to encapsulate what V is would be as follows, the possibility that $a = b$ being allowed.

7.6.3. Definition. *For a real-valued function $f : [a, b] \rightarrow \mathbb{R}$, the **total variation** from a to x, where $x \in [a, b]$, is*

$$V_a^x(f) = \sup \sum_{j=1}^{n} |f(x_j) - f(x_{j-1})|$$

*over all possible partitions $P : a = x_0 < x_1 < \cdots < x_n = x$ of $[a, x]$. The function f is said to be of **bounded variation** over $[a, x]$ if $V_a^x(f) < \infty$. The sum on the right side above will be denoted by $S(f, P)$.*

It is clear that any monotone function $f : [a, b] \to \mathbb{R}$ is of bounded variation over $[a, x]$ with $V_a^x(f) = \pm(f(x) - f(a))$ according as the function is increasing or decreasing. Also, since the only partition P of $[a, a]$ consists of a alone and $S(f, P) = |f(a) - f(a)| = 0$, we have $V_a^a(f) = 0$, whatever the function f may be.

7.6.4. Example. Let $f : [0, 1] \to \mathbb{R}$ be any function such that $f(0) = 0$ and, for any $k \in \mathbb{N}$, we have $f\left(\frac{1}{k}\right) = \frac{1}{k}$ if k is even and $f\left(\frac{1}{k}\right) = 0$ if k is odd. An example of a continuous function of this kind is given by $f(x) = x \cos^2(\pi/2x)$ for $x \neq 0$ and $f(0) = 0$. To see that f is not of bounded variation over $[0, 1]$, we consider the partitions

$$P : 0 < \frac{1}{2N} < \frac{1}{2N - 1} < \frac{1}{2N - 2} < \cdots < \frac{1}{2} < 1, \quad \text{where } N \in \mathbb{N}.$$

The values of f at these points are respectively $0, \frac{1}{2N}, 0, \frac{1}{2N-2}, \ldots, \frac{1}{2}, 0$. Therefore

$$S(f, P) = \frac{1}{N} + \frac{1}{N - 1} + \cdots + 1.$$

The supremum $V_0^1(f)$ is therefore ∞.

The appearance of partitions in the definition already suggests that some of the arguments regarding total variation will be reminiscent of those encountered in the basic considerations about Riemann integration.

7.6.5. Exercise. Let P_1 be a partition of $[a, x] \subseteq [a, b]$. Then for any refinement P_2 of P_1 and any function $f : [a, b] \to \mathbb{R}$, show that $S(f, P_1) \leq S(f, P_2)$.

Solution: It is sufficient to show that the inequality holds when P_2 contains one more point than P_1. Let P_1 consist of the points $a = x_0 < x_1 < \cdots < x_n = x$ and P_2 contain one more point ξ. Then ξ must lie in one of the intervals (x_{k-1}, x_k). Then

$$S(f, P_1) = \sum_{j=1}^{k-1} |f(x_j) - f(x_{j-1})| + |f(x_k) - f(x_{k-1})| + \sum_{j=k+1}^{n} |f(x_j) - f(x_{j-1})|,$$

with the proviso that in the case when $k = 1$, the first summation, being empty, is 0, and correspondingly in the case when $k = n$. Also,

$$S(f, P_2) = \sum_{j=1}^{k-1} |f(x_j) - f(x_{j-1})| + |f(\xi) - f(x_{k-1})| + |f(x_k) - f(\xi)|$$

$$+ \sum_{j=k+1}^{n} |f(x_j) - f(x_{j-1})|,$$

with a similar proviso. In combination with the simple inequality

$$|f(x_k) - f(x_{k-1})| \leq |f(\xi) - f(x_{k-1})| + |f(x_k) - f(\xi)|,$$

the above expressions for $S(f, P_1)$ and $S(f, P_2)$ lead to the desired inequality at once.

7.6.6. Proposition. *Let* $f : [a, b] \to \mathbb{R}$ *and* $g : [a, b] \to \mathbb{R}$ *be any two functions and* $x \in [a, b]$. *Then*

(a) $V_a^x(f) \geq 0$;
(b) $V_a^x(f + g) \leq V_a^x(f) + V_a^x(g)$;
(c) $V_a^x(cf) = |c| V_a^x(f)$ *whenever* $c \in \mathbb{R}$;
(d) $V_a^z(f) = V_a^y(f) + V_y^z(f)$ *whenever* $a \leq y \leq z \leq b$;
(e) *If* f *is of bounded variation over* $[a, b]$, *then it is of bounded variation over* $[a, x]$.

Proof: (a) Obvious.
(b) This is a consequence of the inequality $S(f + g, P) \leq S(f, P) + S(g, P)$, which follows from the fact that

$$\left| (f + g)(x_j) - (f + g)(x_{j-1}) \right| \leq \left| f(x_j) - f(x_{j-1}) \right| + \left| g(x_j) - g(x_{j-1}) \right|.$$

(c) This is true because $S(cf, P) = |c| S(f, P)$.
(d) First we handle the case when $V_a^y(f)$ is ∞. In this event, we have to deduce that $V_a^z(f) = \infty$. With this in view, consider any $M > 0$. There exists a partition P' of $[a, y]$ such that $S(f, P') > M$. The points of P' and the point z together constitute a partition P of $[a, z]$ and this partition satisfies $S(f, P) \geq S(f, P') > M$. Since this has been proved for any $M > 0$ and some partition P of $[a, z]$, we conclude that $V_a^z(f) = \infty$, as required.
   The case when $V_y^z(f) = \infty$ is entirely analogous.
   Now suppose $V_a^y(f)$ and $V_y^z(f)$ are both finite.
   We shall prove first that $V_a^z(f) \geq V_a^y(f) + V_y^z(f)$. With this in view, consider any $\varepsilon > 0$. There exist partitions Q and R of $[a, y]$ and $[y, z]$ respectively such that $S(f, Q) > V_a^y(f) - \frac{\varepsilon}{2}$ and $S(f, R) > V_y^z(f) - \frac{\varepsilon}{2}$. The points of Q and R together form a partition P of $[a, z]$ and $S(f, P) = S(f, Q) + S(f, R) > V_a^y(f) + V_y^z(f) - \varepsilon$. Since this has been proved for any $\varepsilon > 0$ and some partition P of $[a, z]$, we conclude that $V_a^z(f) \geq V_a^y(f) + V_y^z(f)$, as required.
   It remains to prove the reverse inequality, that is, $V_a^z(f) \leq V_a^y(f) + V_y^z(f)$. Consider an arbitrary partition P of $[a, z]$. If necessary, refine it by adding the point y. Then the refinement P_0 satisfies $S(f, P) \leq S(f, P_0)$ by Exercise 7.6.5. The points of P_0 lying in $[a, y]$ form a partition Q of $[a, y]$ and those lying in $[y, z]$ form a partition R of $[y, z]$. Besides, $S(f, P_0) = S(f, Q) + S(f, R) \leq V_a^y(f) + V_y^z(f)$. It follows that $S(f, P) \leq S(f, P_0) \leq V_a^y(f) + V_y^z(f)$. Since this has been proved for any partition P of $[a, z]$, we conclude that $V_a^z(f) \leq V_a^y(f) + V_y^z(f)$, as required.
(e) (e) It follows from parts (a) and (d) that $V_a^x(f) \leq V_a^b(f)$ for every $x \in [a, b]$. $\square$

If $f: [a, b] \to \mathbb{R}$ is of bounded variation, $V_a^x(f)$ describes a nonnegative real-valued function on $[a, b]$ in view of Proposition 7.6.6(e). It will be convenient to introduce the notation $V_a f$ or $V_a\langle f \rangle$ for this function. Thus, $(V_a f)(\xi)$, $V_a\langle f \rangle(\xi)$ and $V_a f(\xi)$ all mean the same thing as $V_a^\xi(f)$.

It is clear from the definition of total variation that $V_a^x(-f) = V_a^x(f)$ for all x, which is to say, $V_a\langle -f \rangle = V_a f$.

7.6.7. Exercise. Show that if $f : [a, b] \to \mathbb{R}$ is of bounded variation, the non-negative real-valued function $V_a f$ is increasing and that $V_a^x(V_a f) = V_a^x(f)$ for all $x \in [a, b]$.

Solution: Let $a \leq x \leq y \leq b$. By Proposition 7.6.6(d), $V_a^y(f) = V_a^x(f) + V_x^y(f)$. Since $V_a f$ is nonnegative-valued, it follows that $V_a^y(f) \geq V_a^x(f)$. Therefore $V_a f$ is increasing. Hence

$$V_a^x(V_a f) = (V_a f)(x) - (V_a f)(a) = V_a^x(f) - V_a^a(f) = V_a^x(f) \quad \text{for all } x \in [a, b].$$

7.6.8. Theorem. *A real-valued function on a closed bounded interval is of bounded variation if and only if it is the sum of an increasing function and a decreasing function.*

Proof: Since a monotone function is always of bounded variation, the "if" part is trivial by Proposition 7.6.6(b).

For the converse, consider any function $f : [a, b] \to \mathbb{R}$ such that $V_a^b(f) < \infty$. Put $g = \frac{1}{2}(V_a f + f)$, which is to say, $g(x) = \frac{1}{2}(V_a^x(f) + f(x))$. Then $g : [a, b] \to \mathbb{R}$ is real-valued. If $a \leq y \leq z \leq b$, then

$$g(z) - g(y) = \frac{1}{2}\left(V_a^z(f) - V_a^y(f) + f(z) - f(y)\right)$$

$$= \frac{1}{2}\left(V_y^z(f) + f(z) - f(y)\right) \quad \text{by Proposition 7.6.6(d)}$$

$$\geq 0 \quad \text{by Definition 7.6.3.}$$

Thus g is an increasing function. As this has to hold even if we replace f by $-f$, the function $\frac{1}{2}(V_a\langle -f \rangle - f) = \frac{1}{2}(V_a f - f)$ is also increasing, so that the function $h = \frac{1}{2}(f - V_a f)$ is decreasing. Besides, $g + h = \frac{1}{2}(V_a f + f) + \frac{1}{2}(f - V_a f) = f$. □

7.6.9. Remark. It now follows from the Lebesgue Differentiability Theorem 7.4.1 that a function of bounded variation is differentiable a.e. Moreover, by Proposition 7.4.2, its derivative agrees a.e. with a Borel function.

The first paragraph in the proof of Proposition 7.5.1 begins by arguing that $F(x) = \int_{[a,x]} f$ describes a function of bounded variation.

Problem Set 7.6
7.6.P1. Show that the product of two functions of bounded variation is of bounded variation.

7.6.P2. Suppose $a < b$ and the bounded function $f : [a, b] \to \mathbb{R}$ is decreasing on $(a, b]$, so that the right limit $f(a+)$ exists. Show that

$$V_a^b(f) = f(a+) - f(b) + |f(a+) - f(a)|.$$

7.6.P3. If $f : [a, b] \to \mathbb{R}$ has a bounded derivative on the open interval (a, b), show that it is of bounded variation on the closed interval $[a, b]$.

7.6.P4. If $f : [a, b] \to \mathbb{R}$ is of bounded variation, show that it has one sided limits at each point and that, for any $c \in [a, b)$, the equality

$$V_a(f)(c+) - V_a(f)(c) = |f(c+) - f(c)|$$

holds. (In particular, f is right continuous at c if and only if $V_a(f)$ is.)

7.7 Absolute Continuity

A theorem about the integral of the derivative can be expected to say something like $f(x) - f(a) = \int_{[a,x]} f'$ for $a \leq x \leq b$, albeit under appropriate conditions on the derivative f'.

For the equality to make sense, the derivative should exist almost everywhere and the function resulting from extending the derivative function to be 0 elsewhere should be integrable. So, this could be one appropriate condition to assume. Additionally, properties that the indefinite integral of an integrable function must necessarily possess, such as continuity, could also be appropriate to assume about f. Actually, an indefinite integral always possesses a stronger property than continuity, called "absolute continuity". If we take f to be absolutely continuous, it turns out that it also has to be of bounded variation, and the Lebesgue Differentiability Theorem 7.4.1 renders it redundant to assume the existence of the derivative almost everywhere.

Now, assuming absolute continuity, the strategy could be to show that the left side of the equality we seek to establish has the same derivative as the right side. The hurdle in this strategy is that equality of derivatives can be expected to hold only almost everywhere, thereby making the Mean Value Theorem inapplicable. However, absolute continuity comes to the rescue, because vanishing of the derivative almost everywhere does indeed imply that the function is constant provided that the function is absolutely continuous.

It may be pertinent to note here as an aside that even the requirement that the function f be increasing does not warrant the desired equality. The best that can be extracted from the assumption that the function is increasing will be given later in Theorem 7.8.1.

7.7.1. Definition. *A real-valued function $f : [a, b] \to \mathbb{R}$ is said to be* **absolutely continuous** *if, for every $\varepsilon > 0$, there exists a $\delta > 0$ such that whenever $(a_1, b_1), (a_2, b_2), \ldots, (a_n, b_n)$ are disjoint subintervals of $[a, b]$,*

$$\sum_{j=1}^{n}(b_j - a_j) < \delta \;\; implies \;\; \sum_{j=1}^{n}\left|f(b_j)-f(a_j)\right| < \varepsilon.$$

It is clear what it means for a function $f:[a,b]\to\mathbb{R}$ to be absolutely continuous on a closed subinterval of the domain.

7.7.2. Exercise. Let $(a_1,b_1),(a_2,b_2),\ldots,(a_n,b_n)$ be subintervals of $[a, b]$ and k be an integer such that $1\le k \le n$. Suppose the intervals $(\alpha_1,\beta_1),(\alpha_2,\beta_2),\ldots,(\alpha_{n+1},\beta_{n+1})$ are obtained by replacing (a_k,b_k) by the two intervals (a_k,c) and (c,b_k), while retaining all the intervals (a_i,b_i) for which $i\neq k$. Show that $\sum_{i=1}^{n}|f(b_i)-f(a_i)| \le \sum_{j=1}^{n}|f(\beta_j)-f(\alpha_j)|$.

Solution: In view of the manner in which the intervals (α_j,β_j) are obtained from the intervals (a_i,b_i),

$$\sum_{j=1}^{n+1}\left|f(\beta_j)-f(\alpha_j)\right| - \sum_{i=1}^{n}\left|f(b_i)-f(a_i)\right|$$
$$= |f(b_k)-f(c)| + |f(c)-f(a_k)| - |f(b_k)-f(a_k)|$$
$$\ge 0 \quad \text{by the triangle inequality.}$$

7.7.3. Theorem. *An absolutely continuous function $f:[a,b]\to\mathbb{R}$ is continuous and of bounded variation. In particular, it is differentiable a.e. and its derivative agrees a.e. with a Borel function.*

Proof: That it is continuous is trivial from the definition of absolute continuity.

Corresponding to $\varepsilon=1$, choose $\delta>0$ satisfying the condition in the definition of absolute continuity. Let N be some positive integer such that $N\delta>b-a$. We shall show that $V_a^b(f)\le N$.

Observe that by choice of δ, any subinterval $[\alpha,\beta]\subseteq[a,b]$ of length less than δ satisfies $V_\alpha^\beta(f)\le\varepsilon=1$.

If $N=1$, then $\delta>b-a$ and it is immediate from the observation in the preceding paragraph that $V_a^b(f)\le 1 = N$. So, we need consider only $N>1$.

Let δ' be the number such that $N\delta'=b-a$. Then $\delta'>0$ and $N\delta'=b-a<N\delta$, so that $\delta'<\delta$. Thus, $0<\delta'<\delta$. Now consider the points $x_0<x_1<\ldots<x_N$ in $[a,b]$ given by $x_r=a+r\delta', 0\le r\le N$. Obviously, $x_0=a$ and $x_N=b$. Since each interval $[x_r,x_{r+1}]$ has length $\delta'<\delta$, it follows by the observation above that each $V_{x_r}^{x_{r+1}}(f)\le 1$. Therefore, on applying Proposition 7.6.6(d) repeatedly $N-1$ times, we obtain

$$V_a^b(f) = \sum_{r=0}^{N-1}V_{x_r}^{x_{r+1}}(f) \le N,$$

as promised.

The rest follows from Remark 7.6.9.   □

In Example 7.6.4, we have exhibited a continuous function that is not of bounded variation. By Theorem 7.7.3, it cannot be absolutely continuous; continuity therefore does not imply absolute continuity.

It will be seen later in Example 7.8.2 that there exists a function of bounded variation, in fact a strictly increasing one, which is continuous but not absolutely continuous.

Now for the result that lets us hurtle over the hurdle of the Mean Value Theorem not being available.

7.7.4. Theorem. *If an absolutely continuous function $f : [a, b] \to \mathbb{R}$ has the property that $f' = 0$ a.e. on $[a, b]$, then it is constant on $[a, b]$.*

Proof: It is sufficient to show that $f(a) = f(b)$. Given any $\varepsilon > 0$, we shall prove that

$$|f(b) - f(a)| \leq \varepsilon(1 + (b - a)).$$

Choose $\delta > 0$ satisfying the condition in the definition of absolute continuity.

Let $N \subseteq [a, b]$ consist of b and all the points where f' is either nonzero or does not exist. Then N has measure 0. To each $\alpha \in [a, b] \backslash N$ correspond intervals $[\alpha, \beta] \subseteq [a, b]$ with $\beta - \alpha$ arbitrarily small such that

$$|f(\beta) - f(\alpha)| < \varepsilon(\beta - \alpha).$$

Since this holds for each $\alpha \in [a, b] \backslash N$, the intervals constitute a Vitali covering of the set. By the Vitali Covering Theorem 7.2.7, there exist finitely many disjoint intervals $[\alpha_j, \beta_j]$, $1 \leq j \leq n$, in the Vitali covering such that

$$m(N_0) < \delta, \quad \text{where } N_0 = ([a, b] \backslash N) \backslash \bigcup_{j=1}^{n} [\alpha_j, \beta_j].$$

Since $N_0 \cup N \supseteq [a, b] \backslash \cup_{j=1}^{n} [\alpha_j, \beta_j]$, we obtain

$$m([a, b] \backslash \bigcup_{j=1}^{n} [\alpha_j, \beta_j]) \leq m(N_0 \cup N) \leq m(N_0) + m(N) = m(N_0) < \delta.$$

By renumbering the intervals $[\alpha_j, \beta_j]$ if necessary, we may assume that their centers are in increasing order. Set $\beta_0 = a$ and $\alpha_{n+1} = b$. Then we have

$$a = \beta_0 \leq \alpha_1 < \beta_1 \leq \cdots < \beta_n \leq \alpha_{n+1} = b.$$

It follows that

$$|f(b) - f(a)| \leq \sum_{j=1}^{n} |f(\beta_j) - f(\alpha_j)| + \sum_{j=1}^{n+1} |f(\alpha_j) - f(\beta_{j-1})|.$$

The choice of the Vitali covering ensures that

$$\sum_{j=1}^{n} |f(\beta_j) - f(\alpha_j)| \leq \sum_{j=1}^{n} \varepsilon(\beta_j - \alpha_j) \leq \varepsilon(b - a).$$

Now the intervals (β_{j-1}, α_j) are all disjoint from $\cup_{j=1}^{n} [\alpha_j, \beta_j]$ and therefore their union is a subset of $[a, b] \backslash \cup_{j=1}^{n} [\alpha_j, \beta_j]$, which has been shown to have measure less than δ. As they are disjoint from each other, their total length is the same as the measure of their union and hence is less than δ. Consequently, in view of the condition that δ was chosen to satisfy with reference to the definition of absolute continuity,

$$\sum_{j=1}^{n+1} |f(\alpha_j) - f(\beta_{j-1})| < \varepsilon.$$

When this is combined with the preceding two inequalities, the desired inequality springs forth. $\square$

7.7.5. Proposition. *Suppose $f : [a, b] \rightarrow \mathbb{R}$ is integrable and $F : [a, b] \rightarrow \mathbb{R}$ is the function given by*

$$F(x) = \int_{[a,x]} f \quad \text{for all } x \in [a, b].$$

Then F is absolutely continuous.

Proof: Let $\varepsilon > 0$. By Problem 3.2.P4, there exists a $\delta > 0$ such that for any measurable A with $m(A) < \delta$, we have $\int_A f < \varepsilon$. Now consider disjoint subintervals $(a_1, b_1), (a_2, b_2), \ldots, (a_n, b_n)$ of $[a, b]$ such that $\sum_{j=1}^{n} (b_j - a_j) < \delta$. Put $A = \cup_{j=1}^{n} [a_j, b_j]$. Then $m(A) = \sum_{j=1}^{n} (b_j - a_j) < \delta$, with the consequence that $\int_A f < \varepsilon$. But

$$\int_A f = \sum_{j=1}^{n} \int_{[a_j, b_j]} f = \sum_{j=1}^{n} (F(b_j) - F(a_j)).$$

$\square$

Problem Set 7.7

7.7.P1 (a) Show that a continuous function on $[0, 1]$ which is absolutely continuous on $[\eta, 1]$ for every positive η less than 1 need not be absolutely continuous on $[0, 1]$.

(b) Show that a continuous function on $[0, 1]$ which is absolutely continuous on $[\eta, 1]$ for every positive η less than 1 and is of bounded variation must be absolutely continuous on $[0, 1]$.

7.7.P2. (a) Show that the function $g : [0, 1] \to \mathbb{R}$ given by $g(x) = x^{\frac{1}{2}}$ is absolutely continuous.

(b) Show that the function $f : [0, 1] \to \mathbb{R}$ given by $f(x) = x^2 \cos^4(\pi/2x)$ for $x \neq 0$ and $f(0) = 0$ is absolutely continuous.

(c) Show that a composition of absolutely continuous functions need not be absolutely continuous.

7.7.P3. (a) Suppose $f : [a, b] \to [\alpha, \beta]$ and $g : [\alpha, \beta] \to \mathbb{R}$ are absolutely continuous. If f is increasing, show that the composition $g \circ f : [a, b] \to \mathbb{R}$ is absolutely continuous. Does the same conclusion hold if the function f is decreasing instead?

(b) Give an example of a function that has an unbounded derivative, is not monotone, but is absolutely continuous.

7.7.P4. If $f : [a, b] \to \mathbb{R}$ is absolutely continuous on each of the subintervals $[a, c]$ and $[c, b]$, where $a < c < b$, show that it is absolutely continuous on $[a, b]$.

7.7.P5. Given a finite number of points in (a, b), show that there exists an absolutely continuous function on $[a, b]$ that fails to be differentiable at the given points.

7.8 The Integral of the Derivative

As recorded just before the statement of Theorem 7.5.2, a Borel function that agrees a.e. with the derivative of a function f is denoted by f' and we speak of it as though it is unique. By Proposition 7.4.2, this may be done with a monotone function, as we shall do in Theorem 7.8.1. By virtue of Theorem 7.7.3, Theorem 7.6.8 and Proposition 7.4.2, the same can also be done with an absolutely continuous function, and we shall do so in Theorem 7.8.3.

The reader is reminded that, for a real-valued monotone function $f : [a, b] \to \mathbb{R}$ one sided limits $f(x+)$ and $f(x-)$ exist at every interior point $x \in (a, b)$ and so do $f(a+)$ and $f(b-)$.

7.8.1. Theorem. *If $f : [a, b] \to \mathbb{R}$ is increasing, then f' is integrable on $[a, b]$ and*

$$f(b-) - f(a) \geq \int_{[a,b]} f'.$$

Proof: The result will follow by the Dominated Convergence Theorem 3.2.6 if we prove it with b replaced by an arbitrary number $c \in (a, b)$ on the smaller side of the inequality. The integral $\int_{[a,c+\frac{1}{n}]} f$ then makes sense when $\frac{1}{b-c} < n \in \mathbb{N}$.

Let the functions $f_n : [a, c] \to \mathbb{R}$ be given by

$$f_n(x) = \frac{f(x + \frac{1}{n}) - f(x)}{\frac{1}{n}}, \quad \frac{1}{b-c} < n \in \mathbb{N}.$$

Then $\lim_{n\to\infty} f_n = f'$ a.e. on $[a, c]$. In tandem with Remark 6.5.4(c) (a consequence of Fatou's Lemma), this leads to

$$\int_{[a,c]} f' \le \liminf_{n\to\infty} \int_{[a,c]} f_n = \liminf_{n\to\infty} \left(n \int_{[a,c]} f(x + \frac{1}{n}) - f(x)dx \right)$$

$$= \liminf_{n\to\infty} \left(n \int_{[c,c+\frac{1}{n}]} f(x)dx - n \int_{[a,a+\frac{1}{n}]} f(x)dx \right)$$

$$\le \liminf_{n\to\infty} \left(n \int_{[c,c+\frac{1}{n}]} f(c-)dx - n \int_{[a,a+\frac{1}{n}]} f(a)dx \right)$$

$$\le f(c-) - f(a). \quad \square$$

The inequality in the above theorem cannot be replaced by equality. There actually exists a continuous function from $[0, 1]$ to $[0, 1]$, which is surjective, has derivative 0 a.e., and is also *strictly* increasing to boot.

7.8.2. Example. Define a sequence $\{f_n\}_{n\ge0}$ of functions on the domain $[0, 1]$ as follows. The first one is $f_0(x) = x$, a strictly increasing function whose values on any interval $\left[\frac{k}{2^0}, \frac{k+1}{2^0}\right]$ with $0 \le k \le 2^0 - 1$ (yes, there is only one!) are linear interpolations of its values at the endpoints (straight line graph on the interval). Suppose f_n is a strictly increasing function on $[0, 1]$ whose values on any interval $\left[\frac{k}{2^n}, \frac{k+1}{2^n}\right]$ with $0 \le k \le 2^n - 1$ are linear interpolations of its values at the endpoints. It is pertinent to note for later purposes that linear interpolation implies that the value at the midpoint is the mean of the values at the endpoints:

$$f_n\left(\frac{2k+1}{2^{n+1}}\right) = \frac{1}{2}f_n\left(\frac{k}{2^n}\right) + \frac{1}{2}f_n\left(\frac{k+1}{2^n}\right). \tag{7.36}$$

The function f_{n+1} is obtained from f_n by (see figure below) first setting its values at the $2^{n+1} + 1$ points $\frac{k}{2^{n+1}}$, where $0 \le k \le 2^{n+1}$, to be

$$f_{n+1}\left(\frac{2k}{2^{n+1}}\right) = f_{n+1}\left(\frac{k}{2^n}\right) = f_n\left(\frac{k}{2^n}\right) \quad \text{for } 0 \le k \le 2^n, \tag{7.37}$$

$$f_{n+1}\left(\frac{2k+1}{2^{n+1}}\right) = \frac{1}{3}f_n\left(\frac{k}{2^n}\right) + \frac{2}{3}f_n\left(\frac{k+1}{2^n}\right) \quad \text{for } 0 \le k \le 2^n - 1, \tag{7.38}$$

and then extending it by linear interpolation to the interiors of the 2^{n+1} intervals formed by the points. Note that the 2^{n+1} intervals are $\left[\frac{k}{2^{n+1}}, \frac{k+1}{2^{n+1}}\right], 0 \leq k \leq 2^{n+1} - 1$.

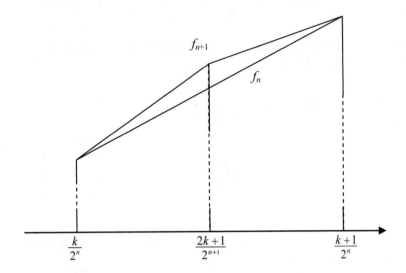

Equation (7.38) mandates the value of f_{n+1} at the midpoint of $\left[\frac{k}{2^n}, \frac{k+1}{2^n}\right]$ to be a convex combination of the values of f_n at the endpoints but "biased" in favor of the value at the right endpoint, which is the larger one. Therefore, intuition would lead one to expect that the convex combination must work out to be greater than the mean. The expectation is confirmed by the computation

$$\left(\frac{1}{3}\alpha + \frac{2}{3}\beta\right) - \left(\frac{1}{2}\alpha + \frac{1}{2}\beta\right) = \frac{1}{6}(\beta - \alpha).$$

Accordingly,

$$f_{n+1}\left(\frac{2k+1}{2^{n+1}}\right) > \frac{1}{2}f_n\left(\frac{k}{2^n}\right) + \frac{1}{2}f_n\left(\frac{k+1}{2^n}\right),$$

and hence by (7.36),

$$f_{n+1}\left(\frac{2k+1}{2^{n+1}}\right) > f_n\left(\frac{2k+1}{2^{n+1}}\right).$$

The stipulation about linear interpolation now implies that $f_{n+1} > f_n$ on $\left(\frac{k}{2^n}, \frac{k+1}{2^n}\right)$ and hence that $f_{n+1} \geq f_n$ on all of $[0, 1]$. Since these functions have values lying between 0 and 1, there is a function F on $[0, 1]$ with values lying between 0 and 1 such that $F = \lim_{p\to\infty} f_p \geq f_n$ for every n. From (7.37), we infer that

$$F\left(\frac{k}{2^n}\right) = f_n\left(\frac{k}{2^n}\right) \quad \text{for } 0 \leq k \leq 2^n \text{ and } n \geq 0. \tag{7.39}$$

In particular, $F(0) = 0$ and $F(1) = 1$.
  The computations

$$\left(\frac{1}{3}\alpha + \frac{2}{3}\beta\right) - \alpha = \frac{2}{3}(\beta - \alpha) \quad \text{and} \quad \beta - \left(\frac{1}{3}\alpha + \frac{2}{3}\beta\right) = \frac{1}{3}(\beta - \alpha),$$

when viewed in the light of (7.37) and (7.38), confirm that

$$f_{n+1}\left(\frac{k}{2^n}\right) < f_{n+1}\left(\frac{2k+1}{2^{n+1}}\right) < f_{n+1}\left(\frac{k+1}{2^n}\right).$$

The stipulation about linear interpolation now implies that f_{n+1} is strictly increasing. Consequently, every function f_n in the inductively constructed sequence is strictly increasing. It follows that the limit function F is increasing. We shall show that it is in fact *strictly* increasing. To this end, suppose $x < y$, where $x, y \in [0, 1]$. Some k and n must satisfy $x < \frac{k}{2^n} < y$. Using the already established facts that (a) F is increasing, (b) $F\left(\frac{k}{2^n}\right) = f_n\left(\frac{k}{2^n}\right)$ always, (c) every f_n is strictly increasing, and (d) $F \geq f_n$ always, in that order, we find that

$$F(x) \leq F\left(\frac{k}{2^n}\right) = f_n\left(\frac{k}{2^n}\right) < f_n(y) \leq F(y).$$

The computations at the beginning of the preceding paragraph also confirm that

$$f_{n+1}\left(\frac{2k+1}{2^{n+1}}\right) - f_{n+1}\left(\frac{k}{2^n}\right) = \frac{2}{3}\left(f_n\left(\frac{k+1}{2^n}\right) - f_n\left(\frac{k}{2^n}\right)\right) \tag{7.40}$$

and

$$f_{n+1}\left(\frac{k+1}{2^n}\right) - f_{n+1}\left(\frac{2k+1}{2^{n+1}}\right) = \frac{1}{3}\left(f_n\left(\frac{k+1}{2^n}\right) - f_n\left(\frac{k}{2^n}\right)\right). \tag{7.41}$$

Now consider an arbitrary $x \in [0, 1]$. There exists a nested sequence of intervals $[\alpha_n, \beta_n] = \left[\frac{k_n}{2^n}, \frac{k_n+1}{2^n}\right]$, each containing x. From (7.39), (7.40) and (7.41), it follows for $n \geq 0$ that

$$F(\beta_{n+1}) - F(\alpha_{n+1}) = \lambda_n(F(\beta_n) - F(\alpha_n)), \quad \text{where each } \lambda_n = \frac{1}{3} \text{ or } \frac{2}{3}.$$

Repeated application of this equality leads to

$$F(\beta_{n+1}) - F(\alpha_{n+1}) = \prod_{k=0}^{n} \lambda_k (F(\beta_0) - F(\alpha_0)) = \prod_{k=0}^{n} \lambda_k.$$

Since each $|\lambda_k| \leq \frac{2}{3}$, the right side here has limit 0 as $n \to \infty$. This proves that F is continuous at x.

Since $\beta_{n+1} - \alpha_{n+1} = \frac{1}{2^{n+1}}$, the derivative $F'(x)$, if it exists, must be the limit of

$$2^{n+1}(F(\beta_{n+1}) - F(\alpha_{n+1})) = \prod_{k=0}^{n} \tau_k, \quad \text{where each } \tau_k = \frac{2}{3} \text{ or } \frac{4}{3}.$$

We shall demonstrate that the product on the right side cannot have a finite positive limit, no matter which among the τ_k are $\frac{2}{3}$ and which ones are $\frac{4}{3}$. Suppose to the contrary that it has a finite limit $\tau > 0$. Then there exists an $n \in \mathbb{N}$ such that

$$\frac{14}{15}\tau < \prod_{k=0}^{n} \tau_k < \frac{16}{15}\tau \quad \text{as well as} \quad \frac{14}{15}\tau < \prod_{k=0}^{n+1} \tau_k < \frac{16}{15}\tau.$$

If $\tau_{n+1} = \frac{2}{3}$, the second part of the first double inequality implies

$$\prod_{k=0}^{n+1} \tau_k < \frac{2}{3}\frac{16}{15}\tau < \frac{14}{15}\tau,$$

contradicting the first part of the second double inequality, whereas if $\tau_{n+1} = \frac{4}{3}$, the first part of the first double inequality implies

$$\prod_{k=0}^{n+1} \tau_k > \frac{4}{3}\frac{14}{15}\tau > \frac{16}{15}\tau,$$

contradicting the second part of the second double inequality. The possibility that $F'(x)$ is finite and nonzero is thereby ruled out. Since F must have a derivative a.e. by the Lebesgue Differentiability Theorem 7.4.1, we conclude that the strictly increasing continuous function F has derivative 0 a.e.

   Thus $F(1-) - F(0) = F(1) - F(0) = 1 > 0 = \int_{[a,b]} F'$, showing that equality need not hold in Theorem 7.8.1. Furthermore, by the next theorem, F cannot be absolutely continuous, although it is continuous and strictly increasing, in particular, of bounded variation.

   The above function F is surprising because it is a continuous surjection from [0, 1] to itself with derivative 0 a.e. but is strictly increasing nonetheless. At the opposite extreme, there are continuous surjections from [0, 1] to itself not only with derivative 0 a.e. but also constant on all the middle thirds that are removed in the process of forming the Cantor set (see Problem 4.1.P4). The standard example of such a continuous surjection, called the *Cantor function*, will be presented in the next chapter. These continuous surjections illustrate the same thing regarding Theorem 7.8.1 as F does.

7.8.3. Theorem (Integral of the Derivative). *If $f: [a, b] \to \mathbb{R}$ is absolutely continuous, then f' is integrable on $[a, b]$ and*

$$f(x) - f(a) = \int_{[a,x]} f' \quad \text{for all} x \in [a, b].$$

Proof: Since f is absolutely continuous, it is of bounded variation and therefore a sum of an increasing function g and a decreasing function h. By Theorem 7.8.1, both g'

and h' are integrable on $[a, b]$. Since $f' = g' - h'$, the derivative f' is integrable. The right side of the equality sought to be proved has derivative f' a.e. by Theorem 7.5.2 (Derivative of the Integral) and is absolutely continuous by Proposition 7.7.5. The same is true of the left side. Upon invoking Theorem 7.7.4, we find that the left side and right side differ by a constant. As they agree at $x = a$, we can assert that they are equal everywhere.   □

7.8.4. Corollary. *A function* $f : [a, b] \to \mathbb{R}$ *is absolutely continuous if and only if there exists an integrable function* $g : [a, b] \to \mathbb{R}$ *such that*

$$f(x) - f(a) = \int_{[a,x]} g \quad \text{for all } x \in [a, b].$$

Proof: Immediate from Proposition 7.7.5 and Theorem 7.8.3.   □

Problem Set 7.8

7.8.P1. (a) Let f be a real-valued function on $[a, b]$ such that its values on (a, b) are linear interpolations between its values at the endpoints. If one among $f(a)$ and $f(b)$ is positive and the other nonnegative, show that $f(x)$ is positive for all $x \in (a, b)$.

(b) Let f and g be real-valued functions on $[a, b]$ such that $f(a) \geq g(a), f(b) \geq g(b)$, one of the inequalities being strict, and let the values of both functions on (a, b) be linear interpolations between their values at the endpoints. Show that $f(x) > g(x)$ for all $x \in (a, b)$.

7.8.P2. (a) Let f be a real-valued function on $[a, b]$ such that its values on (a, b) are linear interpolations between its values at the endpoints and let $a < c < b$. Show that the function has the same interpolation property over the intervals (a, c) and (c, b).

(b) Let f and g be real-valued functions on $[a, b]$ such that $f(a) \geq g(a), f(b) \geq g(b)$, and the values of g on (a, b) are linear interpolations between its values at the endpoints. Also, let $a < c < b$ and the values of f on each of (a, c) and (c, b) be linear interpolations between its values at the endpoints of the respective intervals. If $f(c) > g(c)$, show that $f(x) > g(x)$ for all $x \in (a, b)$.

7.8.P3. Suppose $0 < \alpha < 1 < \beta$ and $\{\tau_j\}_{j \geq 1}$ is a sequence of numbers such that, for each j, either $0 < \tau_j \leq \alpha$ or $\tau_j \geq \beta$. Let $\sigma_n = \prod_{j=1}^{n} \tau_j$. Show that σ_n cannot have a finite positive limit.

7.8.P4. Given that $f : [a, b] \to \mathbb{R}$ is absolutely continuous, show that the function $V_a f : [a, b] \to \mathbb{R}$ defined by $(V_a f)(\xi) = V_a^\xi(f)$ for all $\xi \in [a, b]$ is also absolutely continuous. Can one draw the inference that an absolutely continuous function is the sum of an absolutely continuous increasing function and an absolutely continuous decreasing function?

Chapter 8
The Cantor Set and Function

8.1   A Surjection from the Cantor Set to [0, 1]

Let b be an integer greater than 1. If $\{\alpha_k\}_{k\geq 1}$ is a sequence of integers such that $0 \leq \alpha_k \leq b - 1$ for each k, it is easy to show that the series $\sum_{k=1}^{\infty} \alpha_k b^{-k}$ converges to a sum $x \in [0, 1]$ by comparing it with the geometric series with common ratio b^{-1}. It is well known that every $x \in [0, 1]$ has at least one representation as $x = \sum_{k=1}^{\infty} \alpha_k b^{-k}$, where the α_k are integers satisfying $0 \leq \alpha_k \leq b - 1$ (see Theorem 4.1.9 in Shirali and Vasudeva [9]). We shall refer to any such representation as a **b-expansion** (or **b-representation**) of x. The easily verified fact that 0 and 1 have unique b-expansions will matter but lack of uniqueness in general is an inconvenience. However, the inconvenience can be managed.

The integer b in this context is called the **base** of the expansion. The standard name for a 2-expansion is **binary** expansion (representation). The terms of the sequence $\{\alpha_k\}_{k\geq 1}$ are called the **digits** in the b-expansion.

One learns in early encounters with decimal expansions (base 10) how to compare two numbers by spotting the first decimal place where they differ. The procedure needs to be formulated with due care to adjust for absence of uniqueness, and then validated. We shall do so for an arbitrary base b although we shall use it, with one exception, only for base 2 afterwards. The exception is Problem 8.1.P3, which concerns base 3 and is a standalone with no purpose to serve in anything else relating to the Cantor set that we discuss.

If $\sum_{k=1}^{\infty} \alpha_k b^{-k}$ and $\sum_{k=1}^{\infty} \beta_k b^{-k}$ are b-expansions, then we have $0 \leq \alpha_k \leq b - 1$ as well as $0 \leq \beta_k \leq b - 1$ for each k. Two simple consequences of these inequalities will be used in the next proof: For each k, (1) $|\alpha_k - \beta_k| \leq b - 1$ and (2) $\alpha_k - \beta_k = -(b - 1) \Leftrightarrow \alpha_k = 0, \beta_k = b - 1$.

8.1.1. Proposition. *Suppose* $x = \sum_{k=1}^{\infty} \alpha_k b^{-k}$ *and* $y = \sum_{k=1}^{\infty} \beta_k b^{-k}$ *are b-expansions and there exists a positive integer j such that $\alpha_j > \beta_j$ and $\alpha_k = \beta_k$ for $k < j$ (vacuously true in case $j = 1$). Then (1) $x \geq y$ and (2) $x = y \Leftrightarrow$ not only $\alpha_j - \beta_j = 1$ but also every $k > j$ satisfies $\alpha_k = 0$, $\beta_k = b - 1$.*

© Springer Nature Switzerland AG 2018
S. Shirali, *A Concise Introduction to Measure Theory*,
https://doi.org/10.1007/978-3-030-03241-8_8

Proof: First note that

$$\sum_{k=j}^{\infty} (\alpha_k - \beta_k)b^{-k} \geq b^{-j} + \sum_{k=j+1}^{\infty} (\alpha_k - \beta_k)b^{-k} \text{ because } \alpha_j - \beta_j \geq 1$$

$$\geq b^{-j} - (b-1)\sum_{k=j+1}^{\infty} b^{-k} \text{ because } \alpha_k - \beta_k \geq -(b-1)\text{ for each } k.$$

Now, equality holds in the first line above $\Leftrightarrow \alpha_j - \beta_j = 1$ and holds in the second line $\Leftrightarrow$ every $k > j$ satisfies $\alpha_k - \beta_k = -(b-1) \Leftrightarrow$ every $k > j$ satisfies $\alpha_k = 0, \beta_k = b - 1$. So, equality holds in both lines $\Leftrightarrow$ not only $\alpha_j - \beta_j = 1$ but also every $k > j$ satisfies $\alpha_k = 0, \beta_k = b - 1$. And now the sockdolager: $\sum_{k=j}^{\infty} (\alpha_k - \beta_k)b^{-k} = x - y$ and $b^{-j} - (b-1)\sum_{k=j+1}^{\infty} b^{-k} = 0$. □

In the above proposition, the integer j is the smallest k such that $\alpha_k \neq \beta_k$ and is unique if it exists, which it certainly does unless $\alpha_k = \beta_k$ for all k.

8.1.2. Example. Let $b = 2$ and

$$x = 1 \cdot 2^{-1} + 0 \cdot 2^{-2} + 0 \cdot 2^{-2} + \cdots, \quad y = 0 \cdot 2^{-1} + 1 \cdot 2^{-2} + 1 \cdot 2^{-2} + \cdots.$$

Then the smallest k such that $\alpha_k \neq \beta_k$ is $k = 1$, and $\alpha_1 > \beta_1$. From Proposition 8.1.1, we obtain $x \geq y$, which can independently be seen to be true because $x = \frac{1}{2} = y$. Besides, $\alpha_1 - \beta_1 = 1 - 0 = 1$ and every $k > 1$ satisfies $\alpha_k = 0, \beta_k = 1 = b - 1$.

8.1.3. Remarks. (a) When the open middle third of a closed bounded interval is removed, what remains is a union of two disjoint closed subintervals, which we may distinguish as the left and right thirds. If two numbers in the closed interval belong to different ones among the left and right thirds, then the bigger number must be in the right third (and the smaller in the left).

(b) When the open middle third of a closed bounded interval is removed, the left endpoint not only belongs to the left third but is also its left endpoint. In particular, when the open middle third of the left third is removed, it again belongs to the left third of the left third. Analogously for the right endpoint.

Recall that the Cantor set is defined as $\bigcap_{n=1}^{\infty} C_n$, where $C_1 = \left[0, \frac{1}{3}\right] \cup \left[\frac{2}{3}, 1\right]$, and C_{n+1} is obtained from C_n by removing the open middle thirds of its closed intervals. Thus, $C_2 = \left[0, \frac{1}{9}\right] \cup \left[\frac{2}{9}, \frac{1}{3}\right] \cup \left[\frac{2}{3}, \frac{7}{9}\right] \cup \left[\frac{8}{9}, 1\right]$, and so on.

In general, C_{n+1} is the union of the closed intervals obtained by removing the open middle thirds from each of the disjoint closed intervals whose union is C_n. Thus C_1 is the union of the left third $\left[0, \frac{1}{3}\right]$ of $[0, 1]$ and the right third $\left[\frac{2}{3}, 1\right]$. By induction, C_n is the union of 2^n closed intervals, each of length $\left(\frac{1}{3}\right)^n$. So the total length of the closed intervals in C_n is $\left(\frac{2}{3}\right)^n$. It will be convenient to refer to the 2^n intervals as "comprising" C_n or as "components" of C_n and distinguish them as left and right components. They are left and right thirds respectively of the components of C_{n-1} if

$n > 1$ and of [0, 1] if $n = 1$. In this terminology, half of the 2^n intervals comprising C_n are left components and half are right components and they are all disjoint from each other. The disjointness ensures that, for each n, a given number in the Cantor set belongs to precisely one among the components of C_n. The interval may be a left component or a right component. Accordingly, we can map each number u in the Cantor set into a uniquely determined sequence $\{\alpha_k\}_{k\geq 1}$ of 0s and 1s by setting $\alpha_k = 0$ or 1 according as the number is in a left component or a right component of C_k.

We have thus described a mapping from the Cantor set into the set of all sequences $\{\alpha_k\}_{k\geq 1}$ of 0s and 1s.

8.1.4. Example. The number $\frac{1}{3}$ belongs to the left component $\left[0, \frac{1}{3}\right]$ among the two components of C_1. On the other hand, it is the right endpoint of that left component and therefore by Remark 8.1.3(b), it is the right endpoint of the right third of $\left[0, \frac{1}{3}\right]$ when the open middle third of $\left[0, \frac{1}{3}\right]$ is removed. In fact, $\left[0, \frac{1}{3}\right] = \left[0, \frac{3}{9}\right]$ and its right third is $\left[\frac{2}{9}, \frac{1}{3}\right]$. So, $\frac{1}{3}$ is the right endpoint of a right component of C_2. By repeated application of Remark 8.1.3(b), we find that $\frac{1}{3}$ is in a right component of C_n for all $n > 1$. This shows not only that $\frac{1}{3}$ is in the Cantor set but also that the sequence $\{\alpha_k\}_{k\geq 1}$ obtained by applying the above mapping to $\frac{1}{3}$ is 01111.... In contrast, the number $\frac{2}{3}$ is in the right component $\left[\frac{2}{3}, 1\right]$ among the two components of C_1 but is the left endpoint of it and therefore belongs to a left component of C_n for all $n > 1$. Hence not only is $\frac{2}{3}$ in the Cantor set but also the sequence $\{\alpha_k\}_{k\geq 1}$ obtained by applying the above mapping to $\frac{2}{3}$ is 10000....

Let m be a positive integer. The interval $\left[0, \frac{1}{3^{m+1}}\right]$ is the left third of $\left[0, \frac{1}{3^m}\right]$, which shows by induction that the latter is a component interval of C_m. On the other hand, $\frac{1}{3^{m+1}}$ is the right endpoint of that left component and therefore by Remark 8.1.3(b), it is the right endpoint of the right third of $\left[0, \frac{1}{3^{m+1}}\right]$ when the open middle third of $\left[0, \frac{1}{3^{m+1}}\right]$ is removed. Applying Remark 8.1.3(b) repeatedly as in the preceding paragraph, we deduce that $\frac{1}{3^{m+1}}$ is in a right component of C_n for all $n > m + 1$. This shows that $\frac{1}{3^{m+1}}$ is in the Cantor set and hence that the Cantor set contains arbitrarily small positive numbers.

We note certain properties of the mapping that we shall need.

It is manifest from the Nested Interval Theorem (see Theorem 3.3.6 of Shirali and Vasudeva [9]) that the mapping is surjective.

That it is injective will follow from something stronger that we need and are about to prove. The inverse of the mapping will subsequently be denoted by ϕ.

8.1.5. Proposition. *Suppose $u \in C$ is mapped into $\{\alpha_k\}_{k\geq 1}$ and $v \in C$ is mapped into $\{\beta_k\}_{k\geq 1}$. If $\alpha_k = \beta_k$ for $1 \leq k \leq n$, then u and v belong to the same component of C_n. If $u > v$, then there is some k for which $\alpha_k \neq \beta_k$ and if j denotes the smallest such k, we have $\alpha_j > \beta_j$. In particular, the mapping is injective.*

Proof: If $n = 1$, this is obvious. Assume it true for some n and let $\alpha_k = \beta_k f$ for $1 \leq k \leq n+1$. Then u and v belong to the same component I of C_n. Since $\alpha_{n+1} = \beta_{n+1}$, the numbers u and v both belong to a left third among the components of C_{n+1} or both belong to a right third among the components of C_{n+1}. However, I contains only one left third and one right third. Therefore u and v belong to the same component of C_{n+1}.

Now suppose $u > v$. If we were to have $\alpha_k = \beta_k$ for all k, then u and v would be in the same component of C_k for every k, so that the distance between them would be at most $\left(\frac{1}{3}\right)^k$ for every k. This is not possible because $u \neq v$. Therefore $\alpha_k \neq \beta_k$ for some k and there must be a smallest such k. Denote it by j. If $j = 1$, then u and v belong to different subintervals among $\left[0, \frac{1}{3}\right]$ and $\left[\frac{2}{3}, 1\right]$ and the fact that $u > v$ implies that $u \in \left[\frac{2}{3}, 1\right]$ and $v \in \left[0, \frac{1}{3}\right]$, which further implies that $\alpha_1 = 1$ and $\beta_1 = 0$, so that $\alpha_j > \beta_j$, as desired. Now suppose $j > 1$. Then $\alpha_k = \beta_k$ for $1 \leq k \leq j-1$ and therefore u and v are in the same component of C_{j-1}. Since $\alpha_j \neq \beta_j$, one among u and v belongs to the left third of a component of C_{j-1} and the other belongs to the right third. The fact that $u > v$ implies by Remark 8.1.3(a) that u must belong to the right third and v to the left third. So, $\alpha_j = 1$ and $\beta_j = 0$; this confirms that $\alpha_j > \beta_j$, as desired.   □

8.1.6. Remark. It is trivial to deduce from the second assertion of Proposition 8.1.5 that its converse is also true. Thus if the sequences $\{\alpha_k\}_{k \geq 1}$, $\{\beta_k\}_{k \geq 1}$ fulfill the condition that there is a k for which $\alpha_k \neq \beta_k$ and $\alpha_j > \beta_j$, where j is the smallest k with $\alpha_k \neq \beta_k$, then $\phi(\{\alpha_k\}_{k \geq 1}) > \phi(\{\beta_k\}_{k \geq 1})$.

Now, the mapping which takes a sequence $\{\alpha_k\}_{k \geq 1}$ into $\sum_{k=1}^{\infty} \alpha_k 2^{-k}$ has range $[0, 1]$ because every number in $[0, 1]$ has at least one binary expansion. We make no claim that it is injective (see Example 8.1.2.). By composing this mapping with the one discussed in Proposition 8.1.5 above, namely ϕ^{-1}, we obtain a surjective map κ_0 from the Cantor set to $[0, 1]$. Although an application of the Bernstein–Schröder Theorem immediately yields the conclusion that there is a bijection between the Cantor set and $[0, 1]$, we shall derive this conclusion independently of that theorem in the next section (see Theorem 8.2.3), wherein we shall use the function κ_0.

A more formal definition of κ_0 is that $\kappa_0(u) = \sum_{k=1}^{\infty} \alpha_k 2^{-k}$, where $\{\alpha_k\}_{k \geq 1} = \phi^{-1}(u)$. It is worth bearing in mind that $\kappa_0(u)$ is the number in $[0, 1]$ having $\phi^{-1}(u)$ as its sequence of digits in at least one binary expansion of it.

A rephrasing of Example 8.1.4 in terms of ϕ^{-1} is that the sequence $\phi^{-1}\left(\frac{1}{3}\right)$ has first term 0 and all subsequent terms 1, whereas $\phi^{-1}\left(\frac{2}{3}\right)$ has first term 1 and all subsequent terms 0. This leads to $\kappa_0\left(\frac{1}{3}\right) = 0 \cdot 2^{-1} + 1 \cdot 2^{-2} + 1 \cdot 2^{-2} + \cdots$ and $\kappa_0\left(\frac{2}{3}\right) = 1 \cdot 2^{-1} + 0 \cdot 2^{-2} + 0 \cdot 2^{-2} + \cdots$. So, $\kappa_0\left(\frac{1}{3}\right) = \frac{1}{2} = \kappa_0\left(\frac{1}{2}\right)$, showing that κ_0 is not injective.

8.1.7. Proposition. *The function κ_0 is increasing* (not strictly of course) *and $\kappa_0(u) = 0$ implies $u = 0$.*

Proof: Consider $u > v$ and let $\phi^{-1}(u) = \{\alpha_k\}_{k \geq 1}$, $\phi^{-1}(v) = \{\beta_k\}_{k \geq 1}$. By definition of the function κ_0, we have $\kappa_0(u) = \sum_{k=1}^{\infty} \alpha_k 2^{-k}$ and $\kappa_0(v) = \sum_{k=1}^{\infty} \beta_k 2^{-k}$. We want to show that

$$\kappa_0(u) \geq \kappa_0(v), \text{ i.e. } \sum_{k=1}^{\infty} \alpha_k 2^{-k} \geq \sum_{k=1}^{\infty} \beta_k 2^{-k}.$$

From Proposition 8.1.5 and the inequality $u > v$, we know that there is a k for which $\alpha_k \neq \beta_k$ and that $\alpha_j > \beta_j$, where j is the smallest such k; hence (1) of Proposition 8.1.1 yields just what we want.

Suppose $\kappa_0(u) = 0$. Then 0 has $\phi^{-1}(u)$ as its sequence of digits in at least one binary expansion of it. However, 0 has only one binary expansion and every digit in it is 0; therefore $\phi^{-1}(u)$ is the sequence with every term equal to 0, which is to say, $\phi^{-1}(u) = \phi^{-1}(0)$. Consequently, $u = 0$.   □

Problem Set 8.1

8.1.P1. Suppose $x = \sum_{k=1}^{\infty} \alpha_k b^{-k}$ and $y = \sum_{k=1}^{\infty} \beta_k b^{-k}$ are b-expansions and there exists a positive integer j such that $\alpha_j > \beta_j$ and $\alpha_k = \beta_k$ for $k < j$ (vacuously true in case $j = 1$). If there does not exist any positive integer J such that $\alpha_k = 0$ for every $k > J$, show that $x > y$.

8.1.P2. If $y, z \in [a, b]$ and $|y-z| = b - a$, show that one among y and z is a and the other is b. (Ignore the trivial case when $a = b$.)

8.1.P3. Let $\{\beta_k\}_{k\geq 1}$ be a sequence in which each term is either 0 or 1.

(a) Then $\sum_{k=1}^{\infty} 2 \cdot \beta_k 3^{-k} \in [0, 1]$ and the series is a 3-expansion (called *ternary*) of the sum.

(b) If the first n terms of the sequence $\{\beta_k\}_{k\geq 1}$ of digits in any one ternary expansion $x = \sum_{k=1}^{\infty} 2 \cdot \beta_k 3^{-k}$ of the number x agree with the first n terms of $\phi^{-1}(u)$, where $u \in C$, then x is in the same component of C_n as u is.

(c) If $\phi^{-1}(u) = \{\alpha_k\}_{k\geq 1}$, then $\sum_{k=1}^{\infty} 2 \cdot \alpha_k 3^{-k}$ is a ternary expansion of u.

(d) A number in [0, 1] belongs to C if and only if it has a ternary expansion in which every digit is either 0 or 2.

(e) If a number in [0, 1] has a ternary expansion in which every digit is either 0 or 2, then it has no other ternary expansion of the same kind.

8.2   A Bijection from the Cantor Set to [0, 1] and the Cantor Function

For the purposes of this section, a sequence $\{\xi_k\}_{k\geq 1}$ of any kind of terms will be called *recurring* if all the terms from some k onward are equal to each other. It could also have been called 'eventually constant'. A b-expansion will be called recurring if the sequence of digits in it is recurring.

8.2.1. Proposition. *If a number $x \in [0, 1]$ has two distinct b-expansions, then both are recurring.*

Proof: Left to the reader as Problem 8.2.P1(a).   □

8.2.2. Proposition. *If a number $x \in [0, 1]$ has a recurring b-expansion, then it is rational.*

Proof: Left to the reader as Problem 8.2.P1(b).   □

8.2.3. Theorem. *There exists a bijection from the Cantor set to $[0, 1]$.*

Proof: Let C_0 be the subset of the Cantor set C consisting of those u for which the associated sequence $\phi^{-1}(u) = \{\alpha_k\}_{k \geq 1}$ is recurring. Such sequences are countably infinite in number, and therefore C_0 is countably infinite. Consider also the infinite subset I_0 of $[0, 1]$ consisting of numbers having a recurring binary expansion. Any number having such an expansion is rational by Proposition 8.2.2 and therefore I_0 too is countably infinite. Therefore there is a bijection from C_0 to I_0.

Suppose $u, v \in C \backslash C_0$, where $u > v$ and let $\phi^{-1}(u) = \{\alpha_k\}_{k \geq 1}$ and $\phi^{-1}(v) = \{\beta_k\}_{k \geq 1}$. Both these sequences are nonrecurring. By definition of the function κ_0 (see Sect. 8.1), the numbers $x = \kappa_0(u)$ and $y = \kappa_0(v)$ are the ones that satisfy $x = \sum_{k=1}^{\infty} \alpha_k b^{-k}$ and $y = \sum_{k=1}^{\infty} \beta_k b^{-k}$. By Proposition 8.1.7, $x \geq y$. Now, by Proposition 8.1.5, there is a smallest k such that $\alpha_k \neq \beta_k$ and if j denotes this smallest k, then $\alpha_j > \beta_j$. Since x has a nonrecurring binary expansion, it is a consequence of Proposition 8.2.1 that $x > y$, i.e. $\kappa_0(u) > \kappa_0(v)$. This shows that κ_0 is strictly increasing on $C \backslash C_0$ and is therefore injective on $C \backslash C_0$.

We shall show that κ_0 maps $C \backslash C_0$ into $[0, 1] \backslash I_0$ surjectively. For any $u \in C \backslash C_0$, the sequence $\phi^{-1}(u)$ is nonrecurring and therefore $\kappa_0(u)$ has a nonrecurring binary expansion. By Proposition 8.2.1, it has no other binary expansion and therefore $\kappa_0(u) \in [0, 1] \backslash I_0$, showing that κ_0 maps $C \backslash C_0$ into $[0, 1] \backslash I_0$. It is trivial to see that it does so surjectively.

Since we have already shown that κ_0 is injective on $C \backslash C_0$, it follows that it is a bijection from $C \backslash C_0$ to $[0, 1] \backslash I_0$. But there is also a bijection from C_0 to I_0, and the two bijections together provide a bijection from C to $[0, 1]$.   □

8.2.4. Remark. Let g be an increasing real-valued function on a closed interval. At an interior point x of the closed interval, either $g(x-) = g(x+)$, in which case the function is continuous at x, or $g(x-) < g(x+)$. In the latter case, at least one of the intervals $(g(x-), g(x))$ and $(g(x), g(x+))$ is nonempty and forms a gap in the range of the function. Consequently, an increasing real-valued function on a closed interval whose range is an interval must be continuous in the interior of its domain. Similar reasoning shows that continuity holds at an endpoint as well.

8.2.5. Proposition. *Let the function $\kappa_0 : C \to [0, 1]$ be as in Sect. 8.1. Define the function $\kappa : [0, 1] \to \mathbb{R}$ by setting*

$$\kappa(x) = \sup\{\kappa_0(u) : u \in C, u \leq x\}.$$

Then κ_0 is the restriction of κ to C. Moreover, κ is a continuous increasing function that maps $[0, 1]$ surjectively to $[0, 1]$ and is constant on any interval disjoint from C.

Proof: Since $0 \in C$ and the range of κ_0 is $[0, 1]$, the set $\{\kappa_0(u): u \in C, u \leq x\}$ is always a nonempty subset of $[0, 1]$ and hence its supremum always exists and belongs to $[0, 1]$. Thus κ maps $[0, 1]$ to itself. By Proposition 8.1.7, the function κ_0 is increasing and therefore $\kappa(x) = \kappa_0(x)$ whenever $x \in C$. This means κ_0 is the restriction of κ to C.

Now, the range of κ_0 is $[0, 1]$ and therefore the range of κ is also $[0, 1]$. Thus κ is surjective.

For $x, y \in [0, 1]$ such that $x < y$, we have $\{\kappa_0(u): u \in C, u \leq x\} \subseteq \{\kappa_0(u): u \in C, u \leq y\}$ and hence $\kappa(x) \leq \kappa(y)$. Consequently, κ is increasing.

Since κ is an increasing real-valued function whose domain and range are both intervals, it follows by Remark 8.2.4 that it is continuous.

Lastly, let I be a subinterval of $[0, 1]$ that is disjoint from C. We consider only nonempty I, because it is trivial that κ is constant on $\emptyset$. Suppose $x, y \in I$ and $x \leq y$. Since κ has been shown to be increasing, we know that $\kappa(x) \leq \kappa(y)$. Consider any $u \in C$ such that $u \leq y$. Since $[x, y] \subseteq I$ and is therefore disjoint from C, it follows that $u < x$. This shows that $\{\kappa_0(u): u \in C, u \leq y\} \subseteq \{\kappa_0(u): u \in C, u \leq x\}$ and hence $\kappa(y) \leq \kappa(x)$. Consequently, $\kappa(x) = \kappa(y)$. ☐

The function κ of Proposition 8.2.5 is called the **Cantor function**.

8.2.6. Remark. As seen in Example 8.1.4 (second paragraph), the Cantor set contains arbitrarily small positive numbers. It therefore follows from Proposition 8.1.7 that $0 < u \leq 1$ implies $\kappa(u) > 0$.

It is possible to use the Cantor function to set up, though less explicitly, a function of the kind in Example 7.8.2: strictly increasing and continuous with derivative 0 almost everywhere. To do so, we shall need another well known theorem of Fubini, towards which we first discuss a preliminary result.

Lemma. 8.2.7 *Let $\{\phi_k\}_{k \geq 1}$ be a sequence of increasing functions on $[a, b]$ such that the series $\sum_{k=1}^{\infty} \phi_k$ converges everywhere. Then the differentiated series $\sum_{k=1}^{\infty} \phi_k'$ is defined a.e. and its partial sums form an increasing sequence bounded above by the derivative $\left(\sum_{k=1}^{\infty} \phi_k\right)'$ of the sum; in particular, the differentiated series is convergent a.e.*

Proof: Let s denote the sum of the undifferentiated series and s_n its partial sum $\sum_{k=1}^{n} \phi_k$. All of them are increasing functions on $[a, b]$. By the Lebesgue Differentiability Theorem 7.4.1, the sum s and all ϕ_k are differentiable a.e. and therefore there is a single set $E \subseteq [a, b]$ on which s, all s_n and all ϕ_k are differentiable and whose complement in $[a, b]$ has measure 0. It follows that the differentiated series is defined on all of E and hence almost everywhere.

As for convergence, we note that the increasing nature of the functions involved implies regarding derivatives that they satisfy $s_n' \leq s_{n+1}' \leq s'$ everywhere on E. Consequently, the partial sums of the differentiated series, at any point of E, form a bounded increasing sequence, so that its convergence is ensured. Thus the differentiated series converges everywhere on E, hence almost everywhere. ☐

8.2.8. Fubini Series Theorem. *Let $\{\phi_k\}_{k\geq 1}$ be a sequence of increasing functions on $[a, b]$ such that the series $\sum_{k=1}^{\infty}\phi_k$ converges everywhere. Then the differentiated series $\sum_{k=1}^{\infty}\phi_k'$ is converges a.e. to the derivative $\left(\sum_{k=1}^{\infty}\phi_k\right)'$ of the sum.*

Proof: It is sufficient to prove the theorem with the additional hypothesis that $\phi_k(a)=0$ for each k.

Let s denote the sum of the undifferentiated series and s_n its partial sum $\sum_{k=1}^{n}\phi_k$. All of them are increasing functions on $[a, b]$. By Lemma 8.2.7, there exists an $E \subseteq[a, b]$ such that the complement of E in $[a, b]$ has measure 0 and the partial sums s_n' form an increasing sequence on E with s' as an upper bound. The assertion of the theorem is that $s_n' \rightarrow s$ on some subset of E whose complement in $[a, b]$ has measure 0. Since the s_n' form an increasing sequence, the assertion will follow if we construct a subsequence s_{n_k}' such that $s' - s_{n_k}' \rightarrow 0$ on some subset of E whose complement in $[a, b]$ has measure 0.

Taking into account the additional hypothesis that $\phi_k(a)=0$ for each k, we know for every n and every $x \in [a, b]$ that

$$0 = s(a) - s_n(a) \leq s(x) - s_n(x) \leq s(b) - s_n(b).$$

Since $s_n(b) \rightarrow s(b)$, there exists a sequence $\{n_k\}_{k\geq 1}$ such that $s(b) - s_{n_k}(b) < 2^{-k}$ for each k. Taken in conjunction with the inequality displayed above, this implies that the series $\sum_{k=1}^{\infty}(s - s_{n_k})$ converges everywhere. Now, the terms $s - s_{n_k}$ of this series are increasing functions. Therefore, by appealing to Lemma 8.2.7 once again, we deduce that the differentiated series $\sum_{k=1}^{\infty}(s' - s_{n_k}')$ converges on some subset of $[a, b]$ whose complement has measure 0. The intersection of this subset with E is a subset of E with the property that its complement has measure 0 and the series $\sum_{k=1}^{\infty}(s' - s_{n_k}')$ converges on it, so that the kth term $(s' - s_{n_k}')$ tends to 0 on it. □

8.2.9. Example. We now present another example of a continuous strictly increasing function on $[0, 1]$ with derivative 0 almost everywhere taking the value 0 at 0 and 1 at 1.

Let g be the extension to $\mathbb{R}$ of the Cantor function κ obtained by setting it equal to 1 for $x >1$ and equal to 0 for $x <0$. That is,

$$g(x) = \begin{cases} 0 & \text{if } x < 0 \\ \kappa(x) & \text{if } 0 \leq x \leq 1 \\ 1 & \text{if } 1 < x. \end{cases}$$

Then g is continuous and increasing, and it has derivative 0 almost everywhere. Let $\{r_k\}_{k\geq 1}$ be a sequence, the range of which is the set of all rational numbers in $[0, 1]$, and set

$$F(x) = \sum_{k=1}^{\infty} \frac{1}{2^k} g(x - r_k).$$

Since $0 \leq g \leq 1$ everywhere, the series is uniformly convergent and therefore F is continuous. Moreover, $F' = 0$ a.e. by the Fubini Series Theorem 8.2.8. To see why the function F is strictly increasing on $[0, 1]$, consider any $x, y \in [0, 1]$ satisfying $x < y$. Then $x - r_k < y - r_k$ and therefore

$$g(x - r_k) \leq g(y - r_k) \quad \text{for every } k.$$

As the range of the sequence $\{r_k\}_{k \geq 1}$ is the set of all rational numbers in $[0, 1]$, there exists a j such that $x < r_j < y$. This implies $x - r_j < 0 < y - r_j < 1$. The definition of g therefore yields $g(x - r_j) = 0$ and $g(y - r_j) = \kappa(y - r_j) \geq 0$. But $\kappa(y - r_j) > 0$ by Remark 8.2.6, so that $g(x - r_j) < g(y - r_j)$. This inequality, when combined with the fact that $g(x - r_{sk}) \leq g(y - r_k)$ for every k, implies via the definition of F that $F(x) < F(y)$.

Problem Set 8.2

8.2.P1. (a)   Prove Proposition 8.2.1.

(b)   Prove Proposition 8.2.2.

Ordinarily, the "post inverse" (or "left inverse") g of a function G is understood to have the range of G as its domain and to satisfy the condition that $g(G(x)) = x$ for every x in the domain of G. However, for the present purpose, it will be necessary to broaden the concept somewhat. If $G : [a, b] \to \mathbb{R}$ is strictly increasing, then any function g defined on $[G(a), G(b)]$ and satisfying the aforementioned condition will be called a post inverse. Its values at points of $[G(a), G(b)]$ that are outside the range of G are not uniquely determined unless some additional condition is imposed.

8.2.P2. (Rubel's Lemma [6]) Let $G : [a, b] \to \mathbb{R}$ be a strictly increasing function. Then there is a continuous increasing function $g: [G(a), G(b)] \to \mathbb{R}$ which is a "post inverse" of G, i.e. satisfies $g(G(x)) = x$ for each $x \in [a, b]$, and which is constant on any open subinterval of $[G(a), G(b)]$ contained in the complement of the range of G.

8.2.P3. Let G be a strictly increasing real-valued function on the domain $[a, b]$ and g_0 be an increasing function with domain containing the range of G and satisfying $g_0(G(x)) = x$ for each $x \in [a, b]$. (In particular, g_0 could be an increasing post inverse of G.) Show the following.

(a)   For each $y \in [G(a), G(b)]$ that belongs to the domain of g_0,

$$g_0(y) = \sup\{u \in [a, b] : G(u) \leq y\}.$$

(b)   The increasing post inverse is unique and g_0 agrees with the unique increasing post inverse at points of $[G(a), G(b)]$ that belong to its domain.

Hint: Argue for any $y \in [G(a), G(b)]$ that belongs to the domain of g_0 that $g_0(y)$ lies between the numbers $\sup\{u \in [a, b]: G(u) \leq y\}$ and $\inf\{u \in [a, b]: G(u) \geq y\}$ and that these two numbers are equal.

8.2.P4. Let G be a strictly increasing real-valued function on the domain $[a, b]$. Show that the function $g : [G(a), G(b)] \to \mathbb{R}$ defined by

$$g(y) = \inf\{u \in [a, b] : G(u) \geq y\}$$

is an increasing post inverse of G.

8.2.P5. Let G_1 and G_2 be strictly increasing real-valued functions on the domain $[a, b]$ and let g_1 and g_2 be their respective (unique) increasing post inverses. If $G_1 = G_2$ almost everywhere, show that $g_1 = g_2$ everywhere on the intersection of their domains, i.e. on $[G_1(a), G_1(b)] \cap [G_2(a), G_2(b)]$.

8.2.P6. In the course of proving Theorem 8.2.3, it was established that the mapping κ_0 is strictly increasing on $C \backslash C_0$, where C_0 is the subset of the Cantor set C consisting of those u for which the associated sequence $\phi^{-1}(u) = \{\alpha_k\}_{k \geq 1}$ is recurring. Use this fact in conjunction with Problem 6.6.P3 to show that there exists a measurable set that is not a Borel set.

Hint: Consider the inverse image by κ_0 of the intersection of a nonmeasurable subset of $[0, 1]$ with the range of κ_0.

Solutions

1.1 The Riemann Integral Revisited

1.1.P1. For each $t \geq 0$, find $S = \{x \in [0,4]: f(x) > t\}$, where $f: [0,4] \to \mathbb{R}$ is the function given by:

$$f(x) = 6 \text{ if } 0 \leq x < 1, f(1) = 5, f(x) = 2 \text{ if } 1 < x < 3, f(3) = 1, f(x) = 6 \text{ if } 3 < x \leq 4.$$

Solution:

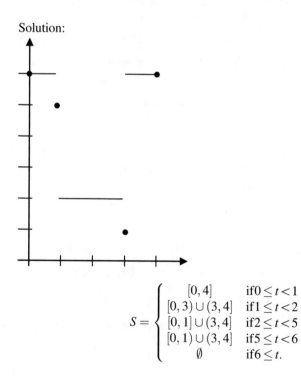

$$S = \begin{cases} [0,4] & \text{if } 0 \leq t < 1 \\ [0,3) \cup (3,4] & \text{if } 1 \leq t < 2 \\ [0,1] \cup (3,4] & \text{if } 2 \leq t < 5 \\ [0,1) \cup (3,4] & \text{if } 5 \leq t < 6 \\ \emptyset & \text{if } 6 \leq t. \end{cases}$$

© Springer Nature Switzerland AG 2018
S. Shirali, *A Concise Introduction to Measure Theory*,
https://doi.org/10.1007/978-3-030-03241-8

1.1.P2. For $f\colon [1,2] \to \mathbb{R}$ given by $f(x) = e^x$, find the distribution $\tilde{f}\colon [0, \sup f] \to \mathbb{R}$ and show that $\int_0^{\sup f} \tilde{f}$ agrees with $\int_1^2 f$.

Solution: The set $\{x \in [1,2]\colon f(x) > t\}$ is $[1,2]$ if $0 \le t < e$, $(\ln t, 2]$ if $e \le t < e^2$ and is $\emptyset$ if $e^2 \le t$. Also, $\sup f = e^2$. Therefore

$$\tilde{f}(t) = \begin{cases} 1 & \text{if } 0 \le t < e \\ 2 - \ln t & \text{if } e \le t < e^2. \end{cases}$$

So,

$$\int_0^{\sup f} \tilde{f} = 1 \cdot (e - 0) + \int_e^{e^2} (2 - \ln t)dt = e + \left((3e^2 - e^2 \ln e^2) - (3e - e \ln e)\right)$$

$$= e^2 - e = \int_1^2 e^x dx.$$

1.1.P3. Let $A > 0$. For the function $f\colon [0, A] \to \mathbb{R}$ given by $f(x) = e^{-x^2}$, find the distribution $\tilde{f}\colon [0, \sup f] \to \mathbb{R}$ and show that $\int_0^{\sup f} \tilde{f}$ agrees with $\int_0^A f$.

Hint: Use an appropriate substitution and integrate by parts.

Solution: Here $\sup f = 1$ and $\inf f = e^{-A^2}$. Also, $\{x \in [0, A]\colon f(x) > t\}$ is $[0, A]$ if $t = 0 \le t < e^{-A^2}$ and is $\left[0, \sqrt{(\ln \frac{1}{t})}\right]$ if $e^{-A^2} \le t \le 1$. Consequently,

$$\tilde{f}(t) = \begin{cases} A & \text{if } 0 \le t < e^{-A^2} \\ \sqrt{\ln \frac{1}{t}} & \text{if } e^{-A^2} \le t \le 1. \end{cases}$$

So, $\int_0^{\sup f} \tilde{f} = Ae^{-A^2} + \int_{\exp(-A^2)}^1 \sqrt{(\ln \frac{1}{t})}dt$. Upon substituting for $\sqrt{(\ln \frac{1}{t})}$ and then integrating by parts, we find that this agrees with $\int_0^A e^{-x^2}dx$.

1.1.P4. Abel summation formula. If $a_1, \ldots, a_n$ and $b_1, \ldots, b_n$ are two finite sequences of n terms each ($n \ge 2$) and $S_k = b_1 + \cdots + b_k$, then $b_k = S_k - S_{k-1}$ for $k = 2, \ldots, n$ and therefore

$$a_1 b_1 + \cdots + a_n b_n = a_1 S_1 + a_2(S_2 - S_1) + a_3(S_3 - S_2) + \cdots + a_n(S_n - S_{n-1})$$
$$= (a_1 - a_2)S_1 + (a_2 - a_3)S_2 + (a_3 - a_4)S_3 + \cdots + (a_{n-1} - a_n)S_{n-1} + a_n S_n.$$

In Σ notation:

Let $\{a_k\}_{k=1}^n$ and $\{b_k\}_{k=1}^n$ be finite sequences of numbers, where $n \ge 2$, and let $S_k = \sum_{j=1}^k b_j$ for $1 \le k \le n$. Then

$$\sum_{k=1}^{n} a_k b_k = \sum_{k=1}^{n-1}(a_k - a_{k+1})S_k + a_n S_n.$$

Prove this in Σ notation without resorting to algebraic manipulations concealed behind an ellipsis $\cdots$. How should T_k be defined in order to obtain the alternative version $\sum_{k=1}^{n} a_k b_k = a_1 T_1 + \sum_{k=2}^{n}(a_k - a_{k-1})T_k$?

Solution: For an induction proof, see Shirali and Vasudeva [9; pp. 133–4].

Here is a proof without induction. Set $S_0 = 0 = a_{n+1}$. Then $S_k - S_{k-1} = b_k$ for $1 \le k \le n$. Also,

$$\sum_{k=1}^{n} a_k S_{k-1} = \sum_{k=0}^{n-1} a_{k+1} S_k$$

$$= \sum_{k=1}^{n-1} a_{k+1} S_k \quad \text{because } S_0 = 0$$

$$= \sum_{k=1}^{n} a_{k+1} S_k \quad \text{because } a_{n+1} = 0.$$

Subtracting this from $\sum_{k=1}^{n} a_k S_k$, we get

$$\sum_{k=1}^{n} a_k(S_k - S_{k-1}) = \sum_{k=1}^{n}(a_k - a_{k+1})S_k.$$

In view of the observation that $S_k - S_{k-1} = b_k$ for $1 \le k \le n$, the above equality becomes

$$\sum_{k=1}^{n} a_k b_k = \sum_{k=1}^{n}(a_k - a_{k+1})S_k = \sum_{k=1}^{n-1}(a_k - a_{k+1})S_k + (a_n - a_{n+1})S_n$$

$$= \sum_{k=1}^{n-1}(a_k - a_{k+1})S_k + a_n S_n \quad \text{because } a_{n+1} = 0.$$

For the alternative version, $T_k = \sum_{j=k}^{n} b_j$.

1.1.P5. (Relevant for Problem 1.1.P6 and to be used in Theorem 4.3.1) If one visualizes two disjoint intervals on the real line, one of them lies entirely on the left of the other. Prove the following precise formulation of this idea for bounded intervals: If two bounded nonempty intervals are disjoint, then the right endpoint of one of them is less than or equal to the left endpoint of the other.

Deduce that, if the union of two disjoint intervals is an interval, then the sum of their lengths is equal to the length of the union.

Solution: Let the left and right endpoints of an interval I be a and b and those of J be α and β respectively; suppose $I \cap J = \emptyset$. We must prove that either $b \leq \alpha$ or $\beta \leq a$. If $a = b$, the interval I consists of a $(= b)$ alone and, because of the disjointness, the inequality $\alpha < a < \beta$ cannot hold; therefore either $\beta \leq a$ or $a \leq \alpha$, which in this case means $b \leq \alpha$. The case $\alpha = \beta$ is similar and we need argue only the case when $a < b$ as well as $\alpha < \beta$. Suppose it is not true that either $b \leq \alpha$ or $\beta \leq a$. Then $\alpha < b$ as well as $a < \beta$. Thus we have the four inequalities

$$a < b, \quad a < \beta, \quad \alpha < \beta, \quad \alpha < b.$$

Therefore $a < \min\{b, \beta\}$ and $\alpha < \min\{b, \beta\}$, so that there exist x and y such that $a < x < \min\{b, \beta\}$ and $\alpha < y < \min\{b, \beta\}$. If $\alpha \leq a$, then $x \in I \cap J$, while if $a \leq \alpha$, then $y \in I \cap J$, thus contradicting the hypothesis that I and J are disjoint. This contradiction shows that either $b \leq \alpha$ or $\beta \leq a$.

If one of the intervals is empty, there is nothing to prove. Continuing with the notation of the above paragraph, suppose $b \leq \alpha$. If $I \cup J$ is an interval, we claim that $b = \alpha$. If not, then any number between b and α fails to belong to $I \cup J$, although it lies between the midpoints of I and J which are in $I \cup J$. This cannot happen when $I \cup J$ is an interval. Consequently, $b = \alpha$ as claimed. It follows that $I \cup J$ has length $\beta - a = (\beta - \alpha) + (b - a)$.

1.1.P6. Suppose $\mathcal{F}$ is a finite family of disjoint bounded intervals. Show that there exists a nonempty finite family $\mathcal{G}$ of bounded intervals such that

- (a) a union of two distinct intervals in $\mathcal{G}$ is never an interval;
- (b) $\cup \mathcal{G} = \cup \mathcal{F}$ (Here "$\cup \mathcal{H}$" means union of all the intervals in $\mathcal{H}$);
- (c) the total length of the intervals in $\mathcal{G}$ is the same as for $\mathcal{F}$.

(Note that the intervals in $\mathcal{G}$ are necessarily disjoint in view of (a).)

Solution: Induction on the number of intervals in $\mathcal{F}$. If there is only one interval in $\mathcal{F}$, then $\mathcal{G} = \mathcal{F}$ serves the purpose; if $\mathcal{F}$ is empty, then take $\mathcal{G} = \{\emptyset\}$. Assume that such a family $\mathcal{G}$ exists whenever $\mathcal{F}$ is a family of k disjoint bounded intervals (induction hypothesis). Consider a family $\mathcal{F}$ of $(k + 1)$ disjoint bounded intervals. If it is the case that a union of two distinct intervals in $\mathcal{F}$ is never an interval, then $\mathcal{G} = \mathcal{F}$ serves the purpose. Otherwise there exist two distinct intervals in $\mathcal{F}$ whose union is an interval; if one of them is empty, then the length of the union is the sum of the lengths of the two intervals; if neither is empty, then the same holds in view of Problem 1.1.P5. Let $\mathcal{F}'$ be the family obtained from $\mathcal{F}$ by replacing the two intervals by their union. Then first, $\cup \mathcal{F}' = \cup \mathcal{F}$; second, $\mathcal{F}'$ is a family of k disjoint bounded intervals, and third, the total length of the intervals in $\mathcal{F}'$ is the same as for $\mathcal{F}$. Therefore by the induction hypothesis, there is a family $\mathcal{G}$ of bounded intervals such that (a) a union of two distinct intervals in $\mathcal{G}$ is never an interval (b) $\cup \mathcal{G} = \cup \mathcal{F}' = \cup \mathcal{F}$ and (c) the total length of the intervals in $\mathcal{G}$ is the same as for $\mathcal{F}'$, which is the same as for $\mathcal{F}$.

1.1.P7. If an interval I has an endpoint that belongs to an interval J, show that the union $I \cup J$ is an interval. (Only a left endpoint need be considered.)

Solution: Let $a \in J$ be the left endpoint of I. Consider any x, u, y such that $x < u < y$, where $x, y \in I \cup J$. All that need be shown is that $u \in I \cup J$.

First suppose $u \leq a$. Then $x < a$. Since a is the left endpoint of I, it follows that $x \notin I$, so that $x \in J$. Thus $x < u \leq a$ with $x \in J$ and $a \in J$, which implies $u \in J$.

Next, suppose instead that $a < u$. If $y \in J$, we have $a < u < y$ with $a \in J$ and $y \in J$, which implies $u \in J$. If $y \notin J$, we have $y \in I$ and hence $a < u < y$ with a the left endpoint of I and $y \in I$, which implies $u \in I$.

1.1.P8. Let $\mathcal{G}$ be a finite nonempty family of bounded intervals such that a union of two distinct intervals of $\mathcal{G}$ is never an interval. Suppose $\mathcal{F}$ is a finite family of disjoint bounded intervals such that $\cup \mathcal{F} = \cup \mathcal{G}$. Show that every interval of $\mathcal{F}$ is contained in some interval of $\mathcal{G}$. What more can be said if $\mathcal{F}$ too is nonempty and has the property that a union of two distinct intervals of it is never an interval?

Solution: If $\mathcal{G} = \{\emptyset\}$, necessarily $\mathcal{F} = \emptyset$ or $\{\emptyset\}$. If $\mathcal{G}$ contains a nonempty interval, then the hypothesis implies that $\emptyset \notin \mathcal{G}$. So we may assume that all the intervals in $\mathcal{G}$ are nonempty, and therefore use results about intervals from elementary Analysis, wherein $\emptyset$ is usually not considered to be an interval.

Let I_1 be any nonempty interval in $\mathcal{F}$ and $x_1 \in I_1$. Then $x_1 \in \cup \mathcal{F} = \cup \mathcal{G}$ and hence $x_1 \in I$ for some interval I belonging to $\mathcal{G}$. We shall show that $I_1 \subseteq I$.

Let the left and right endpoints of I be a and b respectively (the possibility that $a = b$, in which case $I = \{a\} = [a, a]$ is *not* excluded). Then

$$x_1 \in I_1 \text{ and } x_1 \geq a.$$

We shall establish that the left endpoint of I_1 is greater than or equal to a and that, if $a \notin I$, then $a \notin I_1$. A similar argument can then establish the corresponding assertion about right endpoints and it will follow that $I_1 \subseteq I$.

Case 1. $a \notin I$. Consider any interval J of $\mathcal{G}$ other than I, if there is any. By hypothesis, $I \cup J$ is not an interval. It follows by Problem 1.1.P7 that $a \notin J$. This is true for every J belonging to $\mathcal{G}$ (including for I) and therefore $a \notin \cup \mathcal{G}$. But $\cup \mathcal{G} = \cup \mathcal{F}$ and therefore $a \notin \cup \mathcal{F}$. It follows that $a \notin I_1$. Since $x_1 \in I_1$ and $x_1 > a \notin I_1$, the left endpoint of I_1 must be greater than or equal to a. This establishes what we want in this case.

Case 2. $a \in I$. In the subcase that every $x \in \cup \mathcal{G}$ satisfies $x \geq a$, we have $x \in I_1 \Rightarrow x \geq a$, because $I_1 \subseteq \cup \mathcal{F} = \cup \mathcal{G}$. Hence the left endpoint of I_1 is greater than or equal to a. In the contrary subcase, $\cup \mathcal{G}$ does contains a number less than a and hence there must be an interval in $\mathcal{G}$ which contains that number. By hypothesis, whenever J is an interval of this kind, $I \cup J$ is not an interval. It follows by Problem 1.1.P7 that neither I nor J can contain an endpoint of the other and hence that $a \notin J$ and that a cannot be an endpoint of J. Combined with the fact that J contains a number less than a, this implies that the right endpoint of J must be less than a. But there are only finitely many such intervals ($\mathcal{G}$ is given to be a finite) and there must therefore be a greatest right endpoint c for them. We claim that

$c<x<a \Rightarrow x \notin \cup\mathcal{G}$. Indeed, suppose there is some $x \in \cup\mathcal{G}$ satisfying $c<x<a$; then $x \in J_0$ for some J_0 belonging to $\mathcal{G}$ and it therefore follows as before that the right endpoint of J_0 is less than a while being greater than c, thereby contradicting the maximality of c. Since $c<x<a \Rightarrow x \notin \cup\mathcal{G}$ and $I_1 \subseteq \cup\mathcal{F} = \cup\mathcal{G}$, it follows that I_1 cannot contain numbers between c and a. But I_1 contains the number $x_1 \geq a$. Consequently, the left endpoint of I_1 is greater than or equal to a.

When $\mathcal{F}$ too is nonempty and has the property that a union of two distinct intervals of it is never an interval, it follows from what has been proved above that every interval of $\mathcal{G}$ is also contained in some interval of $\mathcal{F}$. Since the intervals in each family are disjoint among themselves, it follows that $\mathcal{F} = \mathcal{G}$.

1.1.P9. Let $\mathcal{F}_1$ and $\mathcal{F}_2$ be finite families of bounded intervals, each having the property that its intervals are disjoint. If $\cup\mathcal{F}_1 = \cup\mathcal{F}_2$, show that the total length of the intervals of each family is the same.

Solution: In accordance with Problem 1.1.P6, let $\mathcal{G}_i$ $(i = 1, 2)$ be a nonempty finite family of bounded intervals such that

- (a) a union of two distinct intervals in $\mathcal{G}_i$ is never an interval;
- (b) $\cup\mathcal{G}_i = \cup\mathcal{F}_i$;
- (c) the total length of the intervals in $\mathcal{G}_i$ is the same as for $\mathcal{F}_i$.

By (b) and the hypothesis that $\cup\mathcal{F}_1 = \cup\mathcal{F}_2$, we have

$$\bigcup\mathcal{G}_1 = \bigcup\mathcal{F}_1 = \bigcup\mathcal{F}_2 = \bigcup\mathcal{G}_2.$$

By (a) and the last part of Problem 1.1.P8, we have $\mathcal{G}_1 = \mathcal{G}_2$. The required conclusion now follows from (c).

1.2 Improper Integrals

1.2.P1. For the double sequence $a: \mathbb{N} \times \mathbb{N} \to \mathbb{R}$ given by

$$a_{n,k} = 1 \text{ if } n = k, \quad a_{n,k} = -1 \text{ if } k = n+1 \quad \text{and} \quad a_{n,k} = 0 \text{ in all other cases,}$$

show that the repeated sums $\sum_{n=1}^{\infty}\sum_{k=1}^{\infty} a_{n,k}$ and $\sum_{k=1}^{\infty}\sum_{n=1}^{\infty} a_{n,k}$ are unequal, although $\sum_{n=1}^{\infty}\left|\sum_{k=1}^{\infty} a_{n,k}\right|$ and $\sum_{k=1}^{\infty}\left|\sum_{n=1}^{\infty} a_{n,k}\right|$ are convergent.

Hint: One can think of $a_{n,k}$ as representing an infinite matrix with first three rows:

$$
\begin{array}{ccccccc}
1 & -1 & 0 & 0 & & \cdots & \\
0 & 1 & -1 & 0 & 0 & \cdots & \\
0 & 0 & 1 & -1 & 0 & 0 & \cdots
\end{array}
$$

Solution: It is clear from the given description of $a_{n,k}$ that, for each n, there always exists a k such that $k = n+1$, and therefore $\sum_{k=1}^{\infty} a_{n,k} = 0$. For the other sum, the possibility that $k = n + 1$ arises if and only if $k > 1$ and therefore $\sum_{n=1}^{\infty} a_{n,k} = 1$ when $k = 1$ and $\sum_{n=1}^{\infty} a_{n,k} = 1$ when $k > 1$. Consequently,

$$\sum_{n=1}^{\infty} \sum_{k=1}^{\infty} a_{n,k} = \sum_{n=1}^{\infty} \left| \sum_{k=1}^{\infty} a_{n,k} \right| = 0 \quad \text{but} \quad \sum_{k=1}^{\infty} \sum_{n=1}^{\infty} a_{n,k} = \sum_{k=1}^{\infty} \left| \sum_{n=1}^{\infty} a_{n,k} \right| = 1.$$

1.2.P2. Let $b > 0$ and g be a decreasing nonnegative function on $[0, b]$ that is real-valued except perhaps at 0. Denote by f the decreasing nonnegative function on $[0, \infty)$ obtained by extending g to be equal to 0 to the right of b. Show that

$$\int\limits_{0}^{\infty} f = \lim_{x \to 0} \int\limits_{x}^{b} g.$$

Note: If $g(0) < \infty$, the function g is Riemann integrable on $[0, b]$ and the limit here is equal to $\int_{0}^{b} g$ (see Problem 9.3.P2 of Shirali and Vasudeva [9]).

Solution: Consider any $x \in (0, b)$. On the interval $[x, b]$, the function g is Riemann integrable and satisfies $0 \leq g \leq f$. It follows by Definition 1.2.2 that $\int_{x}^{b} g \leq \int_{0}^{\infty} f$. Consequently, $\lim_{x \to 0} \int_{x}^{b} g \leq \int_{0}^{\infty} f$. We proceed to prove the reverse inequality.

Let $0 < \alpha < \beta < \infty$ and g_0 be any Riemann integrable function on $[\alpha, \beta]$ such that $0 \leq g_0 \leq f$. Keeping in mind how f is obtained from g, we see that $g_0 \leq g$ on $[0, b]$ and g_0 vanishes on (b, ∞). We claim that $\int_{\alpha}^{\beta} g_0 \leq \lim_{x \to 0} \int_{x}^{b} g$. For $b < \alpha$, the inequality claimed must hold because the left side is 0. Suppose $\alpha \leq b$. If $\beta \leq b$, the validity of the inequality $g_0 \leq g$ on $[0, b]$ leads to $\int_{\alpha}^{\beta} g_0 \leq \int_{\alpha}^{\beta} g$ and the nonnegativity of g leads to $\int_{\alpha}^{\beta} g \leq \lim_{x \to 0} \int_{x}^{b} g$. Thus the inequality claimed holds if $\beta \leq b$. In the contrary case $b < \beta$, the vanishing of g_0 on (b, ∞) leads to $\int_{\alpha}^{\beta} g_0 = \int_{\alpha}^{b} g_0$, the validity of the inequality $g_0 \leq g$ on $[0, b]$ leads to $\int_{\alpha}^{b} g_0 \leq \int_{\alpha}^{b} g$ and the nonnegativity of g leads to $\int_{\alpha}^{b} g \leq \lim_{x \to 0} \int_{x}^{b} g$. Thus the inequality claimed holds also if $b < \beta$. This completes the justification of the claim in all cases. Since g_0 is an arbitrary Riemann integrable function on $[\alpha, \beta] \subseteq (0, \infty)$ satisfying $0 \leq g_0 \leq f$, Definition 1.2.2 now yields the desired reverse inequality.

1.2.P3. In Lemma 1.2.3, show that the condition $g(a) < \infty$ does not follow from the remaining conditions in the hypothesis even if we impose the additional condition that g is real-valued on $(a, b]$. If $g(a) = \infty$ but g (which must be decreasing) is real-valued on $(a, b]$, then the improper integral $\int_{a}^{b} g$ in the usual sense is $\lim_{x \to 0} \int_{x}^{b} g$, though it may be ∞; show that $\int_{a}^{b} g = \lim_{n \to \infty} \int_{a}^{b} g_n$. If g is not real-valued on $(a, b]$, which to say, $g(\alpha) = \infty$ for some $\alpha > a$, show that $\lim_{n \to \infty} \int_{a}^{b} g_n = \infty$.

Solution: Let $g_n(0) = n$ and $g_n(x) = \min\{n, (x - a)^{-1}\}$ for $x \in (a, b]$. Also, let $g(a) = \infty$ and $g(x) = (x - a)^{-1}$ for $x \in (a, b]$. Then all other conditions in the hypothesis are fulfilled but $g(a) = \infty$ and g is real-valued on $(a, b]$.

Without loss of generality, we may assume $a = 0$. Since $g(x) \geq g_n(x) \geq g_n(b) \geq g_1(b)$ for all x and all n, we may, by subtracting $g_1(b)$ from every function, assume that all functions are nonnegative-valued. (This lays the groundwork for applying Proposition 1.2.4.). Extend all functions g_n and g to the right of b by setting them equal to 0 there and call them f_n and f. Then the hypothesis of Proposition 1.2.4. holds and hence we can infer that $\int_0^\infty f = \lim_{n\to\infty} \int_0^\infty f_n$. It remains only to note that by virtue of Problem 1.2.P2,

$$\int_0^\infty f_n = \lim_{x\to 0} \int_x^b g_n = \int_0^b g_n \quad \text{and} \quad \int_0^\infty f = \lim_{x\to 0} \int_x^b g = \int_0^b g.$$

For the last part, suppose $g(\alpha) = \infty$, where $\alpha > 0$. Then for any real $M > 0$, there exists an integer $N \in \mathbb{N}$ such that $n \geq N \Rightarrow g_n(\alpha) \geq M \Rightarrow g_n \geq M$ on $[0, \alpha] \Rightarrow \int_a^b g_n \geq \alpha M$. Consequently, $\lim_{n\to\infty} \int_a^b g_n = \infty$.

2.1 Measure and Measurability

2.1.P1. Let $\{S_n\}_{n\geq 1}$ be a sequence of subsets of a set X, such that $S_n \subseteq S_{n+1}$ for every $n \in \mathbb{N}$ (no σ-algebra intended) and let $S = \cup_{n=1}^\infty S_n$. Denote the characteristic functions of S_n and of S respectively by f_n and f. For each $x \in X$, show that

$$\text{(i) } f_n(x) \leq f_{n+1}(x) \quad \text{and} \quad \text{(ii) } f(x) = \lim_{n\to\infty} f_n(x).$$

Solution: (i) If $x \in S_n$, then $f_n(x) = 1$ and $x \in S_{n+1}$, so that $f_{n+1}(x) = 1$ and consequently, $f_n(x) \leq f_{n+1}(x)$. If $x \notin S_n$, then $f_n(x) = 0 \leq f_{n+1}(x)$.
(ii) Let $x \in S$. Then $f(x) = 1$. Also, since $S = \cup_{n=1}^\infty S_n$, there exists some N such that $x \in S_N$. But $S_n \subseteq S_{n+1}$ for every $n \in \mathbb{N}$, and therefore $n \geq N \Rightarrow S_n \supseteq S_N \Rightarrow f_n(x) = 1$, so that $\lim_{n\to\infty} f_n(x) = 1 = f(x)$. Now let $x \notin S$. In this case, $f(x) = 0$. Also, since $S = \cup_{n=1}^\infty S_n$, we have $x \notin S_n$ for every $n \in \mathbb{N}$, which has the consequence that $f_n(x) = 0$, so that $\lim_{n\to\infty} f_n(x) = 0 = f(x)$.
2.1.P2. If f and g are functions such that $f \leq g$ everywhere, show for any real t that the t-cut of f is a subset of the t-cut of g. If f and g are also measurable and μ is a measure on their domain, show that $\tilde{f}(t) \leq \tilde{g}(t)$ for every t.
Solution: $f(x) > t \Rightarrow g(x) > t$; so, $X(f > t) \subseteq X(g > t)$. By monotonicity of measure, $\mu(X(f > t)) \leq \mu(X(g > t))$, i.e. $\tilde{f}(t) \leq \tilde{g}(t)$.

2.1.P3. Let $\{f_n\}_{n\geq 1}$ be a sequence of functions on X such that, for every $x \in X$ and $n \in \mathbb{N}$, the inequality $f_n(x) \leq f_{n+1}(x)$ holds. For any $t \in \mathbb{R}$, denote the t-cut of f_n by $S_n(t)$. Show that, $S_n(t) \subseteq S_{n+1}(t)$ for every $n \in \mathbb{N}$. If furthermore f is a function on X such that $f(x) = \lim_{n\to\infty} f_n(x)$ for every $x \in X$, show that $X(f > t) = \cup_{n=1}^{\infty} S_n(t)$. Hence conclude that if each f_n is measurable, then f is measurable. (The latter part is true without the hypothesis that $f_n(x) \leq f_{n+1}(x)$, as is to be proved in Problem 2.2.P4.)

Solution: $x \in S_n(t) \Rightarrow f_n(x) > t \Rightarrow f_{n+1}(x) > t \Rightarrow x \in S_{n+1}(t)$. This proves $S_n(t) \subseteq S_{n+1}(t)$ for every $n \in \mathbb{N}$.

For the next part, first note that, since $f(x) = \lim_{n\to\infty} f_n(x)$ and also $f_n(x) \leq f_{n+1}(x)$ for every $n \in \mathbb{N}$, it follows when $f(x) < \infty$ that every $N \in \mathbb{N}$ satisfies $0 \leq f(x) - f_N(x)$ and hence satisfies $|f(x) - f_N(x)| = f(x) - f_N(x)$.

Now, let $f(x) > t$. Since $f(x) = \lim_{n\to\infty} f_n(x)$, there exists some $N \in \mathbb{N}$ such that either $|f(x) - f_N(x)| < f(x) - t$ or $f_N(x) > t$ according as $f(x) < \infty$ or $f(x) = \infty$. In the former case, as noted in the preceding paragraph, $|f(x) - f_N(x)| = f(x) - f_N(x)$ and therefore $f(x) - f_N(x) < f(x) - t$ and hence $f_N(x) > t$, just as in the latter case. Thus $x \in S_N(t)$ in either case. Hence $X(f > t) \subseteq \cup_{n=1}^{\infty} S_n(t)$. Conversely, consider $x \in \cup_{n=1}^{\infty} S_n(t)$. There exists some $N \in \mathbb{N}$ such that $x \in S_N(t)$, which means $f_N(x) > t$. Since $f_N(x) \leq f(x)$, it follows that $f(x) > t$. Thus $\cup_{n=1}^{\infty} S_n(t) \subseteq X(f > t)$.

Finally, if each f_n is measurable, then $S_n(t)$ is measurable for every n and every t. By definition of σ-algebra, it follows that the union $\cup_{n=1}^{\infty} S_n(t)$ is measurable, whereby $X(f > t)$ is measurable. Thus the function f is measurable.

2.1.P4. (Note: This problem is of interest in Combinatorics rather than Measure Theory, because it relates to the *Inclusion-Exclusion Principle* there; see Problem 3.1.P1.) Let $A_1, \ldots, A_n$ be subsets of X. Using the symbol $\chi[B]$ to denote the characteristic function of a set B, show that

$$\chi[A_1 \cup \cdots \cup A_n] = \sum_{1\leq r\leq n} \left((-1)^{r-1} \sum_{1\leq j_1 < \cdots < j_r \leq n} \chi[A_{j_1} \cap \cdots \cap A_{j_r}]\right).$$

Solution: First we note the two (obvious) equalities

$$\chi[B_1]\cdots\chi[B_n] = \chi[B_1 \cap \cdots \cap B_n] \quad \text{and} \quad 1 - \chi[B] = \chi[B^c].$$

Now we invoke the purely algebraic identity (justified for the sake of the skeptical reader at the end of this solution),

$$1 - \prod_{1\leq j\leq n}(1-a_j) = \sum_{1\leq r\leq n}\left((-1)^{r-1}\sum_{1\leq j_1<\cdots<j_r\leq n} a_{j_1}\cdots a_{j_r}\right).$$

Upon substituting $a_j = \chi[A_j]$ and using the first of the two equalities noted above, we find that the right side becomes

$$\sum_{1 \le r \le n} ((-1)^{r-1} \sum_{1 \le j_1 < \cdots < j_r \le n} \chi[A_{j_1} \cap \cdots \cap A_{j_r}]).$$

Also, each factor $1 - a_j$ on the left side is $1 - \chi[A_j]$, which is $\chi[A_j^c]$ by the second of the two equalities noted above. Using the first of the two equalities once again, we find that the product on the left side equals the characteristic function of the intersection $A_1^c \cap \cdots \cap A_n^c$. Therefore the left side equals the characteristic function of the complement of $A_1^c \cap \cdots \cap A_n^c$. However, this complement is nothing but the union $A_1 \cup \cdots \cup A_n$.

For the justification of the algebraic identity used above, it is sufficient to demonstrate that

$$\prod_{1 \le j \le n} (1 + a_j) = 1 + \sum_{1 \le r \le n} \left(\sum_{1 \le j_1 < \cdots < j_r \le n} a_{j_1} \cdots a_{j_r} \right).$$

For $n = 2$, the equality becomes $(1 + a_1)(1 + a_2) = 1 + a_1 + a_2 + a_1 a_2$, which is true. Assume it to hold for some $n \ge 2$ and we shall show that it holds for $n + 1$. Since it has been assumed to hold for n, we deduce that

$$\prod_{1 \le j \le n+1} (1 + a_j) = [\prod_{1 \le j \le n} (1 + a_j)] \cdot (1 + a_{n+1}) = S + T,$$

where

$$S = 1 + \sum_{1 \le r \le n} \left(\sum_{1 \le j_1 < \cdots < j_r \le n} a_{j_1} \cdots a_{j_r} \right)$$

and

$$T = a_{n+1} + \sum_{1 \le r \le n} \left(\sum_{1 \le j_1 < \cdots < j_r \le n} a_{j_1} \cdots a_{j_r} a_{n+1} \right).$$

It is therefore sufficient to prove

$$S + T = 1 + \sum_{1 \le r \le n+1} \left(\sum_{1 \le j_1 < \cdots < j_r \le n+1} a_{j_1} \cdots a_{j_r} \right). \tag{A}$$

We rewrite S by separating out the term with $r = 1$ as

$$S = 1 + (a_1 + \cdots + a_n) + U, \quad \text{where} \quad U = \sum_{2 \le r \le n} \left(\sum_{1 \le j_1 < \cdots < j_r \le n} a_{j_1} \cdots a_{j_r} \right),$$

and also rewrite T by separating out the term with $r = n$ as

$$T = a_{n+1} + V + a_1 \cdots a_{n+1},$$

$$\text{where } V = \sum_{1 \le r \le n-1} \left(\sum_{1 \le j_1 < \cdots < j_r \le n} a_{j_1} \cdots a_{j_r} a_{n+1} \right).$$

Now U can be expressed as

$$U = \sum_{2 \le r \le n} \left(\sum_{1 \le j_1 < \cdots < j_r \le n+1, j_r \ne n+1} a_{j_1} \cdots a_{j_r} \right)$$

and V can be expressed as

$$V = \sum_{2 \le r \le n} \left(\sum_{1 \le j_1 < \cdots < j_{r-1} \le n} a_{j_1} \cdots a_{j_{r-1}} a_{n+1} \right)$$

$$= \sum_{2 \le r \le n} \left(\sum_{1 \le j_1 < \cdots < j_r \le n+1, j_r = n+1} a_{j_1} \cdots a_{j_r} \right).$$

These expressions for U and V show that

$$U + V = \sum_{2 \le r \le n} \left(\sum_{1 \le j_1 < \cdots < j_r \le n+1} a_{j_1} \cdots a_{j_r} \right).$$

It follows that

$$S + T = 1 + (a_1 + \cdots + a_{n+1}) + \sum_{2 \le r \le n} \left(\sum_{1 \le j_1 < \cdots < j_r \le n+1} a_{j_1} \cdots a_{j_r} \right) + a_1 \cdots a_{n+1}.$$

But this is the same as (A) with the terms corresponding to $r = 1$ and $r = n + 1$ separated out.

2.1.P5. If $\mu \colon \mathcal{F} \to \mathbb{R}$ is monotone as well as finitely additive, then show that it is also *finitely subadditive* in the sense that $\mu(A \cup B) \le \mu(A) + \mu(B)$ for measurable sets A and B, disjoint or not.
Solution: Whether the measurable sets A and B are disjoint or not, $\mu(A \cup B) = \mu(A) + \mu(B \cap A^c)$ because the sets A and $B \cap A^c$ are disjoint and measurable, with union equal to $A \cup B$, and $\mu(B \cap A^c) \le \mu(B)$ by virtue of monotonicity.

2.1.P6. Let μ be as in Example 2.1.3(b) and the functions f, g be defined on $X = \{a, b, c\}$ as

$$f(a) = 1, f(b) = \sqrt{3}, f(c) = 2, \quad g(a) = \sqrt{2}, g(b) = 1, g(c) = 2.$$

Show that

$$\sqrt{\int_0^{\sup (f+g)^2} \widetilde{(f+g)^2}} < \sqrt{\int_0^{\sup f^2} \widetilde{f^2}} + \sqrt{\int_0^{\sup g^2} \widetilde{g^2}},$$

but

$$\sqrt{\int_0^{\sup (f+g)^2} \widetilde{(f+g)^2}} > \sqrt{\int_0^{\sup f^2} \widetilde{f^2}} + \sqrt{\int_0^{\sup g^2} \widetilde{g^2}}.$$

Solution: $(f+g)(a) = 1 + \sqrt{2}$, $(f+g)(b) = 1 + \sqrt{3}$, $(f+g)(c) = 4$. Also,

$$\widetilde{f^2}(t) = \begin{cases} 8 & 0 \le t < 1 \\ 6 & 1 \le t < 3 \\ 4 & 3 \le t < 4 \\ 0 & 4 \le t < \infty, \end{cases} \qquad \widetilde{g^2}(t) = \begin{cases} 8 & 0 \le t < 1 \\ 5 & 1 \le t < 2 \\ 4 & 2 \le t < 4 \\ 0 & 4 \le t < \infty \end{cases}$$

and $\widetilde{(f+g)^2}(t) = \begin{cases} 8 & 0 \le t < 3 + 2\sqrt{2} \\ 6 & 3 + 2\sqrt{2} \le t < 4 + 2\sqrt{3} \\ 4 & 4 + 2\sqrt{3} \le t < 16 \\ 0 & 16 \le t < \infty. \end{cases}$

Hence

$$\int_0^{\sup f^2} \widetilde{f^2} = 8 + 12 + 4 = 24, \qquad \int_0^{\sup g^2} \widetilde{g^2} = 8 + 5 + 8 = 21,$$

and

$$\int_0^{\sup(f+g)^2} \widetilde{(f+g)^2} = 8(3 + 2\sqrt{2}) + 6(1 + 2\sqrt{3} - 2\sqrt{2}) + 4(12 - 2\sqrt{3})$$

$$= 78 + 4\sqrt{2} + 4\sqrt{3}.$$

It can be verified (without abject subservience to a calculator) that

$$78 + 4\sqrt{2} + 4\sqrt{3} < (24^{1/3} + 21^{1/3})^3 \text{ but } 78 + 4\sqrt{2} + 4\sqrt{3} > (24^{1/2} + 21^{1/2})^2$$

as follows:

$78 + 4\sqrt{2} + 4\sqrt{3} < 78 + 4(2+3) = 98$. On the other hand, $21 > 15.625 = 2.5^3$. Therefore $21^{1/3} > 2.5$ and hence $24^{1/3} > 2.5$. It follows that $24^{1/3} + 21^{1/3} > 5$, which leads to $(24^{1/3} + 21^{1/3})^3 > 125$. Thus $78 + 4\sqrt{2} + 4\sqrt{3} < 98 < 125 < (24^{1/3} + 21^{1/3})^3$.

$78 + 4\sqrt{2} + 4\sqrt{3} > 78 + 4(\sqrt{1.96} + \sqrt{2.89}) = 78 + (1.4 + 1.7) > 78 + 4(3) = 90$. On the other hand, $(24^{1/2} + 21^{1/2})^2 = 45 + 2\sqrt{504} < 90$, because $2\sqrt{504} < 45$, as can be seen from the inequality $4(504) < 45^2$, which is the same as $2016 < 2025$.

2.1.P7. Show that a measure is *countably subadditive* in the following sense: If $(X, \mathcal{F}, \mu)$ is a measure space and $\{A_j\}_{j \in \mathbb{N}}$ is a sequence of measurable sets, then $\mu(\cup_{j=1}^{\infty} A_j) \leq \sum_{j=1}^{\infty} \mu(A_j)$. Is the corresponding statement true regarding a finite sequence of measurable sets?

Solution: Let $B_1 = A_1$ and, for $k > 1$, let $B_k = A_k \cap (\cup_{j=1}^{k-1} A_j)^c$. In other words, $B_k = A_k \setminus (A_1 \cup \cdots \cup A_{k-1})$. Then each $B_k \subseteq A_k$ and is measurable (by definition of σ-algebra), so that $\mu(B_k) \leq \mu(A_k)$. Furthermore, the sets B_k are all disjoint and therefore by countable additivity of measure, we have

$$\mu\left(\bigcup_{j=1}^{\infty} B_j\right) = \sum_{j=1}^{\infty} \mu(B_j) \leq \sum_{j=1}^{\infty} \mu(A_j).$$

But the definition of B_k also implies that $\cup_{j=1}^{n} A_j = \cup_{j=1}^{n} B_j$ for all $n \in \mathbb{N}$ and hence $\cup_{j=1}^{\infty} A_j = \cup_{j=1}^{\infty} B_j$.

The same argument shows that the corresponding statement is true for a finite sequence.

2.1.P8. Let $\mathcal{F}$ be a σ-algebra of subsets of a set X. Suppose $\mu \colon \mathcal{F} \to \mathbb{R}^{*+}$ is finitely additive and vanishes at $\emptyset$. If it is also countably subadditive (as defined in Problem 2.1.P7), show that it is a measure.

Solution: We need show only that μ is countably additive. Since it is nonnegative and finitely additive, it is monotone. Let $\{A_k\}_{k \geq 1}$ be a sequence of disjoint sets, each belonging to $\mathcal{F}$. Then $\mu(\cup_{k=1}^{\infty} A_k) \geq \mu(\cup_{k=1}^{m} A_k)$ for each $m \in \mathbb{N}$ by monotonicity. But $\mu(\cup_{k=1}^{m} A_k) = \sum_{k=1}^{m} \mu(A_k)$. Therefore $\mu(\cup_{k=1}^{\infty} A_k) \geq \sum_{k=1}^{m} \mu(A_k)$ for each m. It follows that $\mu(\cup_{k=1}^{\infty} A_k) \geq \sum_{k=1}^{\infty} \mu(A_k)$. The reverse inequality is assured by the given countable subadditivity.

2.1.P9. Suppose that, for each $j \in \mathbb{N}$, $\mathcal{F}_j$ is a σ-algebra of subsets of X_j. Define the family $\mathcal{F}$ of sets to consist of those subsets A of the union $X = \cup_{j=1}^{\infty} X_j$ for which $A \cap X_j \in \mathcal{F}_j$ for each $j \in \mathbb{N}$. Show that $\mathcal{F}$ is a σ-algebra.

Solution: Clearly, $\emptyset, X \in \mathcal{F}$. For any $A \subseteq X$, the intersection $A^c \cap X_j$ is the complement in X_j of the intersection $A \cap X_j$. Therefore $A \in \mathcal{F} \Rightarrow A^c \in \mathcal{F}$. Finally, let $\{A_k\}_{k \geq 1}$ be a sequence of sets in $\mathcal{F}$; then for any j, we have $(\cup_{k=1}^{\infty} A_k) \cap X_j = \cup_{k=1}^{\infty} (A_k \cap X_j) \in \mathcal{F}_j$.

2.1.P10. For each $j \in \mathbb{N}$, let $(X_j, \mathcal{F}_j, \mu_j)$ be a measure space and denote the union $\cup_{j=1}^{\infty} X_j$ by X. Set $\mathcal{F} = \{A \subseteq X : A \cap X_j \in \mathcal{F}_j \text{ for each } j \in \mathbb{N}\}$ and for each $A \in \mathcal{F}$, set $\mu(A) = \sum_{j=1}^{\infty} \mu_j(A \cap X_j)$. Show that μ is a measure.

Solution: Obviously, $\mu(A)$ is nonnegative and vanishes at $\emptyset$. From Problem 2.1.P9, we know that $\mathcal{F}$ is a σ-algebra. We shall first prove that μ is finitely additive. It is sufficient to show that $\mu(A \cup B) = \mu(A) + \mu(B)$ for any two disjoint sets $A, B \in \mathcal{F}$. In view of the fact that each μ_j is finitely additive, we have $\mu_j((A \cup B) \cap X_j) = \mu_j(A \cap X_j) + \mu_j(B \cap X_j)$. Therefore

$$\mu(A \cup B) = \sum_{j=1}^{\infty} \mu_j((A \cup B) \cap X_j) = \sum_{j=1}^{\infty} (\mu_j(A \cap X_j) + \mu_j(B \cap X_j))$$

$$= \sum_{j=1}^{\infty} \mu_j(A \cap X_j) + \sum_{j=1}^{\infty} \mu_j(B \cap X_j) = \mu(A) + \mu(B).$$

This proves the finite additivity of μ, and therefore by Problem 2.1.P8, it remains only to prove its countable subadditivity.

Consider a sequence $\{A_k\}_{k \geq 1}$ of sets, each belonging to $\mathcal{F}$. Thus $A_k \cap X_j \in \mathcal{F}_j$ for each $j \in \mathbb{N}$. Since $\mu(A_k) = \sum_{j=1}^{\infty} \mu_j(A_k \cap X_j)$, what we want to prove is that $\mu(\cup_{k=1}^{\infty} A_k) \leq \sum_{k=1}^{\infty}(\sum_{j=1}^{\infty} \mu_j(A_k \cap X_j))$. Since each μ_j is countably subadditive (by Problem 2.1.P7), we have

$$\mu\left(\bigcup_{k=1}^{\infty} A_k\right) = \sum_{j=1}^{\infty} \mu_j\left(\left(\bigcup_{k=1}^{\infty} A_k\right) \cap X_j\right) = \sum_{j=1}^{\infty} \mu_j\left(\bigcup_{k=1}^{\infty} (A_k \cap X_j)\right)$$

$$\leq \sum_{j=1}^{\infty} \left(\sum_{k=1}^{\infty} \mu_j(A_k \cap X_j)\right).$$

When the order of summation is interchanged, the required inequality emerges.

2.1.P11. Let $\mathcal{F}$ be a σ-algebra of subsets of a nonempty set X and Y any set with a map $F: X \rightarrow Y$. Take $\mathcal{G}$ to be the family of subsets $\{B \subseteq Y : F^{-1}(B) \in \mathcal{F}\}$ of Y. Here, $F^{-1}(B)$ means $\{x \in X : F(x) \in B\}$, as usual. Show that $\mathcal{G}$ is a σ-algebra. If μ is a measure on $\mathcal{F}$, show that the map $\nu(B): \mathcal{G} \rightarrow \mathbb{R}^{*+}$ defined by $\nu(B) = \mu(F^{-1}(B))$ is a measure on $\mathcal{G}$.

Solution: It is trivial why $\emptyset$, $Y \in \mathcal{G}$. Suppose $B \subseteq Y$ and B^c is the complement of B in Y. Then

$$x \in F^{-1}(B^c) \Leftrightarrow F(x) \in B^c \Leftrightarrow F(x) \notin B \Leftrightarrow x \notin F^{-1}(B) \Leftrightarrow x \in F^{-1}(B)^c$$

and hence

$$F^{-1}(B^c) = F^{-1}(B)^c.$$

Consequently, $B \in \mathcal{G} \Rightarrow F^{-1}(B) \in \mathcal{F} \Rightarrow F^{-1}(B)^c \in \mathcal{F} \Rightarrow F^{-1}(B^c) \in \mathcal{F} \Rightarrow B^c \in \mathcal{G}$. Thus $\mathcal{G}$ is closed under complements. We shall show that it is also closed under countable unions.

Let $\{B_k\}_{k \geq 1}$ be a sequence of subsets of Y. Then

$$x \in F^{-1}(\bigcup_{k=1}^{\infty} B_k) \Leftrightarrow F(x) \in \bigcup_{k=1}^{\infty} B_k \Leftrightarrow F(x) \in B_k \text{ for some } k \Leftrightarrow x \in F^{-1}(B_k) \text{ for some}$$

$$k \Leftrightarrow x \in \bigcup_{k=1}^{\infty} F^{-1}(B_k).$$

Consequently, $F^{-1}(\cup_{k=1}^{\infty} B_k) = \cup_{k=1}^{\infty} F^{-1}(B_k)$. Now suppose each $B_k \in \mathcal{G}$. This means each $F^{-1}(B_k) \in \mathcal{F}$ and it follows that $\cup_{k=1}^{\infty} F^{-1}(B_k) \in \mathcal{F}$, and hence that $F^{-1}(\cup_{k=1}^{\infty} B_k) \in \mathcal{F}$, i.e. $\cup_{k=1}^{\infty} B_k \in \mathcal{G}$.

We proceed to argue that v is a measure. Certainly, v is nonnegative and vanishes at $\emptyset$. We shall show that it is countably additive. Let $\{B_k\}_{k \geq 1}$ be a sequence of disjoint sets belonging to $\mathcal{G}$. Then the sets $F^{-1}(B_k)$ are also disjoint, because for $j \neq k$, $x \in F^{-1}(B_j) \cap F^{-1}(B_k) \Rightarrow F(x) \in B_j \cap B_k = \emptyset$. Since $F^{-1}(\cup_{k=1}^{\infty} B_k) = \cup_{k=1}^{\infty} F^{-1}(B_k)$ as proved above,

$$v(\bigcup_{k=1}^{\infty} B_k) = \mu\left(F^{-1}(\bigcup_{k=1}^{\infty} B_k)\right) = \mu\left(\bigcup_{k=1}^{\infty} F^{-1}(B_k)\right) = \sum_{k=1}^{\infty} \mu(F^{-1}(B_k)) = \sum_{k=1}^{\infty} v(B_k).$$

2.1.P12. Let $\mathcal{F}$ be a σ-algebra of subsets of a set X, and $E \in \mathcal{F}$ have the property that every subset of it belongs to $\mathcal{F}$.

(a) If $B \subseteq A \cup E$, show that $A^c \cap B \in \mathcal{F}$.
(b) If $B \subseteq A \cup E$, $B \in \mathcal{F}$, show that $A \cap B \in \mathcal{F}$.
(c) If $B \subseteq A \cup E$, $A \subseteq B \cup E$, $B \in \mathcal{F}$, show that $A \in \mathcal{F}$.

Solution:

(a) $A^c \cap B \subseteq A^c \cap (A \cup E) = (A^c \cap A) \cup (A^c \cap E) = \emptyset \cup (A^c \cap E) \subseteq E$. Since every subset of E belongs to $\mathcal{F}$, we have $A^c \cap B \in \mathcal{F}$.
(b) $A \cap B = (A \cap B) \cup (B^c \cap B) = (A \cup B^c) \cap B = (A^c \cap B)^c \cap B$. Since $A^c \cap B \in \mathcal{F}$ by part (a) and $B \in \mathcal{F}$ by hypothesis, we have $A \cap B \in \mathcal{F}$.

(c) $A = A \cap (B^c \cup B) = (A \cap B^c) \cup (A \cap B)$. Since $A \cap B^c \in \mathcal{F}$ by part (a) and $A \cap B \in \mathcal{F}$ by part (b), we get $A \in \mathcal{F}$.

2.1.P13. Prove that the result of Exercise 2.1.5 holds even if the set function $\mu \colon \mathcal{F} \to \mathbb{R}^{*+}$ is assumed to be only finitely additive provided it is monotone and vanishes at $\emptyset$.

Solution: Consider any real t and the t-cuts B and C of f and g respectively. Since μ is monotone and nonnegative the hypothesis $\mu(A^c) = 0$ implies that $\mu(B \cap A^c) = 0$. Now $B = (B \cap A^c) \cup (B \cap A)$ and finite subadditivity of μ implies $\mu(B) \le \mu(B \cap A^c) + \mu(B \cap A) = \mu(B \cap A)$. But $\mu(B \cap A) \le \mu(B)$ by monotonicity, so that $\mu(B \cap A) = \mu(B)$. This means $\tilde{f}(t) = \mu(B \cap A)$. Similarly, $\tilde{g}(t) = \mu(C \cap A)$. By hypothesis, f and g agree on the set A, which implies that the t-cuts B and C satisfy $B \cap A = C \cap A$.

2.2 Measurability and the Integral

2.2.P1. Prove Proposition 2.2.2(b).
Solution: Suppose f is real-valued so that $-f$ is defined. As $X(-f > t) = X(f < -t)$, measurability of f implies that of $-f$ in view of (ii) of part (a) of this proposition. Now consider any $f \colon X \to \mathbb{R} \cup \{\infty\}$. While $-f$ may not be defined, it is true nevertheless that

$$|f(x)| > t \Leftrightarrow f(x) > t \text{ or } f(x) < -t,$$

and hence $X(|f| > t) = X(f > t) \cup X(f < -t)$. It now follows from (ii) of part (a) of this proposition and the fact that a σ-algebra is closed under finite unions that measurability of f implies that of $|f|$.

2.2.P2. If $f \colon X \to \mathbb{R}^{*+}$ has the property that $X(f \ge t)$ is measurable for every $t \in \mathbb{Q}$, show that f is measurable.
Solution: Consider any $s \in \mathbb{R}$. Then $f(x) > s \Leftrightarrow f(x) \ge t$ for some rational $t > s$. This means $X(f > s)$ is the union of all sets $X(f \ge t)$ with $t \in \mathbb{Q}$ and $t > s$. But these sets are measurable and are also countable in number (i.e. can be arranged in a sequence). Therefore their union, which has been shown to be $X(f > s)$, must be measurable.

2.2.P3. Let $\{f_n\}_{n \ge 1}$ be a sequence of extended real-valued measurable functions on X. Then the function f defined on X as $f(x) = \inf f_n(x)$, i.e. $\inf\{f_n(x) \colon n \in \mathbb{N}\}$, is measurable.
Solution: Let $x \in X$ and $n \in \mathbb{N}$. Then $f(x) \ge t \Leftrightarrow f_n(x) \ge t$ for every $n \in \mathbb{N}$. So, $X(f \ge t) = \cap_{n=1}^{\infty} X(f_n \ge t)$. By Proposition 2.2.2, each set $X(f_n \ge t)$ is measurable and hence, so is their intersection. Thus $X(f \ge t)$ is measurable. This has been shown to be the case for all real t. Using Proposition 2.2.2 once again, we find that f is measurable.

2.2.P4. In Problem 2.1.P3, show that the measurability of the limit function f holds without the hypothesis that $f_n(x) \le f_{n+1}(x)$.

Solution: We are given a sequence $\{f_n\}_{n \ge 1}$ of measurable functions and a function f such that $f(x) = \lim_{n \to \infty} f_n(x)$ for every $x \in X$. We must show that the limit function f is measurable. For each $x \in X$,

$$f(x) = \lim_{n \to \infty} \inf f_n(x) = \lim_{n \to \infty} g_n(x),$$

where $g_n(x) = \inf\{f_m(x) : n \le m \in \mathbb{N}\}$. By Problem 2.2.P3, each function g_n is measurable, and obviously, $g_n(x) \le g_{n+1}(x)$. It follows by Problem 2.1.P3 that f is measurable.

2.2.P5. If $f: X \to \mathbb{R}$ is measurable and $\phi: \mathbb{R} \to \mathbb{R}$ is increasing, show that the composition $\phi \circ f: X \to \mathbb{R}$ is measurable.

Solution: $X(\phi \circ f > t) = \{x \in X : f(x) \in \{u \in \mathbb{R} : \phi(u) > t\}\}$. Since ϕ is increasing, $\{u \in \mathbb{R} : \phi(u) > t\}$ is an interval (perhaps empty). For any interval I, the set $\{x \in X : f(x) \in I\}$ is measurable by Proposition 2.2.2. Hence $X(\phi \circ f > t)$ is measurable for any t.

2.2.P6. Suppose μ is a measure on a set X. Prove that

(a) For a measurable function $f: X \to \mathbb{R}^{*+}$ and any real $p > 0$, the nonnegative function $f^p: X \to \mathbb{R}^{*+}$ is measurable; moreover,

$$\int f = 0 \text{ if and only if } \tilde{f}(t) = 0 \text{ for every } t > 0,$$

which, in turn, is equivalent to $\int f^p = 0$ for any $p > 0$.

(b) If $f: X \to \mathbb{R}^{*+}$ and $g: X \to \mathbb{R}^{*+}$ are measurable functions satisfying $\int f = 0 = \int g$, then $\int (f + g) = 0$.

Solution: (a) The measurability of f^p follows from the fact that $X(f^p > t) = X(f > t^{\frac{1}{p}})$. For the rest, the "if" part is clear. For the "only if" part, suppose for some $t > 0$ that $\tilde{f}(t) > 0$. Since $\tilde{f}$ is a decreasing function, we have $\tilde{f}(\tau) \ge \tilde{f}(t)$ for $0 \le \tau \le t$ and therefore $\int f = \int_0^\infty \tilde{f} \ge \int_0^t \tilde{f} \ge t\tilde{f}(t) > 0$. The equivalence of $\int f = 0$ and $\int f^p = 0$ follows from the fact that

$$\tilde{f}(t) = \mu(X(f > t)) = 0 \quad \text{for every } t > 0$$

if and only if

$$\widetilde{f^p}(t) = \mu(X(f^p > t)) = 0 \quad \text{for every } t > 0.$$

(b) Suppose $\int f = 0 = \int g$. Then by (a), we have $\tilde{f}(s) = 0 = \tilde{g}(s)$ for every $s > 0$. By definition of a distribution, this means $\mu(X(f > s)) = 0 = \mu(X(g > s))$ for every $s > 0$. Now, consider an arbitrary $t > 0$. For any $x \in X$,

$$(f+g)(x) > t \quad \Rightarrow \quad \text{either} f(x) > \frac{t}{2} \text{ or } g(x) > \frac{t}{2}.$$

Therefore $X(f+g > t) \subseteq X(f > \frac{t}{2}) \cup X(g > \frac{t}{2})$. It follows by monotonicity and finite subadditivity proved in Problem 2.1.P5 that $\mu(X(f+g > t)) = 0$. This is so for an arbitrary $t > 0$ and hence by (a), $\int (f+g) = 0$.

2.2.P7. (a) Let f, g be nonnegative measurable functions on a space X with measure μ, and $f(x) = g(x)$ for all $x \in A$, where A is a measurable subset of X satisfying $\mu(A^c) = 0$. Show that $\int f = \int g$.

(b) Let f, g be measurable functions on a space X with measure μ, and $f(x) = g(x)$ for all $x \in A$, where A is a measurable subset of X satisfying $\mu(A^c) = 0$. If g is integrable, show that f is also integrable and $\int f = \int g$.

Solution: (a) By Example 2.1.5, $\tilde{f} = \tilde{g}$.

(b) Under the given hypothesis, f^+ and g^+ are nonnegative measurable functions satisfying $f^+(x) = g^+(x)$ for all $x \in A$. It follows from part (a) that $\int f^+ = \int g^+$. Similarly, $\int f^- = \int g^-$.

2.2.P8. Let f, g be nonnegative measurable functions on a space X with measure μ, and $f(x) \leq g(x)$ for all $x \in A$, where A is a measurable subset of X satisfying $\mu(A^c) = 0$. Show that $\int f \leq \int g$.

Solution: Let χ be the characteristic function of A. Then $f(x) = (f\chi)(x)$ for all $x \in A$ and similarly for g. Also, for any $t \geq 0$, the t-cut of $f\chi$ is the intersection of the t-cut of f with A; therefore $f\chi$ is a measurable nonnegative function. Similarly, $g\chi$ is a measurable nonnegative function. It follows by Problem 2.2.P7 that $\int f = \int (f\chi)$ and similarly for g. But $(f\chi)(x) \leq (g\chi)(x)$ for all $x \in X$. Therefore by Proposition 2.2.8, $\int (f\chi) \leq \int (g\chi)$.

2.2.P9. For a function $f: X \to \mathbb{R} \cup \{\infty\}$, let A be the set $X(f < \infty)$. Show that

(a) If $t < 0$, then $X(f\chi_A > t) = A^c \cup (A \cap X(f > t))$.
(b) If $t \geq 0$, then $X(f\chi_A > t) = A \cup X(f > t)$.
(c) The product $f\chi_A$ is real-valued.
(d) If f is measurable, then A is measurable and $f\chi_A$ is measurable.

Solution: (a) First consider any $t \in \mathbb{R}$. Suppose $(f\chi_A)(x) > t$. Either $x \in A^c$ or $x \in A$. If $x \in A$, we have $(f\chi_A)(x) = f(x)$ and therefore also $f(x) > t$, so that $x \in A \cap X(f > t)$. This proves the inclusion $X(f\chi_A > t) \subseteq A^c \cup (A \cap X(f > t))$ for all $t \in \mathbb{R}$. For $t < 0$, we shall prove the reverse inclusion. With this in view, consider any $x \in A^c \cup (A \cap X(f > t))$. If $x \in A^c$, then $(f\chi_A)(x) = 0 > t$. If $x \in A \cap X(f > t)$, then $(f\chi_A)(x) = f(x) > t$. This proves the reverse inclusion for $t < 0$.

(b) Now suppose $t \geq 0$ and $(f\chi_A)(x) > t$. Then $(f\chi_A)(x) > 0$ and therefore $\chi_A(x) \neq 0$. It follows that $\chi_A(x) = 1$, so that $x \in A$, and hence $f(x) = (f\chi_A)(x) > t$. This proves the inclusion $X(f\chi_A > t) \subseteq A \cap X(f > t)$ for $t \geq 0$. To prove the reverse inclusion, consider any $x \in A \cap X(f > t)$. Since $x \in A$, we have $(f\chi_A)(x) = f(x)$ and since $x \in X(f > t)$, we also have $f(x) > t$. Therefore $(f\chi_A)(x) > t$, thereby proving the reverse inclusion.

(c) If $x \in A$, then $(f\chi_A)(x) = f(x)$ and $f(x) < \infty$. If $x \in A^c$, then $(f\chi_A)(x) = 0 < \infty$.

(d) Since $A = (\cap_{n=1}^{\infty} X(f > n))^c$ and f is measurable, the set A is measurable. That $f\chi_A$ is real measurable now follows from parts (a) and (b).

2.2.P10. Let the function $f: X \to \mathbb{R} \cup \{\infty\}$ have the property that $X(t < f < \infty)$ is measurable for every $t \in \mathbb{R}$. Show that f is measurable. (The converse is trivial.)
Solution: It is true for any $x \in X$ that $f(x) < \infty \Leftrightarrow n < f(x) < \infty$ for some $n \in \mathbb{Z}$. This means $X(f < \infty) = \cup_{n \in \mathbb{Z}} X(n < f < \infty)$. From the given property of f, it follows that $X(f < \infty)$ is measurable and therefore its complement $X(f = \infty)$ is also measurable. Since $X(f > t) = X(t < f < \infty) \cup X(f = \infty)$, it further follows that $X(f > t)$ is measurable.

2.2.P11. If $f: X \to \mathbb{R}^{*+}$ and $g: X \to \mathbb{R}^{*+}$ are both measurable, show that their product $fg: X \to \mathbb{R}^{*+}$ is measurable.
Solution: Let $A = X(f < \infty)$ and $B = X(g < \infty)$, both measurable by Problem 2.2.P9 (d). By Problem 2.2.P10, it is sufficient to show that $X(t < fg < \infty)$ is measurable for every $t \in \mathbb{R}$.

First consider the possibility that $t < 0$. Since $fg \geq 0 > t$ everywhere, the set in question is then the same as $X(fg < \infty)$. Now $f(x)g(x) < \infty$ if and only if either $f(x) < \infty$ as well as $g(x) < \infty$ or one among $f(x)$ and $g(x)$ is 0. Consequently,

$$X(fg < \infty) = (A \cap B) \cup X(f = 0) \cup X(g = 0),$$

which is measurable. So, $X(t < fg < \infty)$ is measurable when $t < 0$.

Next consider the possibility that $t \geq 0$. When this is so,

$$X(t < fg < \infty) = X(0 < f < \infty) \cap X(0 < g < \infty) \cap X(f\chi_A g\chi_B > t).$$

The first two sets on the right side are measurable. Since $f\chi_A$ and $g\chi_B$ are real-valued measurable functions by Problem 2.2.P9(c) and (d), their product is measurable by Theorem 2.2.3(c), as a consequence of which the third set on the right is also measurable. It follows that $X(t < fg < \infty)$ is again measurable when $t \geq 0$.

2.2.P12. For an integrable function $f: X \to \mathbb{R}^*$, show that

$$\alpha \cdot \mu(X(|f| > \alpha)) \leq \int |f| \quad \text{for all } \alpha > 0.$$

Solution: It is sufficient to prove the inequality for a nonnegative measurable function. Since $\tilde{f}$ is a decreasing function, we have $\tilde{f}(t) \geq \tilde{f}(\alpha)$ for $0 \leq t \leq \alpha$. This yields $\alpha \cdot \tilde{f}(\alpha) \leq \int_0^\alpha \tilde{f} \leq \int_0^\infty \tilde{f}$ for all $\alpha > 0$. However, $\tilde{f}(\alpha) = \mu(X(f > \alpha))$ and $\int_0^\infty \tilde{f} = \int f$.

2.2.P13. Suppose $f: X \to \mathbb{R}^{*+}$ is a measurable function such that, for some $\alpha > 0$, the function $\min\{f, \alpha\}$ has a finite integral. Show that f has a finite integral if and only if the series $\sum_{k=1}^\infty \tilde{f}(k)$ converges.

Solution: Denote $\min\{f, \alpha\}$ by g. The distribution $\tilde{g}$ agrees with $\tilde{f}$ on the interval $[0, \alpha)$ and is 0 on $[\alpha, \infty)$. Therefore $\int_0^\alpha \tilde{f} = \int_0^\alpha \tilde{g} = \int_0^\infty \tilde{g} = \int g$, which is given to be finite. One consequence of this is that $\int_0^\infty \tilde{f}$ is finite if and only if $\int_\alpha^\infty \tilde{f}$ is finite. Another consequence is that $\tilde{f}(\alpha)$ is finite and hence $\int_1^\alpha \tilde{f} (\text{or } \int_\alpha^1 \tilde{f})$ is finite, so that $\int_\alpha^\infty \tilde{f}$ is finite if and only if $\int_1^\infty \tilde{f}$ is finite. These two consequences together show that $\int_0^\infty \tilde{f}$ is finite if and only if $\int_1^\infty \tilde{f}$ is finite. The equivalence of the finiteness of this integral and the series $\sum_{k=1}^\infty \tilde{f}(k)$ is now just the integral test for the convergence of a series of decreasing nonnegative terms (Theorem 10.2.4 of Shirali and Vasudeva [9] or Example 8 on pp.138–9 of Rudin [7]).

2.3 The Monotone Convergence Theorem

2.3.P1. (Needed in Problem 3.2.P1 and Theorem 4.3.4) Let μ be a measure on X and f a nonnegative measurable function such that $\int f = 0$. Prove $\mu(X(f > 0)) = 0$.

Solution: It follows by Problem 2.2.P6 that $\mu(X(f > t)) = 0$ for every $t > 0$. Now, $f(x) > 0$ if and only if $f(x) > \frac{1}{n}$ for some $n \in \mathbb{N}$, which means $X(f > 0) = \cup_{n=1}^\infty X(f > \frac{1}{n})$. Since $X(f > \frac{1}{n}) \subseteq X(f > \frac{1}{n+1})$, inner continuity yields

$$\mu(X(f > 0)) = \lim_{n \to \infty} \mu\left(X(f > \frac{1}{n})\right).$$

But each $\mu(X(f > \frac{1}{n+1})) = 0$ according to what has been proved earlier.

2.3.P2. Let $\{f_n\}_{n \geq 1}$ be a sequence of real-valued functions on an arbitrary nonempty set X. Suppose f is a real-valued function on X such that, for every $x \in X$, some subsequence of the real sequence $\{f_n(x)\}_{n \geq 1}$ converges to $f(x)$. Show for any real t that $X(f > t) \subseteq \cup_{n=1}^\infty X(f_n > t)$.

Solution: Consider any $x \in X(f > t)$, which means $f(x) > t$. Since the real sequence $\{f_n(x)\}_{n \geq 1}$ has some subsequence converging to the real number $f(x)$, there exists an $n \in \mathbb{N}$ such that $|f_n(x) - f(x)| < f(x) - t$. This inequality implies $f(x) - f_n(x) < f(x) - t$, which in turn implies $f_n(x) > t$. Thus there exists an $n \in \mathbb{N}$ such that $f_n(x) > t$, i.e. $x \in X(f_n > t)$.

2.3.P3. In a measure space, suppose that every set of infinite measure contains a subset of finite positive measure. Show that every set of infinite measure contains a subset of arbitrarily large finite measure.

Hint: For a given set of infinite measure, show that there exists a subset whose measure is the supremum of the measures of all subsets of finite measure. If its measure is finite, then the measure of its complement in the given set must be infinite.

Solution: Denote the measure by μ. Consider any measurable set A with $\mu(A) = \infty$ and set $\alpha = \sup\{\mu(B): B \subseteq A, 0 < \mu(B) < \infty\}$. Then $\alpha > 0$ and we have to prove that $\alpha = \infty$. Assume otherwise. For each $n \in \mathbb{N}$, there exists a $B_n \subseteq A$ such that $\alpha - \frac{1}{n} < \mu(B_n) < \infty$. The set $B = \cup_{n=1}^{\infty} B_n$ is measurable and $\mu(B) = \alpha$ by Proposition 2.3.1. Since $\mu(A) = \mu(B) + \mu(A \backslash B) = \alpha + \mu(A \backslash B)$ and α has been assumed finite we have $\mu(A \backslash B) = \infty$. According to the hypothesis, $A \backslash B$ contains a subset C such that $0 < \mu(C) < \infty$. Since B and C are disjoint, $\mu(C \cup B) = \mu(C) + \mu(B) = \mu(C) + \alpha$. Since $C \cup B \subseteq A$ and $0 < \mu(C) + \alpha < \infty$, this is contrary to the definition of α.

2.3.P4. Let μ be a measure. For a sequence of measurable sets $\{S_n\}_{n \geq 1}$ with $\mu(S_1) < \infty$, show that

$\mu(\cap_{n=1}^{\infty} S_n) = \lim_{n \to \infty} \mu(S_n)$ provided that $S_n \supseteq S_{n+1}$ for every n. (outer continuous)

Solution: Let $T_n = S_1 \cap S_n^c$. Since $S_n \subseteq S_1$, the set S_1 is the disjoint union $S_n \cup T_n$, where T_n is also measurable. Therefore $\mu(S_1) = \mu(T_n) + \mu(S_n)$. Since $\mu(S_1) < \infty$, we can rewrite this as

$$\mu(T_n) = \mu(S_1) - \mu(S_n). \tag{Sol.1}$$

Now,

$$\bigcup_{n=1}^{\infty} T_n = \bigcup_{n=1}^{\infty} (S_1 \cap S_n^c) = S_1 \cap \bigcup_{n=1}^{\infty} S_n^c = S_1 \cap (\bigcap_{n=1}^{\infty} S_n)^c$$

and

$$\bigcap_{n=1}^{\infty} S_n \subseteq S_1.$$

An argument similar to that leading to (Sol.1) now leads to

$$\mu(\bigcup_{n=1}^{\infty} T_n) = \mu(S_1) - \mu(\bigcap_{n=1}^{\infty} S_n). \tag{Sol.2}$$

Moreover, $T_1 \subseteq T_2 \subseteq \cdots$ and hence by inner continuity (Proposition 2.3.1) and (Sol.1),

$$\mu\left(\bigcup_{n=1}^{\infty} T_n\right) = \lim_{n\to\infty} \mu(T_n) = \mu(S_1) - \lim_{n\to\infty} \mu(S_n). \tag{Sol.3}$$

The required equality follows from (Sol.2), (Sol.3) and the finiteness of $\mu(S_1)$.

3.1 Simple Functions

3.1.P1. Let μ be a measure on X and $A_1, \ldots, A_n$ be measurable subsets of X such that $\mu(A_1 \cup \cdots \cup A_n) < \infty$. Show that the *Inclusion-Exclusion Principle* holds:

$$\mu(A_1 \cup \cdots \cup A_n) = \sum_{1 \le r \le n} \left((-1)^{r-1} \sum_{1 \le j_1 < \cdots < j_r \le n} \mu(A_{j_1} \cap \cdots \cap A_{j_r}) \right).$$

If it is not given that $\mu(A_1 \cup \cdots \cup A_n) < \infty$, what is the best one can say?
Solution: In the equality established in Problem 2.1.P4, transpose all terms with even r to other side, so that all powers of -1 can be taken to be 1. Then apply Proposition 3.1.8. Convert the integrals of characteristic functions to measures of the corresponding sets. If $\mu(A_1 \cup \cdots \cup A_n) < \infty$ all terms are finite and one may therefore transpose them back to their original positions.

If it is not given that $\mu(A_1 \cup \cdots \cup A_n) < \infty$, the transposing back is impermissible and one can only say that

$$\mu(A_1 \cup \cdots \cup A_n) + \sum_{1 \le r \le n \text{ and } r \text{ even}} \left(\sum_{1 \le j_1 < \cdots < j_r \le n} \mu(A_{j_1} \cap \cdots \cap A_{j_r}) \right)$$

$$= \sum_{1 \le r \le n \text{ and } r \text{ odd}} \left(\sum_{1 \le j_1 < \cdots < j_r \le n} \mu(A_{j_1} \cap \cdots \cap A_{j_r}) \right).$$

3.1.P2. When $X = \{a, b, c, d\}$ and $A = X, B = \{c, d\}, C = \{d\}$, find the canonical form of the simple function $s = \chi_A + 2\chi_B + \chi_C$.
Solution: Since a and b belong only to A, we have $s(a) = s(b) = 1$. Since c belongs to only A and B, we have $s(c) = 1 + 2 = 3$. Lastly, since d belongs to all three sets, $s(d) = 1 + 2 + 1 = 4$. To summarize,

$$s(a) = s(b) = 1, s(c) = 3 \quad \text{and} \quad s(d) = 4.$$

Thus s takes the three values 1, 3, 4. Now,

$$A_1 = X(s = 1) = \{a, b\}, A_2 = X(s = 3) = \{c\} \quad \text{and} \quad A_3 = X(s = 4) = \{d\}.$$

So, the canonical form is

$$s = 1 \cdot \chi_{A_1} + 3 \cdot \chi_{A_2} + 4 \cdot \chi_{A_3} = 1 \cdot \chi_{\{a,b\}} + 3 \cdot \chi_{\{c\}} + 4 \cdot \chi_{\{d\}}.$$

3.1.P3. Does the data of Exercise 3.1.9 determine the sets A, B, C uniquely?
Solution: No. There are at least two possibilities consistent with the data: (in abbreviated notation) *abe, ace, bde* and also *abe, ade, bce*.

3.1.P4. Each of three junior managers in a company submits a list of three cities to the manager, because the senior manager wants them to. The manager goes through the lists and informs the senior manager that there are five cities in all the three lists taken together, and that two cities appear in two lists each and two others in three lists each. When the senior manager hears this, she is furious—Why?
Solution: Let μ denote counting measure on the set of all five cities and let A, B, C denote the three lists. If we take $s = \chi_A + \chi_B + \chi_C$, then by Proposition 3.1.8, $\int s = \int \chi_A + \int \chi_B + \int \chi_C = \mu(A) + \mu(B) + \mu(C) = 3 + 3 + 3 = 9$. But according to the information provided by the manager, $\mu(X(s = 2)) \geq 2$ and $\mu(X(s = 3)) \geq 2$. Therefore $\int s \geq 2 \cdot 2 + 3 \cdot 2 = 10 > 9$.

3.1.P5. Let $\sum \alpha_j$ be a series of nonnegative terms. Show that there is a measure μ on $\mathbb{N}$, every subset measurable, such that $\mu(A) = \sum_{j \in A} \alpha_j$ and $\int f = \sum_{j=1}^{\infty} \alpha_j f(j)$ for every nonnegative function f on $\mathbb{N}$. (Note that when each α_j is 1, this is the counting measure.)
Solution: Consider the sequence of measure spaces $(X_j, \mathcal{F}_j, \mu_j), j \in \mathbb{N}$, in which $X_j = \{j\} \in \mathbb{N}, \mathcal{F}_j$ is the only possible σ-algebra of subsets of a set containing a single element and $\mu_j(X_j) = \alpha_j$. Then $\cup_{j=1}^{\infty} X_j = \mathbb{N}$. By Problem 2.1.P10, there is a measure μ on $\mathbb{N}$ with domain $\mathcal{F} = \{A \subseteq X : A \cap X_j \in \mathcal{F}_j \text{ for each } j \in \mathbb{N}\}$ and for each $A \in \mathcal{F}, \mu(A) = \sum_{j=1}^{\infty} \mu_j(A \cap X_j)$. However, in the present situation, $\mathcal{F}$ is seen to consist of all subsets of $\mathbb{N}$, and moreover, $\sum_{j=1}^{\infty} \mu_j(A \cap X_j) = \sum_{j=1}^{\infty} \mu_j(A \cap \{j\}) = \sum_{j \in A} \alpha_j$.

Now let f be any nonnegative function on $\mathbb{N}$ but real-valued. For any $n \in \mathbb{N}$, define f_n to agree with f on $\{1, 2, \ldots, n\}$ and vanish outside the set $\{1, 2, \ldots, n\}$. Then each f_n is a simple nonnegative-valued function with integral $\int f_n = \sum_{j=1}^{n} \alpha_j f(j)$ by Proposition 3.1.7 or by Proposition 3.1.8. But we have $f_n(j) \leq f_{n+1}(j)$ and $f(j) = \lim_{n \to \infty} f_n(j)$ for every $j \in \mathbb{N}$. It follows by the Monotone Convergence Theorem that $\int f = \lim_{n \to \infty} \int f_n = \sum_{j=1}^{\infty} \alpha_j f(j)$.

Next, consider the possibility when f takes ∞ as a value. If for some j, we have $f(j) = \infty$ with $\alpha_j > 0$, then $\sum_{j=1}^{\infty} \alpha_j f(j) = \infty$ and also $\int f = \infty$, the latter because $f > K\chi_{\{j\}}$ for any $K > 0$, and $\int K\chi_{\{j\}} = K\mu(\{j\}) = K\alpha_j$. If there is no such j, then $\alpha_j = 0$ whenever $f(j) = \infty$. This means $\mu(A^c) = 0$, where $A \subseteq \mathbb{N}$ is the subset on which f is finite-valued. Define $\phi: \mathbb{N} \to \mathbb{N}$ to agree with f on A and be zero on A^c. Then by Problem 2.2.P7, $\int f = \int \phi$. But ϕ is nonnegative real-valued and therefore $\int \phi = \sum_{j=1}^{\infty} \alpha_j \phi(j)$. We shall argue that $\sum_{j=1}^{\infty} \alpha_j \phi(j) = \sum_{j=1}^{\infty} \alpha_j f(j)$. In fact, it is true that $\alpha_j \phi(j) = \alpha_j f(j)$ or each j, the reason being that, if $f(j) < \infty$, then $\phi(j) = f(j)$, and if $f(j) = \infty$, then $\alpha_j = 0$ and so, $\alpha_j \phi(j) = 0 \cdot \phi(j) = 0 = 0 \cdot \infty = \alpha_j f(j)$. We have thus shown that $\int f = \int \phi = \sum_{j=1}^{\infty} \alpha_j \phi(j) = \sum_{j=1}^{\infty} \alpha_j f(j)$.

3.1.P6. Show that the following sequence in $\mathbb{R}$ diverges:

$$\left\{\sum_{j=1}^{\infty} \frac{n}{j(n+j)}\right\}_{n\in\mathbb{N}}$$

Solution: Let μ be the measure on $\mathbb{N}$ such that $\mu(\{j\}) = \frac{1}{j^2}$. By Problem 3.1.P5, such a measure is possible and the integral $\int \phi$ of any nonnegative function ϕ on $\mathbb{N}$ is $\sum_{j=1}^{\infty} \frac{1}{j^2}\phi(j)$. Therefore the function $f_n(j) = \frac{nj}{n+j}$ has integral $\int f_n = \sum_{j=1}^{\infty} \frac{n}{j(n+j)}$. Now, $f_n(j) \leq f_{n+1}(j)$ and $\lim_{n\to\infty} f_n(j) = j$ for every $j \in \mathbb{N}$, and the limit function $f(j) = j$ has integral $\int f = \sum_{j=1}^{\infty} \frac{1}{j} = \infty$. By the Monotone Convergence Theorem, $\lim_{n\to\infty} \int f_n = \int f = \infty$.

3.1.P7. Let μ be a measure on X and let $A_1, \ldots, A_n$, where $n \geq 5$, be measurable subsets of X such that whenever $1 \leq j_1 < j_2 < j_3 < j_4 < j_5 \leq n$, the intersection $A_{j_1} \cap A_{j_2} \cap A_{j_3} \cap A_{j_4} \cap A_{j_5}$ of the corresponding subsets is empty. (In symbol-free terminology, the intersection of sets with any five distinct indices is empty.) Show that

$$\sum_i \mu(A_i) \leq \mu(A_1 \cup \cdots \cup A_n) + \sum_{1\leq j_1 < j_2 \leq n} \mu(A_{j_1} \cap A_{j_2}).$$

Solution: If $\mu(A_1 \cup \cdots \cup A_n) = \infty$, there is nothing to prove. So, assume the contrary. Then by the Inclusion-Exclusion Principle of Problem 3.1.P1, we have

$$\mu(A_1 \cup \cdots \cup A_n) = \sum_i \mu(A_i) - \sum_{1\leq j_1 < j_2 \leq n} \mu(A_{j_1} \cap A_{j_2})$$

$$+ \sum_{1\leq j_1 < j_2 < j_3 \leq n} \mu(A_{j_1} \cap A_{j_2} \cap A_{j_3}) - \sum_{1\leq j_1 < j_2 < j_3 < j_4 \leq n} \mu(A_{j_1} \cap A_{j_2} \cap A_{j_3} \cap A_{j_4}).$$

The sets $A_{j_1} \cap A_{j_2} \cap A_{j_3} \cap A_{j_4}$ are disjoint and their union is contained in

$$\bigcup_{1\leq j_1 < j_2 < j_3 \leq n} (A_{j_1} \cap A_{j_2} \cap A_{j_3}).$$

It follows that

$$\sum_{1\leq j_1 < j_2 < j_3 < j_4 \leq n} \mu(A_{j_1} \cap A_{j_2} \cap A_{j_3} \cap A_{j_4}) \leq \mu\left(\bigcup_{1\leq j_1 < j_2 < j_3 \leq n} (A_{j_1} \cap A_{j_2} \cap A_{j_3})\right)$$

$$\leq \sum_{1\leq j_1 < j_2 < j_3 \leq n} \mu(A_{j_1} \cap A_{j_2} \cap A_{j_3}).$$

Hence the sum of the last two summations in the above expression for $\mu(A_1 \cup \cdots \cup A_n)$ is greater than or equal to 0. Consequently,

$$\mu(A_1 \cup \cdots \cup A_n) \geq \sum_i \mu(A_i) - \sum_{1 \leq j_1 < j_2 \leq n} \mu\big(A_{j_1} \cap A_{j_2}\big),$$

which leads at once to the inequality sought to be proved.

3.1.P8. (a) If $A_1, \ldots, A_4$ are respectively $\{2, 3, 4\}$, $\{1, 3, 4\}$, $\{1, 2, 4\}$, $\{1, 2, 3\}$, then the intersection of all four sets is obviously empty. With counting measure μ, compute $\sum_{j=1}^{4} \mu(A_j)$ and $\mu(\cup_{j=1}^{4} A_j)$. Do the same when $A_1, \ldots, A_4$ are respectively $\{1, 2, 3, 4\}$, $\{1, 2, 4\}$, $\{1, 3\}$, $\{2, 3, 4\}$, which also have empty intersection.

(b) Give an example of five distinct sets $A_1, \ldots, A_5$ such that the intersection of any four among them is empty and $\sum_{j=1}^{5} \mu(A_j) = 3\mu\big(\cup_{j=1}^{5} A_j\big)$, where μ again means the counting measure.

Solution: (a) When $A_1, \ldots, A_4$ are respectively $\{2, 3, 4\}, \{1, 3, 4\}, \{1, 2, 4\}, \{1, 2, 3\}$, we have $\sum_{j=1}^{4} \mu(A_j) = 3 + 3 + 3 + 3 = 12$ and $\mu(\cup_{j=1}^{4} A_j) = 4$. When $A_1, \ldots, A_4$ are respectively $\{1, 2, 3, 4\}$, $\{1, 2, 4\}$, $\{1, 3\}$, $\{2, 3, 4\}$, we have $\sum_{j=1}^{4} \mu(A_j) = 4 + 3 + 2 + 3 = 12$ and $\mu(\cup_{j=1}^{4} A_j) = 4$.

(b) Let $A_1, \ldots, A_5$ be respectively $\{1, 2, 5\}, \{1, 2, 3\}, \{3, 4, 5\}, \{2, 4, 5\}, \{1, 3, 4\}$. It is a bit laborious to check that the intersection of any four among them is empty. It is simple to verify that $\sum_{j=1}^{5} \mu(A_j) = 3 + 3 + 3 + 3 + 3 = 15$ and $\mu(\cup_{j=1}^{5} A_j) = 5$.

3.1.P9. Let μ be a measure on X and let $A_1, \ldots, A_n$, where $n \geq 4$, be measurable subsets of X such that whenever $1 \leq j_1 < j_2 < j_3 < j_4 \leq n$, the intersection $A_{j_1} \cap A_{j_2} \cap A_{j_3} \cap A_{j_4}$ of the corresponding subsets is empty. (In symbol-free terminology, the intersection of sets with any four distinct indices is empty.) Suppose also that $\mu(A_1 \cup \cdots \cup A_n) < \infty$. Show that

$$\sum_j \mu(A_j) = \mu\big(\bigcup_j A_j\big) + \mu\Bigg(\bigcup_{1 \leq j_1 < j_2 \leq n} \big(A_{j_1} \cap A_{j_2}\big)\Bigg) + \sum_{1 \leq j_1 < j_2 < j_3 \leq n} \mu\big(A_{j_1} \cap A_{j_2} \cap A_{j_3}\big),$$

$$\sum_{1 \leq j_1 < j_2 \leq n} \mu\big(A_{j_1} \cap A_{j_2}\big) = \mu\Bigg(\bigcup_{1 \leq j_1 < j_2 \leq n} \big(A_{j_1} \cap A_{j_2}\big)\Bigg) + 2 \sum_{1 \leq j_1 < j_2 < j_3 \leq n} \mu\big(A_{j_1} \cap A_{j_2} \cap A_{j_3}\big)$$

and that

$$\sum_j \mu(A_j) \leq 3\mu\big(\bigcup_j A_j\big).$$

Solution: All summations, unions and intersections will be understood to be taken over all distinct indices. By the Inclusion-Exclusion Principle of Problem 3.1.P1, we have

$$\mu\left(\bigcup A_j\right) = \sum \mu(A_j) - \sum \mu(A_{j_1} \cap A_{j_2}) + \sum \mu(A_{j_1} \cap A_{j_2} \cap A_{j_3}).$$

The sets $A_{j_1} \cap A_{j_2}$ are indexed by two-element subsets $\{j_1,j_2\} \subseteq \{1,2,\ldots,n\}$ and may be denoted by B_α, where the index α ranges over the set of two-element subsets $\{j_1,j_2\} \subseteq \{1,2,\ldots,n\}$. Then

$$\sum \mu(B_\alpha) \text{ has the same meaning as } \sum \mu(A_{j_1} \cap A_{j_2})$$

and

$$\mu\left(\bigcup B_\alpha\right) \text{ has the same meaning as } \mu\left(\bigcup (A_{j_1} \cap A_{j_2})\right),$$

facts which will be used below. Now, the number of two-element subsets of a set of 3 elements is $^3C_2 = 3$ and therefore the union of four distinct two-element subsets of $\{1,2,\ldots,n\}$ must contain at least four elements. It follows that the intersection of sets B_α with four distinct α must be empty. Therefore, upon applying the Inclusion-Exclusion Principle of Problem 3.1.P1 to the sets $A_{j_1} \cap A_{j_2} = B_\alpha$, we get

$$\mu\left(\bigcup B_\alpha\right) = \sum \mu(B_\alpha) - \sum \mu(B_{\alpha_1} \cap B_{\alpha_2}) + \sum \mu(B_{\alpha_1} \cap B_{\alpha_2} \cap B_{\alpha_3}).$$

This equality may be written as

$$\sum \mu(A_{j_1} \cap A_{j_2}) = \sum \mu(B_\alpha)$$
$$= \mu\left(\bigcup B_\alpha\right) + \sum \mu(B_{\alpha_1} \cap B_{\alpha_2}) - \sum \mu(B_{\alpha_1} \cap B_{\alpha_2} \cap B_{\alpha_3}).$$

The summations on the right will be examined in this paragraph and the next. Consider any one term in the summation $\sum \mu(B_{\alpha_1} \cap B_{\alpha_2})$. We have $\alpha_1 = \{j_1,j_2\}$ and $\alpha_2 = \{j_3,j_4\}$ for some $j_1,j_2,j_3,j_4 \in \{1,2,\ldots,n\}$. If $\{j_1,j_2,j_3,j_4\}$ consists of four distinct elements, then $B_{\alpha_1} \cap B_{\alpha_2}$ is the intersection of sets A_j with four distinct indices and is therefore empty. Therefore the summation need extend only over those α_1, α_2 whose union consists of precisely three distinct indices from $\{1,2,\ldots,n\}$. If $\alpha_1 \cup \alpha_2 = \{j_1,j_2,j_3\}$, then $B_{\alpha_1} \cap B_{\alpha_2} = A_{j_1} \cap A_{j_2} \cap A_{j_3}$. This means that the term $\mu(B_{\alpha_1} \cap B_{\alpha_2})$ equals $\mu(A_{j_1} \cap A_{j_2} \cap A_{j_3})$. Moreover, there are precisely two other pairs of distinct α_1, α_2 whose union is $\{j_1,j_2,j_3\}$. We deduce from this that

$$\sum \mu(B_{\alpha_1} \cap B_{\alpha_2}) = 3 \sum \mu(A_{j_1} \cap A_{j_2} \cap A_{j_3}).$$

Next, consider any one term in the summation $\sum \mu(B_{\alpha_1} \cap B_{\alpha_2} \cap B_{\alpha_3})$. It can be nonzero only if $\alpha_1 \cup \alpha_2 \cup \alpha_3 = \{j_1, j_2, j_3\}$ for some distinct $j_1, j_2, j_3 \in \{1, 2, \ldots, n\}$. When this so, $B_{\alpha_1} \cap B_{\alpha_2} \cap B_{\alpha_3} = A_{j_1} \cap A_{j_2} \cap A_{j_3}$. We need consider only such terms. So, a term in the summation under discussion is equal to a uniquely determined term in the sum $\sum \mu(A_{j_1} \cap A_{j_2} \cap A_{j_3})$. Besides, the converse can be shown to be true. The converse is the assertion is that, given distinct $j_1, j_2, j_3 \in \{1, 2, \ldots, n\}$, the three distinct two-element sets $\alpha_1, \alpha_2, \alpha_3$ such that $\alpha_1 \cup \alpha_2 \cup \alpha_3 = \{j_1, j_2, j_3\}$ are uniquely determined. This is true because, being a three-element set, $\{j_1, j_2, j_3\}$ has only three distinct two-element subsets (i.e. $^3C_2 = 3$). Consequently,

$$\sum \mu(B_{\alpha_1} \cap B_{\alpha_2} \cap B_{\alpha_3}) = \sum \mu(A_{j_1} \cap A_{j_2} \cap A_{j_3}).$$

Combining the preceding three equalities, we obtain

$$\sum \mu(A_{j_1} \cap A_{j_2}) = \mu\left(\bigcup B_\alpha\right) + 2\sum \mu(A_{j_1} \cap A_{j_2} \cap A_{j_3})$$
$$= \mu\left(\bigcup (A_{j_1} \cap A_{j_2})\right) + 2\sum \mu(A_{j_1} \cap A_{j_2} \cap A_{j_3}).$$

This is the second of the inequalities asked for in the problem. Combining it with the very first equality derived above from the Inclusion-Exclusion Principle, we further obtain

$$\sum_j \mu(A_j) = \mu\left(\bigcup A_j\right) + \mu\left(\bigcup (A_{j_1} \cap A_{j_2})\right) + \sum \mu(A_{j_1} \cap A_{j_2} \cap A_{j_3}).$$

This proves the first equality that was asked for in the problem.

To arrive at the inequality, we need only show that the second and third terms on the right side in the equality that has just been proved are each bounded above by the first. For the second term, this is obvious because the union therein is a subset of the union in the first term. The third term is handled as follows. The sets whose measures are being taken are disjoint and therefore

$$\sum \mu(A_{j_1} \cap A_{j_2} \cap A_{j_3}) = \mu\left(\bigcup (A_{j_1} \cap A_{j_2} \cap A_{j_3})\right).$$

However,

$$\bigcup (A_{j_1} \cap A_{j_2} \cap A_{j_3}) \subseteq \bigcup A_j.$$

3.2 Other Measurable Functions

3.2.P1. Let μ be a measure on X and f, g be integrable functions such that $f \geq g$ everywhere and $\int f = \int g$. Prove $\mu(X(f > g)) = 0$.

Solution: The function $\phi = f - g$ is nonnegative and, by Theorem 3.2.5, $\int (f - g) = \int f - \int g$, which is given to be 0. The result now follows by Problem 2.3.P1.

3.2.P2. Let $X = \{a, b\}$ with counting measure. Let $\{f_n\}_{n \geq 1}$ be the sequence of functions on X defined by $f_n = \chi_{\{a\}}$ if n is even and $\chi_{\{b\}}$ if n is odd. Find

$$\int \left(\lim_{n \to \infty} \inf f_n \right) \quad \text{and} \quad \lim_{n \to \infty} \inf \int f_n.$$

Solution: For each $x \in X$, the sequence $\{f_n(x)\}_{n \geq 1}$ alternates between 0 and 1, starting with 0 or 1 depending on whether $x = a$ or $x = b$. Hence $\lim \inf_{n \to \infty} f_n$ is 0 everywhere and therefore has integral 0. But $\int f_n = 1$ for every n and therefore $\lim \inf_{n \to \infty} \int f_n = 1$.

3.2.P3. In this problem, $(X, \mathcal{F}, \mu)$ is a measure space and f, g are measurable real-valued functions on X. Then the nonnegative real-valued function $|fg|$ is measurable by Theorem 2.2.3(c) and Proposition 2.2.2(b); also, for any real $p > 0$ and $q > 0$, the real-valued functions $|f + g|$, $|f + g|^p$, $|f|^p$ and $|g|^p$ are measurable by Theorem 2.2.3(b), Proposition 2.2.2(b) and Problem 2.2.P6(a). If $p > 1$, then $|f + g|^{p-1}$ is also measurable.

(a) Suppose p, q are positive real numbers such that $\frac{1}{p} + \frac{1}{q} = 1$. Show that

$$\int |fg| \leq \left(\int |f|^p \right)^{\frac{1}{p}} \left(\int |g|^q \right)^{\frac{1}{q}}.$$

This is known as **Hölder's Inequality**.

(b) Suppose p is a real number such that $p \geq 1$. Show that

$$\left(\int |f + g|^p \right)^{\frac{1}{p}} \leq \left(\int |f|^p \right)^{\frac{1}{p}} + \left(\int |g|^p \right)^{\frac{1}{p}}.$$

This is known as **Minkowski's Inequality**.

Solution: (a) If one of the integrals on the right side is 0, then by Problem 2.2.P6, the integral on the left side is also 0; therefore we may assume that both the integrals on the right side are positive. Then if one of them is ∞, the right side is ∞, and there is nothing to prove. So, we also assume both of them to be finite.

Set $F = |f|/(\int |f|^p)^{1/p}$ and $G = |g|/(\int |g|^p)^{1/q}$. Then by homogeneity (second part of Theorem 3.2.2), $\int F^p = 1 = \int G^q$. By the inequality

$$ab \le \frac{1}{p}a^p + \frac{1}{q}b^q \text{ for any nonnegative } a \text{ and } b$$

(easily proved by considering the derivative of $\phi(\xi) = 1/q + \xi/p - \xi^{1/p}$ for $\xi \in (0, \infty)$), we have $FG \le F^p/p + G^q/q$. Employing Proposition 2.2.8 and both parts of Theorem 3.2.2, we obtain $\int FG \le \frac{1}{p} + \frac{1}{q} = 1$. This leads to the required inequality upon using the second part of Theorem 3.2.2 once again.

(b) If $p = 1$, then the required inequality is a straightforward consequence of Proposition 2.2.8. So, suppose $p > 1$. We need argue only the case when both integrals on the right side are finite. When this is so, the inequality

$$(a+b)^{p+1} \le 2^p \left(a^{p+1} + b^{p+1} \right) \text{ for all } a, b, p \ge 0$$

(easily proved by considering the derivative of $\phi(x) = (x+b)^{p+1} - 2^p(x^{p+1} + b^{p+1})$, $x \in [0, b]$) shows that $|f+g|^p \le 2^{p-1}(|f|^p + |g|^p)$ and hence by Proposition 2.2.8 and Theorem 3.2.2, $\int |f+g|^p$ is also finite, a fact needed later in this proof. Since

$$|f+g|^p = |f+g|^{p-1}|f+g| \le |f+g|^{p-1}(|f| + |g|),$$

we have $\int |f+g|^p \le \int |f+g|^{p-1}|f| + \int |f+g|^{p-1}|g|$. Upon applying Hölder's Inequality of part (a) here, with $q = \frac{p}{p-1}$, so that $\frac{1}{p} + \frac{1}{q} = 1$ and $pq - q = p$, we get

$$\int |f+g|^{p-1}|f| \le \left(\int |f|^p \right)^{\frac{1}{p}} \left(\int |f+g|^{q(p-1)} \right)^{\frac{1}{q}} = \left(\int |f|^p \right)^{\frac{1}{p}} \left(\int |f+g|^p \right)^{\frac{1}{q}}.$$

Similarly, $\int |f+g|^{p-1}|g| \le \left(\int |g|^p \right)^{\frac{1}{p}} \left(\int |f+g|^p \right)^{\frac{1}{q}}$. These two inequalities jointly imply that $\int |f+g|^p \le \left(\left(\int |f|^p \right)^{\frac{1}{p}} + \left(\int |g|^p \right)^{\frac{1}{p}} \right) \left(\int |f+g|^p \right)^{\frac{1}{q}}$. As we have already noted that the integral on the left side is finite, we may divide both sides by $\left(\int |f+g|^p \right)^{\frac{1}{q}}$, whereupon the desired inequality follows.

3.2.P4. (Needed in Proposition 7.7.5) Let f be an integrable function on X, where $(X, \mathcal{F}, \mu)$ is a measure space. Given any $\varepsilon > 0$, show that there exists a $\delta > 0$ such that every $A \in \mathcal{F}$ with $\mu(A) < \delta$ satisfies $\left| \int (\chi_A f) \right| < \varepsilon$.

Solution: Since $\left| \int f \right| \le \int |f|$, we may assume $f \ge 0$. By Theorem 3.2.1, there exists an increasing sequence $\{s_n\}_{n \ge 1}$ of simple functions such that $s_n(x) \to f(x)$ as $n \to \infty$ for every $x \in X$. By the Monotone Convergence Theorem 2.3.2, $\int f = \lim_{n \to \infty} s_n$. Fix any $N \in \mathbb{N}$ such that $0 \le \int (f - s_N) < \frac{\varepsilon}{2}$. Then

$$0 \le \int (\chi_A f) - \int (\chi_A s_N) \le \int (f - s_N) < \frac{\varepsilon}{2} \quad \text{for any } A \in \mathcal{F},$$

because $0 \le \chi_A(f - s_N) \le f - s_N$ everywhere. Being a simple function, s_N is bounded above by some positive real number K. Set $\delta = \frac{\varepsilon}{2K}$. Then any $A \in \mathcal{F}$ with

$\mu(A) < \delta$ satisfies $0 \le \int (\chi_A s_N) \le K \int \chi_A = K\mu(A) < K\delta < \frac{\varepsilon}{2}$. Together with the inequality displayed above, this leads to $0 \le \int (\chi_A f) < \varepsilon$.

3.2.P5. Show that the "dominated" hypothesis of the Dominated Convergence Theorem 3.2.6, namely that, for some integrable g, the inequality $|f_n(x)| \le g(x)$ holds for every $x \in X$, cannot be dropped.

Solution: The same example as in Example 2.3.3 for the Monotone Convergence Theorem will serve the purpose.

3.2.P6. For each fixed $n \in \mathbb{N}$, the series $\sum_{j=1}^{\infty} \frac{1}{n}\left(\frac{n}{n+1}\right)^{j-1}$ converges to $\frac{n+1}{n}$. Show that any series $\sum_{j=1}^{\infty} a_j$ satisfying $a_j \ge \frac{1}{n}\left(\frac{n}{n+1}\right)^{j-1}$ for every $n \in \mathbb{N}$ and every $j \in \mathbb{N}$ diverges.

Solution: Consider the measure μ on $\mathbb{N}$ such that $\mu(\{j\}) = \frac{1}{j^2}$. By Problem 3.1.P5, such a measure is possible and the integral $\int \phi$ of any nonnegative function ϕ on $\mathbb{N}$ is $\sum_{j=1}^{\infty} \frac{1}{j^2}\phi(j)$. Therefore the function $f_n(j) = j^2 \frac{1}{n}\left(\frac{n}{n+1}\right)^{j-1}$ has integral $\int f_n = \frac{n+1}{n}$ and the function $g(j) = j^2 a_j$ has integral $\int g = \sum_{j=1}^{\infty} a_j$. If this series were to be convergent, then g would be integrable and it would follow from the inequality $a_j \ge \frac{1}{n}\left(\frac{n}{n+1}\right)^{j-1}$ and the Dominated Convergence Theorem 3.2.6 that $\frac{n+1}{n} \to \int \lim_{n \to \infty} f_n$. However, this not the case, because $\lim_{n \to \infty} f_n$ is 0 everywhere. This contradiction shows that the series $\sum_{j=1}^{\infty} a_j$ cannot converge if the given inequality holds.

3.2.P7. Show that the following sequence converges to 0:

$$\left\{ \sum_{j=1}^{\infty} \frac{n}{j(n^2 + j^2)} \right\}_{n \in \mathbb{N}}$$

Solution: Consider the same measure as in the solution of Problem 3.2.P6 but with functions f_n and g given by

$$f_n(j) = j^2 \frac{n}{j(n^2 + j^2)} \quad \text{and} \quad g(j) = 1.$$

Then $\lim_{n \to \infty} f_n$ is 0 everywhere. The integral $\int \phi$ of any nonnegative function ϕ on $\mathbb{N}$ is $\sum_{j=1}^{\infty} \frac{1}{j^2}\phi(j)$. Therefore $\int f_n = \sum_{j=1}^{\infty} \frac{n}{j(n^2+j^2)}$ and $\int g = \sum_{j=1}^{\infty} \frac{1}{j^2}$. In particular, g is integrable. Also, $0 \le f_n(j) \le g(j)$ for all n and j. It follows by the Dominated Convergence Theorem 3.2.6 that the given sequence converges to $\int \lim_{n \to \infty} f_n$, which is 0.

3.2.P8. Given any measure space, show that a nonnegative measurable function f satisfies

$$\int f = \sup \left\{ \int s : 0 \le s \le f, \text{ where } s \text{ is simple} \right\}.$$

Hint: Use Theorem 3.2.1 and the Monotone Convergence Theorem 2.3.2.

Solution: Clearly, $\int f$ cannot be less than the supremum. By Theorem 3.2.1 and the Monotone Convergence Theorem 2.3.2, there exists a sequence of simple functions s_n such that $0 \le s_n \le f$ and $\int f = \lim_{n \to \infty} \int s_n$. Therefore $\int f$ cannot be greater than the supremum.

3.2.P9. Given any measure space $(X, \mathcal{F}, \mu)$, let $\mathcal{S}$ denote the class of nonnegative simple functions s vanishing outside a set of finite measure, i.e. $\mu(X(s > 0)) < \infty$. This is obviously equivalent to $\int s < \infty$.

(a) If every set of infinite measure contains a subset of finite positive measure, show for any nonnegative measurable function f that

$$\int f = \sup\left\{ \int s \colon 0 \le s \le f, \ s \in \mathcal{S}. \right\}$$

(b) Prove the converse of (a).

Hint:
(a) Use Problem 2.3.P3.
(b) Take f to be the characteristic function of any given set of infinite measure; the condition yields $s \in \mathcal{S}$ with $\int s > 0$. Now consider any set of positive measure on which s takes a positive value.

Solution: (a) If $\int f < \infty$, the equality is trivial. So, suppose $\int f = \infty$. We need only show that there exists some $s \in \mathcal{S}$ such that $0 \le s \le f$ and $\int s$ is arbitrarily large.

Consider any $M > 0$. Since $\int f = \infty$, we know from Problem 3.2.P8 that there exists a simple function s satisfying $0 \le s \le f$ and $\int s > M$. If $\int s < \infty$, then $s \in \mathcal{S}$ and we already have the required s. Suppose $\int s = \infty$. Then for some finite $\alpha > 0$, we have $\mu(X(s = \alpha)) = \infty$. By Problem 2.3.P3, the set must contain a subset A for which $\infty > \mu(A) > M/\alpha$. Then $s\chi_A = \alpha\chi_A$ is a nonnegative simple function not exceeding s and hence not exceeding f. Also, $\int s\chi_A = \alpha\mu(A) > M$ and is finite, so that $s\chi_A \in \mathcal{S}$. Thus $s\chi_A$ serves our purpose.

(b) For the converse, consider a measurable set A with $\mu(A) = \infty$. Apply the given condition with χ_A as f. We obtain a function $s \in \mathcal{S}$ such that $0 \le s \le \chi_A$ and $\int s > 0$. Now, s must take a positive value on some set of positive measure and that set must not only be a subset of A because $0 \le s \le \chi_A$ but must also have finite measure because $s \in \mathcal{S}$.

3.2.P10. (a) Let μ be a measure on a set X. Suppose f is a nonnegative real-valued function on X and $\{f_n\}_{n \ge 1}$ a sequence of measurable nonnegative real-valued functions such that

$$\text{(i) } f_n(x) \ge f_{n+1}(x) \quad \text{and} \quad \text{(ii) } f(x) = \lim_{n \to \infty} f_n(x) \quad \text{for every } x \in X.$$

Then f is measurable, and if $\int f_N < \infty$ for some N, then $\int f = \lim_{n \to \infty} \int f_n$.

(b) Show that in part (a), the hypothesis that $\int f_N < \infty$ for some N cannot be dropped even if $\mu(X)$ is finite and the only subset of measure 0 is the empty set.

Solution: (a) Measurability of f is a simple consequence of Problem 2.2.P4. We shall assume, as is legitimate, that $N = 1$. The functions $f_n - f$ are nonnegative real-valued and satisfy $|f_n - f| \le f_1$. Since $\int f_1 < \infty$, the Dominated Convergence Theorem 3.2.6 yields the desired conclusion.

(b) Consider the measure μ on $\mathbb{N}$ such that $\mu(\{j\}) = \frac{1}{j^2}$. By Problem 3.1.P5, such a measure is possible and the integral $\int \phi$ of any nonnegative function ϕ on $\mathbb{N}$ is $\sum_{j=1}^{\infty} \frac{1}{j^2}\phi(j)$. Moreover, $\mu(X)$ is finite and the only subset of measure 0 is the empty set. Define f_n on $\mathbb{N}$ by $f_n(j) = j^2(j^{\frac{1}{n}} - 1)$. Then each f_n is nonnegative real-valued and measurable. Moreover, each j satisfies $f_n(j) \ge f_{n+1}(j)$ for each n as well as $\lim_{n\to\infty} f_n(j) = 0$. So, $f = \lim_{n\to\infty} f_n$ is 0 everywhere and has integral 0. However, the integral of f_n is $\sum_{j=1}^{\infty} (j^{\frac{1}{n}} - 1)$ which is ∞ because, for each n, the jth term $(j^{\frac{1}{n}} - 1)$ of the series tends to ∞ as $j \to \infty$. So, $\lim_{n\to\infty} \int f_n = \infty \ne \int f$.

3.2.P11. Let the function $f: [0,1] \to \mathbb{R}$ be given by $f(x) = x^{\frac{1}{2}}$. In the notation of the proof of Theorem 3.2.1, find the set $E_{3,9}$ and the number $s_3(\frac{3}{4})$.
Solution: Since $2^3 = 8$ and $9 - 1 = 8$, the set $E_{3,9}$ is the inverse image $f^{-1}([\frac{8}{8}, \frac{9}{8}])$, which is $\{1\}$, because $f(1) = 1$ and $f(x) < 1$ when $1 \ne x \in [0,1]$.

To evaluate $s_3(\frac{3}{4})$, we need to know the integer k for which $1 \le k \le 3 \cdot 2^3$ and $\frac{3}{4} \in E_{3,k}$ i.e. $\frac{1}{8}(k-1) \le (\frac{3}{4})^{\frac{1}{2}} < \frac{1}{8}k$. This inequality is equivalent to $k - 1 \le 4 \cdot 3^{\frac{1}{2}} < k$. Therefore $k - 1$ is the integer part of $4 \cdot 3^{\frac{1}{2}}$, which can be determined to be 6. Thus $s_3(\frac{3}{4}) = \frac{6}{8} = \frac{3}{4}$. To determine the integer part of $4 \cdot 3^{\frac{1}{2}}$, we need only note that $(4 \cdot 3^{\frac{1}{2}})^2 = 16 \cdot 3 = 48$, which lies between 6^2 and 7^2.

3.3 Subadditive Fuzzy Measures

3.3.P1. Let $X = \mathbb{N}$ and $\mathcal{F}$ be the σ-algebra of all subsets of X. Take $\tilde{v}$ to be defined by $\tilde{v}(E) = \sum_{i=1}^{k} \beta^i$, where $\beta \in (0,1)$ is fixed and k is the number of elements in E. (If $E = \emptyset$, then $k = 0$ and the summation becomes "empty", in which case its value is understood to be 0 as usual.) Show that $\tilde{v}$ is a subadditive fuzzy measure that is not a measure. Is it inner continuous?
Solution: It is immediate from the definition of $\tilde{v}$ that $\tilde{v}\emptyset = 0$ and $\tilde{v}(E) \ge 0$ for all E. If $E \subseteq F$, then F has at least as many elements in it as E, which implies that $\tilde{v}(E) \le \tilde{v}(F)$. Therefore $\tilde{v}$ is monotone and hence a fuzzy measure. Now consider any two subsets E and F. If one of them contains infinitely many elements, there is nothing to prove. Suppose they contain k and ℓ elements respectively. Then $E \cup F$ contains at most $j = k + \ell$ elements and therefore $\tilde{v}(E \cup F) \le \sum_{i=1}^{j} \beta^i = \sum_{i=1}^{k} \beta^i + \sum_{i=1}^{\ell} \beta^{i+k} \le \sum_{i=1}^{k} \beta^i + \sum_{i=1}^{\ell} \beta^i = \tilde{v}(E) + \tilde{v}(F)$. This shows that $\tilde{v}$ is subadditive.

Let E and F consist of one element each, $E \neq F$. Then $E \cap F = \emptyset$, $\tilde{v}(E) = \beta = \tilde{v}(F)$ and $\tilde{v}(E \cup F) = \beta + \beta^2 \neq \beta + \beta = \tilde{v}(E) + \tilde{v}(F)$. This establishes that $\tilde{v}$ is not a measure.

We shall argue that $\tilde{v}$ is inner continuous. Consider a sequence of subsets E_n of X such that $E_n \subseteq E_{n+1}$ for each n and denote their union by E. If E is finite, then there is some p such that $E_p = E$ and it is apparent that $\tilde{v}(E) = \tilde{v}(E_p) = \lim_{n \to \infty} \tilde{v}(E_n)$. If on the other hand E is infinite, then $\tilde{v}(E) = \sum_{i=1}^{\infty} \beta^i$, while at the same time, $\lim_{n \to \infty} \tilde{v}(E_n) = \sum_{i=1}^{\infty} \beta^i$ because the number of elements in E_n must tend to ∞ as $n \to \infty$.

3.3.P2. (Cf. Problem 3.3.P6) Suppose $\tilde{\mu}$ is a subadditive fuzzy measure on a set X and $f: X \to \mathbb{R}^{*+}$, $g: X \to \mathbb{R}^{*+}$ are measurable functions. If $fg = 0$ everywhere, show that $\int_{\sim} f + g \leq \int_{\sim} f + \int_{\sim} g$.

Solution: We begin as in the proof of Theorem 3.3.5. If either $\int_{\sim} f$ or $\int_{\sim} g$ is ∞, there is nothing to prove. So, we may assume both to be finite. The hypothesis that $fg = 0$ everywhere provides that f and g cannot both be nonzero at the same point, and therefore

$$X(f + g > t) \subseteq X(f > t) \cup X(g > t).$$

The monotonicity and subadditivity of $\tilde{\mu}$ now lead to

$$\widetilde{(f+g)}(t) \leq \tilde{f}(t) + \tilde{g}(t).$$

Upon integrating from 0 to ∞, we arrive at the required inequality.

3.3.P3. Given a subadditive fuzzy measure $\tilde{\mu}$ on a σ-algebra $\mathcal{F}$ of subsets of a set X and a nonnegative measurable function $f: X \to \mathbb{R}^{*+}$, prove that the function $\tilde{v}$ defined on all sets $E \in \mathcal{F}$ by setting

$$\tilde{v}(E) = \int_{\sim} f\chi_E \, d\tilde{\mu}$$

is a subadditive fuzzy measure. Prove also that (i) $\tilde{v}(E) = 0$ whenever $\tilde{\mu}(E) = 0$ and (ii) $\tilde{v}$ is a measure (i.e. is countably additive) whenever $\tilde{\mu}$ is.

Solution: That $\tilde{v}(E) \geq 0$ and $\tilde{v}(\emptyset) = 0$ need no argument. Since $A \subseteq B$ implies $\chi_A \leq \chi_B$, which further implies $f\chi_A \leq f\chi_B$, leading to $\int_{\sim} f\chi_A \, d\tilde{\mu} \leq \int_{\sim} f\chi_B \, d\tilde{\mu}$ when $A, B \in \mathcal{F}$, we find that $\tilde{v}$ is monotone.

As regards subadditivity, first consider disjoint sets $A, B \in \mathcal{F}$. In this case, $\chi_A + \chi_B = \chi_{A \cup B}$ and the product $(f\chi_A)(f\chi_B)$ is 0 everywhere. It follows that

$$\tilde{v}(A \cup B) = \int_{\sim} f\chi_{A \cup B} \, d\tilde{\mu} = \int_{\sim} (f\chi_A + f\chi_B) d\tilde{\mu}$$

$$\leq \int_{\sim} f\chi_A \, d\tilde{\mu} + \int_{\sim} f\chi_B \, d\tilde{\mu} \text{ by Problem 3.3.P2}$$

$$\leq \tilde{v}(A) + \tilde{v}(B).$$

When A, B are not disjoint, the sets $A \backslash B$ and B are surely disjoint and have union $A \cup B$. Therefore we can reason as follows:

$$\tilde{v}(A \cup B) = \tilde{v}((A \backslash B) \cup B) \leq \tilde{v}(A \backslash B) + \tilde{v}(B) \leq \tilde{v}(A) + \tilde{v}(B),$$

where the final step invokes the already established motonicity of $\tilde{v}$.

Since $f\chi_E$ can take a nonzero value only on E, the hypothesis $\tilde{\mu}(E) = 0$ implies that $\widetilde{f\chi_E} = 0$ everywhere on $[0, \infty)$ and hence that $\tilde{v}(E) = 0$. This proves (i).

Now suppose $\tilde{\mu}$ is a measure and consider a sequence $\{A_n\}_{n \geq 1}$ of disjoint sets, each belonging to $\mathcal{F}$. Denote the union $\cup_{n=1}^{\infty} A_n$ by A and $\cup_{n=1}^{N} A_n$ by B_N. To prove (ii), we have to demonstrate that $\tilde{v}(A) = \sum_{n=1}^{\infty} \tilde{v}(A_n)$. On account of the disjointness of the sets A_n, we have $\sum_{n=1}^{N} f\chi_{A_n} = f\chi_{B_N}$ and hence $\int_{\sim} f\chi_{B_N} d\tilde{\mu} = \int_{\sim} (\sum_{n=1}^{N} f\chi_{A_n}) d\tilde{\mu}$. Since $\tilde{\mu}$ is a measure, the integral is additive by Theorem 3.2.2, and therefore we further have

$$\int_{\sim} f\chi_{B_N} \, d\tilde{\mu} = \sum_{n=1}^{N} \int_{\sim} f\chi_{A_n} \, d\tilde{\mu}.$$

The definition of A and B_N also has the consequence that $f\chi_{B_N} \leq f\chi_{B_{N+1}}$ for every N, and that $\lim_{N \to \infty} (f\chi_{B_N})(x) = (f\chi_A)(x)$ for every $x \in X$ (cf. Problem 2.1.P1), so that the Monotone Convergence Theorem 2.2.3 yields

$$\lim_{N \to \infty} \int_{\sim} f\chi_{B_N} \, d\tilde{\mu} = \int_{\sim} f\chi_A \, d\tilde{\mu}.$$

Upon combining this with the equality displayed further above, we obtain $\int_{\sim} f\chi_A d\tilde{\mu} = \sum_{n=1}^{\infty} \int_{\sim} f\chi_{A_n} d\tilde{\mu}$, which, by the definition of $\tilde{v}$, is the same as what we had to demonstrate.

3.3.P4. Let $\tilde{\mu}$ and $\tilde{v}$ be finite subadditive fuzzy measures on a σ-algebra $\mathcal{F}$ of subsets of a set X such that $\tilde{v}(E) = 0$ whenever $\tilde{\mu}(E) = 0$. Must there exist a nonnegative measurable function $f: X \to \mathbb{R}^{*+}$ satisfying $\tilde{v}(E) = \int_{\sim} f\chi_E d\tilde{\mu}$ for each $E \in \mathcal{F}$?

Solution: No. Let $\gamma \in (0, 1)$ be fixed and consider the measure $\tilde{\mu}$ on the σ-algebra of all subsets of $\mathbb{N}$ as in Problem 3.1.P5 with every $\alpha_j = \gamma^j$. This measure satisfies $\tilde{\mu}(\mathbb{N}) = \sum_{j \in \mathbb{N}} \gamma^j < \infty$ and also satisfies $\tilde{\mu}(E) = 0 \Rightarrow E = \emptyset$. It follows for any fuzzy measure $\tilde{\nu}$ that $\tilde{\nu}(E) = 0$ whenever $\tilde{\mu}(E) = 0$. If a function $f: X \to \mathbb{R}^{*+}$ of the kind in question were to exist, it would follow from Problem 3.3.P3 that $\tilde{\nu}$ must be a measure. However, the example in Problem 3.3.P1 shows that $\tilde{\nu}$ need not be a measure.

3.3.P5. Let $\alpha, \beta > 0$ and $s > 1$. For any x such that $\max\{\alpha, \beta\} \leq x \leq \alpha + \beta$, show that

$$\left((1+s)^3 x\right)^{\frac{1}{4}} \leq \left(x + (s^3 - 1)\alpha\right)^{\frac{1}{4}} + \left(x + (s^3 - 1)\beta\right)^{\frac{1}{4}}.$$

Solution: Define $\tilde{\mu}$ on a two element set $\{a, b\}$ as $\tilde{\mu}\emptyset = 0$, $\tilde{\mu}\{a\} = \alpha$, $\tilde{\mu}\{b\} = \beta$ and $\tilde{\mu}\{a, b\} = x$. Then $\tilde{\mu}$ is a subadditive fuzzy measure. Consider the functions f and g on $\{a, b\}$ such that $f(a) = 1, f(b) = s$ and $g(a) = s, g(b) = 1$. Then $f + g$ takes the value $1 + s$ at both points and hence $\int_{\sim} (f + g)^3 = (1 + s)^3 x$. Also,

$$\tilde{f^3}(t) = \begin{cases} x & 0 \leq t < 1 \\ \beta & 1 \leq t < s^3 \\ 0 & s^3 \leq t < \infty \end{cases} \quad \text{and} \quad \tilde{g^3}(t) = \begin{cases} x & 0 \leq t < 1 \\ \alpha & 1 \leq t < s^3 \\ 0 & s^3 \leq t < \infty. \end{cases}$$

Therefore $\int_{\sim} f^3 = x + (s^3 - 1)\beta$ and $\int_{\sim} g^3 = x + (s^3 - 1)\alpha$. On applying Theorem 3.3.5 with $p = 3$, we arrive at the desired inequality.

3.3.P6. (Cf. Problem 3.3.P2) Let $\tilde{\mu}$ be a subadditive fuzzy measure on a two-element set $X = \{a, b\}$, the σ-algebra $\mathcal{F}$ being the family of all four subsets of X. For any functions $f: X \to \mathbb{R}^{*+}$, $g: X \to \mathbb{R}^{*+}$, show that $\int_{\sim} (f + g) \leq \int_{\sim} f + \int_{\sim} g$. (Consider only the nontrivial case that $\tilde{\mu}\{a\} > 0$, $\tilde{\mu}\{b\} > 0$ and $\tilde{\mu}\{a, b\} < \infty$.)

Solution: Since $0 < \tilde{\mu}\{a\} \leq \tilde{\mu}\{a, b\} < \infty$, we may assume $\tilde{\mu}\{a, b\} = 1$. Then the numbers $\alpha = \tilde{\mu}\{a\}$ and $\beta = \tilde{\mu}\{b\}$ satisfy $\alpha > 0$, $\beta > 0$ and $\alpha + \beta \geq 1$ (subadditivity of $\tilde{\mu}$). Denote $f(a)$ by $x \geq 0$ and $f(b)$ by $y \geq 0$. Also, denote $g(a)$ by $\xi \geq 0$ and $g(b)$ by $\eta \geq 0$. Then

$$\tilde{f}(t) = \begin{cases} 1 & 0 \leq t < x \\ \beta & x \leq t < y \\ 0 & y \leq t < \infty \end{cases} \text{if } x \leq y \quad \text{and} \quad \tilde{f}(t) = \begin{cases} 1 & 0 \leq t < y \\ \alpha & y \leq t < x \\ 0 & x \leq t < \infty. \end{cases} \text{if } y \leq x$$

(When $y = x$, both descriptions of $\tilde{f}$ hold, the middle line being vacuous.) Similarly for g, but with x and y replaced by ξ and η respectively. From the above computation of $\tilde{f}$, we obtain

$$\int_{\sim} f = \begin{cases} x + \beta(y - x) & \text{if } x \leq y \\ y + \alpha(x - y) & \text{if } x \geq y \end{cases} \quad \text{and} \quad \text{similarly for} \int_{\sim} g.$$

Suppose $x \leq y$. If also $\xi \leq \eta$ and hence $x + \xi \leq y + \eta$, then it is straightforward that $\int_{\sim} (f + g) = \int_{\sim} f + \int_{\sim} g$. So, suppose $\xi \geq \eta$ instead. Then we have

$$\int_{\sim} f = x + \beta(y - x) \quad \text{and} \quad \int_{\sim} g = \eta + \alpha(\xi - \eta).$$

There are two cases to consider: $x + \xi \le y + \eta$ and $x + \xi \ge y + \eta$.

First consider the case when $x + \xi \le y + \eta$. This inequality has the consequence that

$$\int_{\sim} (f + g) = x + \xi + \beta(y + \eta - x - \xi).$$

Keeping in view that $\alpha + \beta \ge 1$, i.e. $\alpha \ge 1 - \beta$, and that $\xi - \eta \ge 0$, we reason that

$$\int_{\sim} f + \int_{\sim} g = x + \beta(y - x) + \eta + \alpha(\xi - \eta)$$
$$\ge x + \beta(y - x) + \eta + (1 - \beta)(\xi - \eta)$$

and the smaller side of this inequality simplifies to

$$x + \beta(y - x) + \eta + (\xi - \eta) - \beta(\xi - \eta) = x + \xi + \beta((y - x) - (\xi - \eta))$$
$$= x + \xi + \beta(y + \eta - x - \xi)$$
$$= \int_{\sim} (f + g).$$

Next, consider the case when $x + \xi \ge y + \eta$. Analogously to the previous case, this inequality has the consequence that

$$\int_{\sim} (f + g) = y + \eta + \alpha(x + \xi - y - \eta).$$

Keeping in view that $\alpha + \beta \ge 1$, i.e. $\beta \ge 1 - \alpha$, and that $y - x \ge 0$, we reason that

$$\int_{\sim} f + \int_{\sim} g = x + \beta(y - x) + \eta + \alpha(\xi - \eta)$$
$$\ge x + (1 - \alpha)(y - x) + \eta + \alpha(\xi - \eta)$$

and the smaller side of this inequality simplifies to

$$x + (y - x) - \alpha(y - x) + \eta + \alpha(\xi - \eta) = y + \eta + \alpha((\xi - \eta) - (y - x))$$
$$= y + \eta + \alpha(x + \xi - y - \eta)$$
$$= \int_{\sim} (f + g).$$

This proves the required inequality when $x \leq y$. In the event that $x \geq y$, an entirely analogous argument goes through.

4.1 Lebesgue Outer Measure

4.1.P1. The outer measure m^* on $[a, b]$ is defined on the σ-algebra $\mathcal{F}$ of all subsets of $[a, b]$ and therefore all functions are measurable and have a distribution. Find the distribution $\tilde{f}$ of the function $f: [a, b] \to \mathbb{R}$ such that $f(x) = 1$ if x is irrational and 0 otherwise and hence find $\int_0^{\sup f} \tilde{f}$. (The reader is reminded that the Riemann integral $\int_a^b f$ does not exist.)

Solution: The function f is the characteristic function of the set A of all irrational numbers in $[a, b]$. By Exercise 4.1.10, its outer measure is $b - a$. Therefore $\tilde{f}(t) = b - a$ if $0 \leq t < 1$ and 0 if $1 \leq t$. Lastly, $\int_0^{\sup f} \tilde{f} = b - a$.

4.1.P2. For nonempty $A \subseteq \mathbb{R}$, the intersection I_A of all intervals containing A is a nonempty interval. Show that $m^*(A) \leq \sup\{|x - y|: x, y \in A\} = \ell(I_A)$.

Solution: Since $A \subseteq I_A$, we have $m^*(A) \leq m^*(I_A) = \ell(I_A)$. We need show only that $\ell(I_A) = \sup\{|x - y|: x, y \in A\} = s$, say. Observe that $s = 0$ (resp. ∞) if and only if $\ell(I_A) = 0$ (resp. ∞). So we take $0 < s < \infty$. Then I_A contains more than one number and is bounded. Denote its left and right endpoints by a and b respectively. Then $[a, b] \supseteq I_A \supseteq A$. So, we have $\ell(I_A) \leq |x - y|$ for all $x, y \in A$. This yields $\ell(I_A) \geq s$. In order to prove the reverse inequality, consider an arbitrary $\varepsilon > 0$ and let $\eta = \min\{\varepsilon, b - a\} > 0$. The interval $[a + \eta, b]$ does not contain I_A and hence does not contain A, but $[a, b]$ does. Therefore there exists an $x \in A$ such that $a \leq x < a + \eta$. Similarly, there exists a $y \in A$ such that $b - \eta < y \leq b$. Consequently, $s > b - a - 2\eta = \ell(I_A) - 2\eta \geq \ell(I_A) - 2\varepsilon$. Since $\varepsilon > 0$ is arbitrary, this proves the required reverse inequality $s \geq \ell(I_A)$.

4.1.P3. For any subset A of $[a, b]$ and any $\varepsilon > 0$, show that there exists a subset B of $[a, b]$ such that $A \subseteq B = \cup_{j=1}^{\infty} J_j, m^*(B) < m^*(A) + \varepsilon$ and each J_j is the intersection of an open interval with $[a, b]$.

Solution: By Definition 4.1.3 of m^*, there exists a sequence $\{I_j\}_{j \geq 1}$ of open intervals such that $A \subseteq \cup_{j=1}^{\infty} I_j$ and

$$\sum_{j=1}^{\infty} \ell(I_j) < m^*(A) + \varepsilon. \qquad (\text{Sol.4})$$

Let $J_j = I_j \cap [a, b]$; then each J_j is the intersection of an open interval with $[a, b]$. Take $B = \cup_{j=1}^{\infty} J_j$. Since $A \subseteq [a, b]$, therefore $A = A \cap [a, b] \subseteq \left(\cup_{j=1}^{\infty} I_j\right) \cap [a, b] = \cup_{j=1}^{\infty} (I_j \cap [a, b]) = B$. By countable subadditivity of m^* (see Proposition 4.1.9), $m^*(B) \leq \sum_{j=1}^{\infty} m^*(J_j) \leq \sum_{j=1}^{\infty} m^*(I_j)$. But since each I_j is an interval, we have by

Theorem 4.1.8 that $m^*(I_j) = \ell(I_j)$. It follows that $m^*(B) \le \sum_{j=1}^{\infty} \ell(I_j)$. By (Sol.4), we have $m^*(B) < m^*(A) + \varepsilon$.

4.1.P4. Consider the set C, known as the Cantor set, defined as $\bigcap_{n=1}^{\infty} C_n$, where $C_1 = \left[0, \frac{1}{3}\right] \cup \left[\frac{2}{3}, 1\right]$, and C_{n+1} is obtained from C_n by removing the open middle thirds of its closed intervals. Thus, $C_2 = \left[0, \frac{1}{9}\right] \cup \left[\frac{2}{9}, \frac{1}{3}\right] \cup \left[\frac{2}{3}, \frac{7}{9}\right] \cup \left[\frac{8}{9}, 1\right]$, and so on. Show that $m^*(C) = 0$.

Solution: Every C_n is the union of 2^n closed intervals of length $1/3^n$ each. Therefore the total length of the closed intervals comprising C_n is $(2/3)^n$. From Theorem 4.1.8 and Proposition 4.1.9, it follows that $m^*(C_n) \le (2/3)^n$. Considering that $C \subseteq C_n$, we obtain $m^*(C) = 0$.

4.1.P5. Prove the reverse of the inequality established in Remark 4.1.2(b): Suppose that for each $n \in \mathbb{N}$, we have a series $\sum_{k=1}^{\infty} a_{n,k}$ of nonnegative terms in $\mathbb{R}^{*+}$. Let $\phi: \mathbb{N} \to \mathbb{N} \times \mathbb{N}$ be a bijection. Prove that

$$\sum_{p=1}^{\infty} a_{\phi(p)} \ge \sum_{n=1}^{\infty} \sum_{k=1}^{\infty} a_{n,k}.$$

Hint: First show that, given any $N, K \in \mathbb{N}$, there exists a $P \in \mathbb{N}$ such that $\sum_{n=1}^{N} \sum_{k=1}^{K} a_{n,k} \le \sum_{p=1}^{P} a_{\phi(p)}$. Also use the fact that the limit of a finite sum of sequences is the sum of their separate limits.

Solution: Consider any $N, K \in \mathbb{N}$. Let $P = \max\{\phi^{-1}(n,k): 1 \le n \le N, 1 \le k \le K\}$. Then $1 \le n \le N$, $1 \le k \le K \Rightarrow \phi^{-1}(n,k) \le P \Rightarrow (n,k) = \phi(p)$ for some $p \le P \Rightarrow a_{n,k} = a_{\phi(p)}$ for some $p \le P$. Therefore $\sum_{n=1}^{N} \sum_{k=1}^{K} a_{n,k} \le \sum_{p=1}^{P} a_{\phi(p)} \le \sum_{p=1}^{\infty} a_{\phi(p)}$. This inequality has been shown to be true for any $N, K \in \mathbb{N}$. Since the limit of a finite sum of sequences is the sum of their separate limits, for any $N \in \mathbb{N}$, we further have $\sum_{n=1}^{N} \sum_{k=1}^{\infty} a_{n,k} \le \lim_{K \to \infty} \sum_{n=1}^{N} \sum_{k=1}^{K} a_{n,k} \le \sum_{p=1}^{\infty} a_{\phi(p)}$. Now take the limit as $N \to \infty$.

4.2 Measure from Outer Measure

4.2.P1. Let μ^* be a finite outer measure on a set X and $A \subseteq B$ be subsets of X. Show that

(a) if $\mu^*(B \backslash A) = 0$, then $\mu^*(B) = \mu^*(A)$;
(b) if $\mu^*(B) = \mu^*(A)$ and $A \in \mathcal{M}$, then $\mu^*(B \backslash A) = 0$.

Solution: (a) By the subadditivity of outer measure, $\mu^*(B) = \mu^*(A \cup (B \backslash A)) \le \mu^*(A) + \mu^*(B \backslash A) = \mu^*(A)$ since $\mu^*(B \backslash A) = 0$. The reverse inequality follows from the monotonicity of outer measure.

(b) Since A is μ^*-measurable,

$$\mu^*(B) = \mu^*(B \cap A) + \mu^*(B \cap A^c) = \mu^*(A) + \mu^*(B \backslash A).$$

4.2.P2. Show that, if E_1 and E_2 are μ^*-measurable, where μ^* is an outer measure, then

$$\mu^*(E_1 \cup E_2) + \mu^*(E_1 \cap E_2) = \mu^*(E_1) + \mu^*(E_2).$$

State the analog for three sets.

Solution: The restriction μ of μ^* to μ^*-measurable sets is a measure (Theorem 4.2.11) and we only need to prove

$$\mu(E_1 \cup E_2) + \mu(E_1 \cap E_2) = \mu(E_1) + \mu(E_2).$$

Although on proceeding as in the solution of Problem 3.1.P1 with $n = 2$ we get the required equality forthwith, we offer a direct proof.

For $i = 1, 2$, the set E_i is the union of the disjoint sets $E_i \backslash (E_1 \cap E_2)$ and $E_1 \cap E_2$, and therefore $\mu(E_i) = \mu(E_i \backslash (E_1 \cap E_2)) + \mu(E_1 \cap E_2)$. Adding these for $i = 1, 2$ results in

$$\begin{aligned}
\mu(E_1) + \mu(E_2) &= \mu(E_1 \backslash (E_1 \cap E_2)) + \mu(E_2 \backslash (E_1 \cap E_2)) + 2\mu(E_1 \cap E_2) \\
&= (\mu(E_1 \backslash (E_1 \cap E_2)) + \mu(E_2 \backslash (E_1 \cap E_2)) + \mu(E_1 \cap E_2)) + \mu(E_1 \cap E_2) \\
&= \mu(E_1 \cup E_2) + \mu(E_1 \cap E_2),
\end{aligned}$$

the justification of the last step being that $E_1 \cup E_2$ is the union of the disjoint sets $E_1 \cap E_2$ and $E_i \backslash (E_1 \cap E_2), i = 1, 2$.

The analog for three sets is that

$$\begin{aligned}
\mu^*(E_1 \cup E_2 \cup E_3) &+ \mu^*(E_1 \cap E_2) + \mu^*(E_2 \cap E_3) + \mu^*(E_1 \cap E_3) \\
&= \mu^*(E_1) + \mu^*(E_2) + \mu^*(E_3) + \mu^*(E_1 \cap E_2 \cap E_3)
\end{aligned}$$

for μ^*-measurable sets E_1, E_2, E_3. A justification can be given by using the two sets case twice.

4.2.P3. What has been called "μ" in Example 2.1.3(b) is easily seen to be an outer measure μ^*. Which subsets are μ^*-measurable?

Solution: Only $\emptyset$ and X. It is enough to show that $\{a\}$, $\{b\}$ and $\{c\}$ are not μ^*-measurable. It will follow that their complements are also not μ^*-measurable (Remark 4.2.3(b)). For $E = \{a\}$, the set $A = \{c, a\}$ does not satisfy the equality required for μ^*-measurability of E, because $\mu^*(A) = 5$, $\mu^*(A \cap E) = 2$ and $\mu^*(A \cap E^c) = 4$. For $E = \{b\}$, the set $A = \{b, c\}$ does not satisfy the equality required for μ^*-measurability of E, because $\mu^*(A) = 6$, $\mu^*(A \cap E) = 3$ and $\mu^*(A \cap E^c) = 4$. For $E = \{c\}$, the set $A = \{b, c\}$ does not satisfy the equality required for μ^*-measurability of E, because $\mu^*(A) = 6$, $\mu^*(A \cap E) = 4$ and $\mu^*(A \cap E^c) = 3$.

4.2.P4. Let $X = \{a, b, c\}$ and $\mathcal{F}$ be the σ-algebra consisting of the subsets $\{a, b\}, \{c\}$ besides $\emptyset$ and X. Then $\mu: \mathcal{F} \to \mathbb{R}$, where $\mu\emptyset = \mu\{a, b\} = 0, \mu\{c\} = \mu X = 1$, is clearly a measure on X. For an arbitrary subset $A \subseteq X$, define $\mu^* A$ to be the infimum of all numbers $\sum_{n=1}^{\infty} \mu(I_n)$, where $\{I_n\}_{n \geq 1}$ is a sequence of sets in $\mathcal{F}$ that covers A (the infimum is taken over all such sequences, but only finitely many need be used). Find $\mu^* A$ for each of the eight subsets of X. It can be checked easily that μ^* is an outer measure; find the μ^*-measurable subsets.

Solution: $\mu^*\{a\} = \mu^*\{b\} = 0, \mu^*\{b,c\} = \mu^*\{a,c\} = 1$; the other four subsets are in $\mathcal{F}$, and μ^* agrees with μ on each of them. All subsets are μ^*-measurable.

4.2.P5. Let $\mathcal{H}$ be a family of subsets of a nonempty set X such that $\emptyset \in \mathcal{H}$ and suppose $\alpha: \mathcal{H} \to \mathbb{R}$ is nonnegative-valued, satisfying $\alpha\emptyset = 0$. Assume that some sequence $\{H_n\}_{n \geq 1}$ of sets belonging to $\mathcal{H}$ covers X, hence also every subset of X. Show that

$$\alpha^* A = \inf \left\{ \sum_{n=1}^{\infty} \alpha(I_n) : \text{every } I_n \in \mathcal{H} \text{ and } A \subseteq \bigcup_{n \geq 1} I_n \right\}$$

defines an outer measure on X.

Solution: It is obvious that α^* is $\mathbb{R}^{*+}$-valued. Consider $A \subseteq B \subseteq X$. Any sequence $\{I_n\}_{n \geq 1}$ of sets I_n belonging to $\mathcal{H}$ that covers B also covers A. Therefore $\alpha^* A \leq \alpha^* B$. Thus α^* is monotone. Since $\emptyset$ is covered by the sequence $\emptyset, \emptyset, \ldots,$ of sets in $\mathcal{H}$ and $\alpha\emptyset = 0$, we have $\alpha^*\emptyset = 0$. It remains only to prove countable subadditivity.

Let $\{A_n\}_{n \geq 1}$ be a sequence of subsets of X and $\varepsilon > 0$. If $\sum_{n=1}^{\infty} \alpha^*(A_n) = \infty$, there is nothing to prove. So, we may assume this sum to be finite and hence also every $\alpha^*(A_n)$. In accordance with the definition of α^*, for each $n \in \mathbb{N}$, there is a sequence $\{I_{n,k}\}_{k \geq 1}$ of sets belonging to $\mathcal{H}$ such that $A_n \subseteq \bigcup_{k=1}^{\infty} I_{n,k}$ and

$$\sum_{k=1}^{\infty} \alpha(I_{n,k}) \leq \alpha^*(A_n) + \frac{\varepsilon}{2^n}.$$

Then the family $\{I_{n,k} : (n,k) \in \times \mathbb{N} \times \mathbb{N}\}$ covers $\bigcup_{n=1}^{\infty} A_n$ and

$$\sum_{n=1}^{\infty} \sum_{k=1}^{\infty} \alpha(I_{n,k}) \leq \sum_{n=1}^{\infty} \alpha^*(A_n) + \varepsilon. \qquad \text{(Sol.5)}$$

Let $\phi: \mathbb{N} \to \mathbb{N} \times \mathbb{N}$ be a bijection. The sequence $\{I_{\phi(p)}\}_{p \geq 1}$ of sets belonging to $\mathcal{H}$ covers $\bigcup_{n=1}^{\infty} A_n$ and hence

$$\alpha^* \left(\bigcup_{n=1}^{\infty} A_n \right) \leq \sum_{p=1}^{\infty} \alpha(I_{\phi(p)}).$$

It follows from (Sol.5) and Remark 4.1.2(b) that

$$\alpha^*\left(\bigcup_{n=1}^{\infty} A_n\right) \le \sum_{n=1}^{\infty} \alpha^*(A_n) + \varepsilon.$$

Since this holds for every $\varepsilon > 0$, countable subadditivity must hold.

4.2.P6. For $X = \{a, b, c, d\}$, a set of four elements, let

$$\mathcal{H} = \{\emptyset, \{d\}, \{a, b\}, \{a, c\}, \{a, b, c\}, X\}$$

and $\alpha \colon \mathcal{H} \to \mathbb{R}$ be defined as

$$\alpha\emptyset = 0, \ \alpha\{d\} = 3, \ \alpha\{a, b\} = 0, \ \alpha\{a, c\} = 3, \ \alpha\{a, b, c\} = 2 \quad \text{and} \quad \alpha X = 2.$$

Find $\alpha^* A$ for all the sixteen subsets of X and determine which ones are α^*-measurable.

Solution: $\alpha^*\emptyset = 0, \alpha^*\{a\} = \alpha^*\{b\} = \alpha^*\{a, b\} = 0$, all other subsets E of X have $\alpha^* E = 2$. The sets $\emptyset, \{a\}, \{b\}, \{a, b\}$ and their complements are α^*-measurable. The remaining eight sets fail to be α^*-measurable.

The subset $E = \{c\}$ is not α^*-measurable, because when $A = \{c, d\}$, we find that $\alpha^*(A \cap E) = \alpha^*(E) = 2, \alpha^*(A \cap E^c) = \alpha^*(\{d\}) = 2$ and $\alpha^*(A) = 2 \ne 2 + 2$.

The subset $E = \{d\}$ is not α^*-measurable, because when $A = \{c, d\}$, we find that $\alpha^*(A \cap E) = \alpha^*(E) = 2, \alpha^*(A \cap E^c) = \alpha^*(\{c\}) = 2$ and $\alpha^*(A) = 2 \ne 2 + 2$.

The subset $E = \{a, c\}$ is not α^*-measurable, because when $A = \{c, d\}$, we find that $\alpha^*(A \cap E) = \alpha^*\{c\} = 2, \alpha^*(A \cap E^c) = \alpha^*(\{d\}) = 2$ and $\alpha^*(A) = 2 \ne 2 + 2$.

The subset $E = \{a, d\}$ is not α^*-measurable, because when $A = \{c, d\}$, we find that $\alpha^*(A \cap E) = \alpha^*\{d\} = 2, \alpha^*(A \cap E^c) = \alpha^*(\{c\}) = 2$ and $\alpha^*(A) = 2 \ne 2 + 2$.

The subset $E = \{b, c\}$ is not α^*-measurable, because its complement $\{a, d\}$ has been shown not to be α^*-measurable.

The subset $E = \{b, d\}$ is not α^*-measurable, because its complement $\{a, c\}$ has been shown not to be α^*-measurable.

The subset $E = \{a, b, d\}$ is not α^*-measurable, because its complement $\{c\}$ has been shown not to be α^*-measurable.

The subset $E = \{a, b, c\}$ is not α^*-measurable, because its complement $\{d\}$ has been shown not to be α^*-measurable.

4.2.P7. Let $\mathcal{H}$ be a σ-algebra of subsets of X and $(X, \mathcal{H}, \alpha)$ be a measure space. Any sequence of sets in $\mathcal{H}$ having X as one of its terms covers X. Show that for the outer measure α^* obtained as in Problem 4.2.P5, every set in $\mathcal{H}$ is α^*-measurable. Is the converse true?

Solution: Let $E \in \mathcal{H}$ and $A \subseteq X$ be an arbitrary subset. We need prove only the Carathéodory condition that $\alpha^*(A) \ge \alpha^*(A \cap E) + \alpha^*(A \cap E^c)$. So, let $\alpha^*(A) < \infty$ and consider an arbitrary sequence $\{E_n\}_{n \ge 1}$ of sets in $\mathcal{H}$ that covers A. Then $\{E_n \cap E\}_{n \ge 1}$ and $\{E_n \cap E^c\}_{n \ge 1}$ are sequences of sets in $\mathcal{H}$ that cover $A \cap E$ and $A \cap E^c$ respectively. It follows that

$$\sum_{n=1}^{\infty} \alpha(E_n \cap E) \geq \alpha^*(A \cap E) \quad \text{and} \quad \sum_{n=1}^{\infty} \alpha(E_n \cap E^c) \geq \alpha^*(A \cap E^c).$$

Since α is a measure, we have $\alpha(E_n \cap E) + \alpha(E_n \cap E^c) = \alpha(E_n)$ for each n. Therefore

$$\sum_{n=1}^{\infty} \alpha(E_n) = \sum_{n=1}^{\infty} \alpha(E_n \cap E) + \sum_{n=1}^{\infty} \alpha(E_n \cap E^c) \geq \alpha^*(A \cap E) + \alpha^*(A \cap E^c).$$

Since $\{E_n\}_{n \geq 1}$ is an arbitrary sequence of sets in $\mathcal{H}$ that covers A, we may take the infimum on the left side, thereby obtaining the desired Carathéodory condition and completing the proof that every set in $\mathcal{H}$ is α^*-measurable.

The converse is false. Let X consist of three points a, b, c and $\mathcal{H}$ consist of $\emptyset, \{a, b\}, \{c\}$ and X. Consider the measure α on $\mathcal{H}$ satisfying $\alpha(\emptyset) = \alpha\{a, b\} = 0$ and $\alpha\{c\} = \alpha X = 1$. Since $\alpha^*\{a\} = 0$, the set $\{a\}$ is α^*-measurable but $\{a\} \notin \mathcal{H}$.

4.2.P8. For infinite $A \subseteq \mathbb{N}$, let $\alpha(A) = 1$ and for finite $A \subseteq \mathbb{N}$, let $\alpha(A) = \sum_{j=1}^{n} (1/2^j)$, where n is the number of elements in A. (If $A = \emptyset$, the sum becomes "empty" and is to be interpreted as usual to be 0.) Show that α is an outer measure on $\mathbb{N}$ and find all α-measurable subsets of $\mathbb{N}$.

Solution: The set function α is certainly nonnegative real-valued and bounded above by 1. Besides, $\alpha(A) < 1$ whenever A is finite. Suppose $A \subseteq B \subseteq \mathbb{N}$. If B is infinite, then $\alpha(B) = 1$ and hence $\alpha(A) \leq \alpha(B)$. If B is finite, then so is A and $\alpha(A) = \sum_{j=1}^{n} (1/2^j), \alpha(B) = \sum_{j=1}^{p} (1/2^j)$, where n is the number of elements in A and p is the number of elements in B. But $n \leq p$, which implies $\alpha(A) \leq \alpha(B)$. This shows that α is monotone.

For countable subadditivity, consider a sequence $\{A_n\}_{n \geq 1}$ of subsets of $\mathbb{N}$. For the purpose at hand, we may take every A_n to be nonempty. If $\cup_{n=1}^{\infty} A_n$ is infinite, then infinitely many among the sets A_n must contain one or more elements and therefore have α equal to $\frac{1}{2}$ or greater. This implies $\sum_{n=1}^{\infty} \alpha(A_n) = \infty$. If $\cup_{n=1}^{\infty} A_n$ is finite, then it can have only finitely many distinct subsets and therefore $\cup_{n=1}^{\infty} A_n = \cup_{n=1}^{N} A_n$ for some N. So, we need prove only finite subadditivity with finite sets, which will follow by induction if we prove it for two finite sets. With this in view, let A and B be finite subsets having n and p elements respectively. Then $A \cup B$ has at most $n + p$ elements and hence $\alpha(A \cup B) \leq \sum_{j=1}^{n+p} (1/2^j) = \sum_{j=1}^{n} (1/2^j) + \sum_{j=1}^{p} (1/2^{j+n}) \leq \sum_{j=1}^{n} (1/2^j) + \sum_{j=1}^{p} (1/2^j) = \alpha(A) + \alpha(B)$.

Let E be a subset that is neither empty nor $\mathbb{N}$. Let A be a subset consisting of any one element belonging to E and any one element belonging to E^c. Then A contains precisely two elements and each of $A \cap E$ and $A \cap E^c$ contains a single element. Therefore $\alpha(A) = \frac{3}{4}$ and $\alpha(A \cap E) = \alpha(A \cap E^c) = \frac{1}{2}$, so that $\alpha(A) < \alpha(A \cap E) + \alpha(A \cap E^c)$. So E is not α-measurable. This shows that the only α-measurable subsets of $\mathbb{N}$ are $\emptyset$ and $\mathbb{N}$.

4.2.P9. (Cf. Problem 4.3.P7) Let $\mathcal{F}$ be a σ-algebra of subsets of a nonempty set X and suppose the set function $\alpha\colon \mathcal{F} \to \mathbb{R}^{*+}$ vanishes at the empty set, is monotone and countably subadditive. (In other words α has all the properties of an outer measure except that its domain is $\mathcal{F}$.) For any $A \in \mathcal{F}$, put

$$\mu(A) = \sup\left\{\sum_{j=1}^{\infty}\alpha(E_j)\colon \bigcup_{j=1}^{\infty}E_j = A, E_j \cap E_k = \emptyset \text{ whenever } j \neq k, \text{ every } E_j \in \mathcal{F}\right\}.$$

(a) Show that μ is a measure and that, for all $A \in \mathcal{F}$, we have (i) $\mu(A) \geq \alpha(A)$
   (ii) $\mu(A) = 0 \Leftrightarrow \alpha(A) = 0$.
(b) If α is as in Problem 4.2.P8, what is μ?

Solution: (a) Given any $A \in \mathcal{F}$, one possible sequence $\{E_j\}_{j\geq 1}$ such that $\cup_{j=1}^{\infty}E_j = A, E_j \cap E_k = \emptyset$ whenever $j \neq k$ and every $E_j \cap \mathcal{F}$ is given by $E_1 = A, E_j = \emptyset$ for $j > 1$. For this sequence, $\sum_{j=1}^{\infty}\alpha(E_j) = \alpha(A)$ and therefore $\mu(A) \geq \alpha(A)$. This proves (i).

It follows from (i) that μ is nonnegative-valued. For $A = \emptyset$, the only sequence $\{E_j\}_{j\geq 1}$ of the required kind has every $E_j = \emptyset$, and therefore $\mu(\emptyset) = 0$.

Before proceeding to countable additivity, we establish monotonicity. Suppose $A \subseteq B$, both sets being in $\mathcal{F}$, i.e. measurable. Consider an arbitrary sequence $\{E_j\}_{j\geq 1}$ of disjoint measurable sets with $\cup_{j=1}^{\infty}E_j = A$. Then the sequence of sets obtained by prefixing $B\backslash A$ to the sequence $\{E_j\}_{j\geq 1}$ consists of disjoint measurable sets and has union equal to B. So, $\mu(B) \geq \alpha(B\backslash A) + \sum_{j=1}^{\infty}\alpha(E_j) \geq \sum_{j=1}^{\infty}\alpha(E_j)$. Since this holds for an arbitrary sequence $\{E_j\}_{j\geq 1}$ of disjoint measurable sets with union A, we have $\mu(B) \geq \mu(A)$.

Regarding countable additivity, consider $A = \cup_{n=1}^{\infty}A_n$, where $\{A_n\}_{n\geq 1}$ is a sequence of disjoint measurable sets. Let $\{E_j\}_{j\geq 1}$ be an arbitrary sequence of disjoint measurable sets such that $\cup_{j=1}^{\infty}E_j = A$. Then for each n, $\{A_n \cap E_j\}_{j\geq 1}$ is a sequence of disjoint measurable sets such that $\cup_{j=1}^{\infty}(A_n \cap E_j) = A_n$. Hence the definition of μ yields $\sum_{j=1}^{\infty}\alpha(A_n \cap E_j) \leq \mu(A_n)$. Consequently, $\sum_{n=1}^{\infty}\sum_{j=1}^{\infty}\alpha(A_n \cap E_j) \leq \sum_{n=1}^{\infty}\mu(A_n)$. By Remark 4.1.2(b) and Problem 4.1.P5, we can assert that $\sum_{j=1}^{\infty}\sum_{n=1}^{\infty}\alpha(A_n \cap E_j) \leq \sum_{n=1}^{\infty}\mu(A_n)$. However, the sequence $\{A_n \cap E_j\}_{n\geq 1}$ consists of disjoint measurable sets satisfying $\cup_{n=1}^{\infty}(A_n \cap E_j) = E_j$ and therefore countable subadditivity of α leads to $\alpha(E_j) \leq \sum_{n=1}^{\infty}\alpha(A_n \cap E_j)$. In tandem with the preceding inequality, this leads to $\sum_{j=1}^{\infty}\alpha(E_j) \leq \sum_{n=1}^{\infty}\mu(A_n)$. This has been shown to hold for an arbitrary sequence $\{E_j\}_{j\geq 1}$ of disjoint measurable sets having union A. By definition of μ, it now follows that $\mu(A) \leq \sum_{n=1}^{\infty}\mu(A_n)$.

To arrive at the reverse inequality, we may assume $\mu(A) < \infty$. Then monotonicity provides $\mu(A_n) < \infty$ for each n. Take an arbitrary $\varepsilon > 0$. Corresponding to each $n \in \mathbb{N}$, there is a sequence $\{E_{n,j}\}_{j \geq 1}$ of disjoint measurable sets such that $A_n = \cup_{j=1}^{\infty} E_{n,j}$ and $\sum_{j=1}^{\infty} \alpha(E_{n,j}) \geq \mu(A_n) - \frac{\varepsilon}{2^n}$. Then the sets of the family $\{E_{n,j} : (n,j) \in \mathbb{N} \times \mathbb{N}\}$ are measurable, are disjoint and have union $\cup_{n=1}^{\infty} A_n = A$. Besides, they satisfy the inequality $\sum_{n=1}^{\infty} \sum_{j=1}^{\infty} \alpha(E_{n,j}) \geq \sum_{n=1}^{\infty} \mu(A_n) - \varepsilon$. Next, let $\phi : \mathbb{N} \to \mathbb{N} \times \mathbb{N}$ be a bijection. Then the sequence $\{E_{\phi(p)}\}_{p \geq 1}$ of disjoint measurable sets has union A and hence $\sum_{p=1}^{\infty} \alpha(E_{\phi(p)}) \leq \mu(A)$. Upon combining this with the previous inequality and Problem 4.1.P5, we obtain $\sum_{n=1}^{\infty} \mu(A_n) - \varepsilon \leq \mu(A)$. Since $\varepsilon > 0$ is arbitrary, we arrive at the desired reverse inequality.

It remains to prove (ii). It is trivial from (i) that $\mu(A) = 0 \Rightarrow \alpha(A) = 0$. If $\alpha(A) = 0$, the monotonicity of α ensures that every sequence $\{E_j\}_{j \geq 1}$ of measurable sets satisfying $\cup_{j=1}^{\infty} E_j = A$ also satisfies $\sum_{j=1}^{\infty} \alpha(E_j) = 0$, thereby rendering $\mu(A) = 0$.

(b) Consider a subset A consisting of a single point x. In any sequence $\{E_j\}_{j \geq 1}$ of disjoint measurable sets having union A, all sets must be $\emptyset$ except one and the exceptional one must be equal to A. Therefore $\sum_{j=1}^{\infty} \alpha(E_j) = \alpha(A) = \frac{1}{2}$. It follows that $\mu\{x\} = \frac{1}{2}$ for every x and hence that μ is $\frac{1}{2}$ times the counting measure.

4.3 Lebesgue Measure

4.3.P1. Suppose that A is a subset of $\mathbb{R}$ with the property that, for every $\varepsilon > 0$, there exist Lebesgue measurable sets B and C such that $B \subseteq A \subseteq C$ and $m(C \cap B^c) < \varepsilon$. (Here m denotes Lebesgue measure as usual.) Show that A is Lebesgue measurable.
Solution: For $n \in \mathbb{N}$, there exist Lebesgue measurable B_n and C_n such that $B_n \subseteq A \subseteq C_n$ and $m(C_n \cap B_n^c) < \frac{1}{n}$. Let $B_0 = \cup_{n=1}^{\infty} B_n$ and $C_0 = \cap_{n=1}^{\infty} C_n$. Then B_0 and C_0 are Lebesgue measurable and $m(C_0 \cap B_0^c) \leq m(C_n \cap B_n^c) \leq \frac{1}{n}$ for each $n \in \mathbb{N}$. This implies $m(C_0 \cap B_0^c) = 0$. Moreover, $A \cap B_0^c \subseteq C_0 \cap B_0^c$, so that $m^*(A \cap B_0^c) \leq m^*(C_0 \cap B_0^c) = 0$. Hence $A \cap B_0^c \in \mathcal{M}$ and so, $A = (A \cap B_0^c) \cup B_0 \in \mathcal{M}$.
4.3.P2. Let $A \subseteq [a, b]$ and m denote Lebesgue measure. Show that there exists a Lebesgue measurable set $B \subseteq [a, b]$ such that $m(B) = m^*(B) = m^*(A)$ and $A \subseteq B$. The same is true with $[a, b]$ replaced by $\mathbb{R}$.
Solution: First suppose $A \subseteq [a, b]$. By Problem 4.1.P3, for each $n \in \mathbb{N}$, there exists a subset B_n of $[a, b]$ such that $A \subseteq B_n = \cup_{j=1}^{\infty} J_{j,n}$ such that $m^*(B_n) \leq m^*(A) + \frac{1}{n}$ and each $J_{j,n}$ is the intersection of an open interval with $[a, b]$. Applying Theorem 4.3.1 and then Theorem 4.2.9, we find that each B_n is Lebesgue measurable. Appealing to Theorem 4.2.9 again, we deduce that the set $B = \cap_{n=1}^{\infty} B_n \subseteq [a, b]$ is also measurable. Now $A \subseteq B$ and by monotonicity of m^*, we have $m^*(A) \leq m^*(B) \leq m^*(B_n) \leq m^*(A) + \frac{1}{n}$ for each $n \in \mathbb{N}$. Hence $m^*(B) = m^*(A)$. Since B is Lebesgue measurable, it is trivial that $m(B) = m^*(B)$.

When $[a, b]$ is replaced by $\mathbb{R}$, the appeal to Problem 4.1.P3 can be replaced by an appeal to Remark 4.1.4; each $J_{j,n}$ is an open interval and $m^*(B_n)$ can be ∞. (The matter is trivial if $m^*(A) = \infty$.)

4.3.P3. Show that the Lebesgue outer measure m^* is inner continuous. In other words, if $\{S_n\}_{n \geq 1}$ is a sequence of sets such that $S_1 \subseteq S_2 \subseteq \cdots$, then $m^*(\cup_{n=1}^{\infty} S_n) = \lim_{n \to \infty} m^*(S_n)$.

Solution: Since the sequence $\{m^*(S_n)\}_{n \geq 1}$ is increasing, we know that $\lim_{n \to \infty} m^*(S_n)$ exists, perhaps as ∞. Also, $m^*(S_k) \leq m^*(\cup_{n=1}^{\infty} S_n)$ for each k, and hence

$$\lim_{n \to \infty} m^*(S_n) \leq m^*\left(\bigcup_{n=1}^{\infty} S_n\right). \tag{Sol.6}$$

By Problem 4.3.P2, there exist Lebesgue measurable $T_n \supseteq S_n$ such that

$$m(T_n) = m^*(T_n) = m^*(S_n). \tag{Sol.7}$$

Setting $A_n = \cap_{j \geq n} T_j$, we have

$$A_1 \subseteq A_2 \subseteq \cdots, \text{ with each } A_n \text{ Lebesgue measurable} \tag{Sol.8}$$

(by Theorem 4.2.9) and also $S_n \subseteq A_n \subseteq T_n$. Therefore, we have

$$\bigcup_{n=1}^{\infty} S_n \subseteq \bigcup_{n=1}^{\infty} A_n \tag{Sol.9}$$

as well as $m^*(S_n) \leq m^*(A_n) \leq m^*(T_n)$. Hence from (Sol.7), we get

$$m^*(A_n) = m^*(S_n). \tag{Sol.10}$$

From (Sol.8) and Proposition 2.3.1 (inner continuity of measure), we know $m(\cup_{n=1}^{\infty} A_n) = \lim_{n \to \infty} m(A_n)$, i.e.

$$m^*\left(\bigcup_{n=1}^{\infty} A_n\right) = \lim_{n \to \infty} m^*(A_n)$$
$$= \lim_{n \to \infty} m^*(S_n) \text{ by (Sol.10)}$$

Together with (Sol.9), this leads to the reverse of inequality (Sol.6).

4.3.P4. Let $E \subseteq [a, b]$. Show that E is a Lebesgue measurable subset of $[a, b]$ if and only if it is a Lebesgue measurable subset of $\mathbb{R}$.

Solution: We shall denote complements in $\mathbb{R}$ by a superscript 'c' as usual; however, we shall denote the complement of a set $F \subseteq [a, b]$ by $[a, b] \setminus F$.

First suppose E is a Lebesgue measurable subset of $\mathbb{R}$. To show that it is a Lebesgue measurable subset of $[a, b]$, consider an arbitrary subset $A \subseteq [a, b]$. Note that $A \cap E^c = \{x \in A : x \notin E\} = \{x \in A : x \in [a, b], x \notin E\} = A \cap ([a, b] \backslash E)$. Since E is a Lebesgue measurable subset of $\mathbb{R}$, we have

$$m^*(A) = m^*(A \cap E) + m^*(A \cap E^c) = m^*(A \cap E) + m^*(A \cap ([a, b] \backslash E)).$$

So, E is a Lebesgue measurable subset of $[a, b]$.

Next, suppose for the converse that E is a Lebesgue measurable subset of $[a, b]$. To show that it is a Lebesgue measurable subset of $\mathbb{R}$, consider an arbitrary subset $A \subseteq \mathbb{R}$. From the inclusion $E \subseteq [a, b]$, we know that

$$A \cap [a, b] \cap E = A \cap E \quad \text{and} \quad [a, b]^c = E^c \cap [a, b]^c.$$

Now, since $[a, b]$ is a Lebesgue measurable subset of $\mathbb{R}$,

$$\begin{aligned} m^*(A) &= m^*(A \cap [a, b]) + m^*(A \cap [a, b]^c) \\ &= m^*(A \cap [a, b]) + m^*(A \cap E^c \cap [a, b]^c). \end{aligned}$$

Also, since E is a Lebesgue measurable subset of $[a, b]$ and $A \cap [a, b] \subseteq [a, b]$, we have

$$\begin{aligned} m^*(A \cap [a, b]) &= m^*(A \cap [a, b] \cap E) + m^*(A \cap [a, b] \cap ([a, b] \backslash E)) \\ &= m^*(A \cap E) + m^*(A \cap [a, b] \cap ([a, b] \backslash E)), \text{ as } E \subseteq [a, b]. \end{aligned}$$

But $[a, b] \backslash E \subseteq [a, b]$ and hence $A \cap [a, b] \cap ([a, b] \backslash E) = A \cap ([a, b] \backslash E)$. So, the preceding equality can be restated as

$$m^*(A \cap [a, b]) = m^*(A \cap E) + m^*(A \cap ([a, b] \backslash E)).$$

It follows that

$$\begin{aligned} m^*(A) &= m^*(A \cap E) + m^*(A \cap ([a, b] \backslash E)) + m^*(A \cap E^c \cap [a, b]^c) \\ &= m^*(A \cap E) + m^*(A \cap E^c \cap [a, b]) + m^*(A \cap E^c \cap [a, b]^c). \end{aligned}$$

However, $m^*(A \cap E^c \cap [a, b]) + m^*(A \cap E^c \cap [a, b]^c) = m^*(A \cap E^c)$ because $[a, b]$ is a Lebesgue measurable subset of $\mathbb{R}$. Hence

$$m^*(A) = m^*(A \cap E) + m^*(A \cap E^c).$$

4.3.P5. For any $S \subseteq \mathbb{R}$ and any $x \in \mathbb{R}$, let $S + x$ denote the set $\{s + x : s \in S\}$. Prove the following:

(a) If E is a Lebesgue measurable subset of $\mathbb{R}$, then $E + x$ is also a Lebesgue measurable subset of $\mathbb{R}$ and $m(E + x) = m(E)$.

(b) If E is a Lebesgue measurable subset of $[a, b]$ and $E + x \subseteq [a, b]$, then $E + x$ is also a Lebesgue measurable subset of $[a, b]$.

Solution: (a) Note that, for any $A \subseteq \mathbb{R}$, $E \subseteq \mathbb{R}$ and any $x \in \mathbb{R}$, the following equalities hold:

$$A \cap (E + x) = [(A - x) \cap E] + x \quad \text{and} \quad A \cap (E + x)^c = [(A - x) \cap E^c] + x.$$

Moreover, in view of Proposition 4.1.5(e),

$$m^*(S + x) = m^*(S) \text{ for any } S \subseteq \mathbb{R} \text{ and any } x \in \mathbb{R}.$$

Now let $A \subseteq \mathbb{R}$ be arbitrary and E be a Lebesgue measurable subset of $\mathbb{R}$. It follows from the three equalities noted in the above paragraph and the measurability of E that

$$\begin{aligned}
&m^*(A \cap (E + x)) + m^*(A \cap (E + x)^c) \\
&= m^*([(A - x) \cap E] + x) + m^*([(A - x) \cap E^c] + x) \\
&= m^*([(A - x) \cap E]) + m^*([(A - x) \cap E^c]) \\
&= m^*(A - x) = m^*(A).
\end{aligned}$$

The equality $m(E + x) = m(E)$ now follows from the fact that it holds for m^*.

(b) By Problem 4.3.P4, the sets E and $E + x$ are Lebesgue measurable of subsets of $[a, b]$ if and only if they are Lebesgue measurable subsets of $\mathbb{R}$. Therefore the required conclusion is immediate from (a).

4.3.P6. Let $\{r_n\}_{n \geq 1}$ be an enumeration of the rationals in $[0, 1]$. For each $i \in \mathbb{N}$, let g_i be a continuous nonnegative function on $[0, 1]$ such that

$$g_i(r_i) \geq i \quad \text{and} \quad \int_0^1 g_i(x)dx = \frac{1}{2^i}.$$

Define $f_n = \sum_{i=1}^n g_i$. Then each f_n is nonnegative and continuous. Besides, the sequence $\{f_n\}_{n \geq 1}$ is increasing and therefore has a limit function f that is nonnegative extended real-valued and measurable. Show that the function f is unbounded on every subinterval of $[0, 1]$ and satisfies $\int_{[0,1]} f = 1$.

Solution: Consider an arbitrary subinterval $[a, b] \subseteq [0, 1]$, where $a < b$, and an arbitrary $M > 0$. The interval $[a, b]$ must contain infinitely many rational numbers and hence must contain some r_n with $n > M$. The given definition of f_n leads to $f_n(r_n) \geq g_n(r_n) \geq n > M$. By continuity, $[a, b]$ has some subinterval $[\alpha, \beta]$ of positive length on which $f_n > M$. It follows that $f \geq f_n > M$ on $[\alpha, \beta]$. This shows that f is unbounded on every subinterval of $[0, 1]$.

By Theorem 4.3.4, each g_n satisfies $\int_{[0,1]} g_i = \int_0^1 g_i(x)dx = 1/2^i$. It follows that $\int_{[0,1]} f_n = \sum_{i=1}^n (1/2^i) = 1 - (1/2^n)$. Since the sequence $\{f_n\}_{n\geq 1}$ is increasing, we infer from the Monotone Convergence Theorem 2.3.2 that $\int_{[0,1]} f = \lim_{n\to\infty} \int_{[0,1]} f_n = 1$.

4.3.P7. For any interval $I \subseteq \mathbb{R}$ of positive length and any $M > 0$, show that there exists a sequence $\{E_j\}_{j\geq 1}$ of disjoint sets such that $\cup_{j=1}^{\infty} E_j = I$ and $\sum_{i=1}^{\infty} m^*(E_j) > M$.

Solution: In Problem 4.2.P9, choose X to be $\mathbb{R}$, $\mathcal{F}$ to be the family of all subsets of $\mathbb{R}$ and α to be m^*. We get a measure μ defined on all subsets of $\mathbb{R}$ such that the first inequality in (i) of Proposition 4.3.7 is satisfied and (ii) holds. It follows by an application of that proposition that the second inequality in (i) cannot hold. Therefore there exists a bounded interval J such that $\mu(J) = \infty$. Being bounded, J is a subset of the union finitely many translates of any interval I of positive length. Therefore such an interval I must satisfy $\mu(I) = \infty$. Considering the manner in which μ is set up in the aforementioned problem as a supremum, the assertion that $\mu(I) = \infty$ is precisely equivalent to what is to be proved here.

4.3.P8. For an increasing nonnegative real-valued function ϕ on $\mathbb{R}$ satisfying $\phi(0) = 0$ and $\phi(u + v) \leq \phi(u) + \phi(v)$ whenever $u, v \in \mathbb{R}$, show that

$$\phi(1)^{1/2} \leq \left(\int_0^1 \phi(1 - \sqrt{t})dt \right)^{1/2} + \left(\int_0^1 \phi(\sqrt{1 - t})dt \right)^{1/2},$$

where the integrals are understood in the sense of Riemann.

Solution: By setting $\tilde{\mu}A = \phi(m(A))$, where m is Lebesgue measure, we obtain a subadditive fuzzy measure on any Lebesgue measurable subset of real numbers, in particular, on the interval $[0, 1]$. Consider the functions f and g on $[0, 1]$ defined as $f(x) = x^2$ and $g(x) = 1 - x^2$. Their distributions are given by

$$\tilde{f}(t) = \phi(1 - \sqrt{t}) \quad \text{for} \quad 0 \leq t < 1 \quad \text{and } 0 \text{ for } t \geq 1$$

and

$$\tilde{g}(t) = \phi(\sqrt{1 - t}) \quad \text{for} \quad 0 \leq t < 1 \quad \text{and} \quad 0 \quad \text{for } t \geq 1.$$

Therefore

$$\int_{\sim} f = \int_0^1 \phi(1 - \sqrt{t})dt \quad \text{and} \quad \int_{\sim} g = \int_0^1 \phi(\sqrt{1 - t})dt.$$

Since $f + g = 1$ everywhere, $\int_{\sim} (f + g) = \tilde{\mu}[0, 1] = \phi(1)$. On applying Theorem 3.3.5 with $p = 1$, we arrive at the desired inequality.

4.4 Induced Measure and an Application

4.4.P1. (a) Let $g\colon [1,\infty) \to \mathbb{R}$ be given by $g(x) = \sin x$ and I_k be the interval $[2k\pi, (2k+1)\pi]$, where $k \in \mathbb{N}$. Show that $\int_{I_k} g(x)dx = 2$.

(b) Let $f\colon [1,\infty) \to \mathbb{R}$ be given by $f(x) = \frac{1}{x}\sin x$ and I_k be the interval $[2k\pi, (2k+1)\pi]$, where $k \in \mathbb{N}$. Show that $\int_{I_k} f(x)dx \geq \frac{2}{(2k+1)\pi}$.

Solution: (a) $\int_{I_k} g(x)dx = \int_{[1,\infty)} g(x)\chi_{I_k}(x)dx = \int_{I_k} g|_{I_k}(x)dx$ by Proposition 4.4.1. Now, by Proposition 4.3.4,

$$\int_{I_k} g|_{I_k}(x)dx = \int_{2k\pi}^{(2k+1)\pi} g|_{I_k}(x)dx = \int_{2k\pi}^{(2k+1)\pi} \sin x\, dx$$
$$= -[\cos((2k+1)\pi) - \cos(2k\pi)] = 2.$$

(b) As in part (a), $\int_{I_k} f(x)dx = \int_{2k\pi}^{(2k+1)\pi} f|_{I_k}(x)dx$ by Proposition 4.4.1 and Proposition 4.3.4. Now, $\int_{2k\pi}^{(2k+1)\pi} f|_{I_k}(x)dx = \int_{2k\pi}^{(2k+1)\pi} \frac{1}{x}\sin x\, dx \geq \frac{1}{(2k+1)\pi}\int_{2k\pi}^{(2k+1)\pi} \sin x\, dx = \frac{2}{(2k+1)\pi}$.

4.4.P2. Let $(X, \mathcal{F}, \mu)$ be a measure space and $\{f_k\}_{k \geq 1}$ be a sequence of integrable functions on X such that $\sum_{k=1}^{\infty} \int |f_k| < \infty$. Show that there exists an integrable function f on X such that the series $\sum_{k=1}^{\infty} f_k$ converges to f on a measurable subset $H \subseteq X$ satisfying $\mu(H^c) = 0$, and that $\int f = \sum_{k=1}^{\infty} \int f_k$.

Solution: Let $g = \sum_{k=1}^{\infty} |f_k|$. Since $\sum_{k=1}^{\infty} \int |f_k| < \infty$, it follows by the Monotone Convergence Theorem 2.3.2 that $\int g < \infty$. By Proposition 2.2.9, there exists a measurable set $H \subseteq X$ such that g is finite on H and $\mu(H^c) = 0$. Thus the series $\sum_{k=1}^{\infty} f_k$ is convergent on H, and its sum f_0 on H satisfies $|f_0(x)| \leq g(x)$ for all $x \in H$. Since H is measurable and restrictions of the partial sums of $\sum_{k=1}^{\infty} f_k$ to H are measurable with reference to the induced σ-algebra $\mathcal{F}_H$ on H, therefore f_0 is an $\mathcal{F}_H$-measurable function on H. Hence by Proposition 4.4.1, its extension f to X obtained by setting it equal to 0 on H^c is real-valued and $\mathcal{F}$-measurable (considering that $f|_H = f_0$ and $f\chi_H = f$). Moreover, the extension satisfies $|f| \leq g$ on X and is therefore integrable. Now, $|\sum_{k=1}^{n} f_k(x)| \leq g(x)$ and $\lim_{n \to \infty} \sum_{k=1}^{n} f_k(x) = f(x)$ for every $x \in H$. The Dominated Convergence Theorem 3.2.6, the fact that $\mu(H^c) = 0$ and Proposition 4.4.3 together lead to

$$\lim_{n \to \infty} \sum_{k=1}^{n} \int f_k = \lim_{n \to \infty} \sum_{k=1}^{n} \int_H f_k = \lim_{n \to \infty} \int_H \sum_{k=1}^{n} f_k = \int_H f = \int f.$$

4.4.P3. Let $\mathcal{F}$ be a σ-algebra of subsets of X and $A, B \in \mathcal{F}$ be nonempty subsets satisfying $A \cup B = X$. Show that a function $\phi \colon X \to \mathbb{R}^*$ is $\mathcal{F}$-measurable if and only if the restrictions $\phi|_A$ and $\phi|_B$ are $\mathcal{F}_A$-measurable and $\mathcal{F}_B$-measurable respectively. Solution: Proposition 4.4.1 will be used repeatedly without reference. If $\phi \colon X \to \mathbb{R}^*$ is $\mathcal{F}$-measurable, then so are the products $\phi\chi_A$ and $\phi\chi_B$ and it follows that the restrictions $\phi|_A$ and $\phi|_B$ are $\mathcal{F}_A$-measurable and $\mathcal{F}_B$-measurable respectively.

For the converse, suppose the restrictions $\phi|_A$ and $\phi|_B$ are $\mathcal{F}_A$-measurable and $\mathcal{F}_B$-measurable respectively. Then the products $\phi\chi_A$ and $\phi\chi_B$ are $\mathcal{F}$-measurable and it follows that the products $(\phi\chi_A)\chi_B$, $(\phi\chi_A)\chi_{A\backslash B}$ and $(\phi\chi_B)\chi_{B\backslash A}$ are also $\mathcal{F}$-measurable. But

$$(\phi\chi_A)\chi_B = \phi(\chi_A\chi_B) = \phi\chi_{A\cap B}, (\phi\chi_A)\chi_{A\backslash B} = \phi(\chi_A\chi_{A\backslash B}) = \phi\chi_{A\backslash B}$$
$$\text{and } (\phi\chi_B)\chi_{B\backslash A} = \phi(\chi_B\chi_{B\backslash A}) = \phi\chi_{B\backslash A}.$$

Thus the products $\phi\chi_{A\cap B}$, $\phi\chi_{A\backslash B}$ and $\phi\chi_{B\backslash A}$ are $\mathcal{F}$-measurable. The sum of these three products is $\phi\chi_{A\cup B}$, which is the same as ϕ because it is given that $A \cup B = X$.

5.1 Interchanging the Order of Summation

5.1.P1. Let X be any nonempty set. Define $\mathcal{F}$ to consist of those subsets of X that are either countable or have a countable complement. (It is easy to argue that $\mathcal{F}$ is a σ-algebra, sometimes called the "co-countable" σ-algebra). Now let $\phi \colon X \to \mathbb{R}^{*+}$ be any function and set $\mu(A) = \sum_{x\in A} \phi(x)$ for any $A \in \mathcal{F}$, where the sum is understood in the sense of Definition 5.1.7. Show that μ is a measure on $\mathcal{F}$. (Note: If $\phi > 0$ everywhere, then the only subset of measure 0 is the empty set.) Solution: Countable additivity is all that needs to be proved. Let $\{A_n\}_{n\geq 1}$ be a sequence of disjoint sets in $\mathcal{F}$ and A be their union. For any finite subset $B\subseteq A$, each of the intersections $B\cap A_n$ is finite and, in view of the disjointness of the A_n, is empty for all except finitely many n. Hence $\sum_{x\in B} \phi(x) = \sum_n \sum_{x\in B\cap A_n} \phi(x) \leq \sum_n \mu(A_n)$. Since this holds for every finite subset $B\subseteq A$, it follows that $\mu(A) \leq \sum_n \mu(A_n)$. To prove the reverse inequality, we first note that it is trivial when $\mu(A) = \infty$, so that we may assume $\mu(A) < \infty$. Consider an arbitrary $\varepsilon > 0$. For each n, there exists a finite subset $B_n\subseteq A_n$ such that $\sum_{x\in B_n} \phi(x) > \mu(A_n) - \frac{\varepsilon}{2^n}$. For each positive integer N, let $C_N = \cup_{n=1}^{N} B_n$. Then $C_N\subseteq A$ and is finite. Combined with the disjointness of the B_n, this implies $\mu(A) \geq \sum_{x\in C_N} \phi(x) = \sum_{1\leq n\leq N} \sum_{x\in B_n} \phi(x) \geq \sum_{1\leq n\leq N} \mu(A_n) - \varepsilon$. As this holds for every positive integer N, we get $\sum_n \mu(A_n) \leq \mu(A) + \varepsilon$. Since $\varepsilon > 0$ is arbitrary, we conclude that $\sum_n \mu(A_n) \leq \mu(A)$, which is the desired reverse inequality.

5.1.P2. Prove Proposition 5.1.9(b). Solution: Let $A = \{x \in X : a(x) > 0\}$. Then any subset B of X containing A has the property that $a(x) = 0$ whenever $x \notin B$. For each $n \in \mathbb{N}$, consider the set

$A_n = \{x \in X: a(x) > \frac{1}{n}\}$. It contains at most $n \cdot \sum_{x \in X} a(x)$ elements and therefore must be finite. The set $\cup_{n=1}^{\infty} A_n$ is therefore countable. But it is precisely the same set as A.

5.1.P3. Let μ be a measure on a set X. Consider $\mathbb{N}$ with counting measure (all subsets measurable) and let $f: X \times \mathbb{N} \to \mathbb{R}$ be a real-valued function such that, for each $k \in \mathbb{N}$, the function f^k on X described by $f^k(x) = f(x, k)$ is measurable. Suppose also that there is an integrable function $\phi: X \to \mathbb{R}^+$ such that $|\sum_{k=1}^{n} f(x, k)| \leq \phi(x)$ for each $x \in X$ and each $n \in \mathbb{N}$. If the series $\sum_{k=1}^{\infty} f(x, k)$ converges, show that interchanging the order of integration is valid, which is to say,

$$\int \left(\sum_{k=1}^{\infty} f(x, k) \right) d\mu(x) = \sum_{k=1}^{\infty} \left(\int f(x, k) d\mu(x) \right).$$

Solution: For brevity, we shall write $d\mu(x)$ as simply dx. Define a sequence of functions ϕ_n on X by $\phi_n(x) = \sum_{k=1}^{n} f(x, k)$. Then each ϕ_n is real-valued and measurable. Besides, $|\phi_n(x)| \leq \phi(x)$ for every $x \in X$ and every $n \in \mathbb{N}$. By the Dominated Convergence Theorem 3.2.6,

$$\int \left(\sum_{k=1}^{\infty} f(x, k) \right) dx = \int \left(\lim_{n \to \infty} \phi_n(x) \right) dx = \lim_{n \to \infty} \left(\int \phi_n(x) dx \right)$$

$$= \lim_{n \to \infty} \sum_{k=1}^{n} \int f(x, k) dx = \sum_{k=1}^{\infty} \left(\int f(x, k) dx \right).$$

5.1.P4. Suppose the real-valued function $a: \mathbb{N} \times \mathbb{N} \to \mathbb{R}$ has the property that there exists a sequence $\{b_m\}_{m \geq 1}$ such that $\sum_{m=1}^{\infty} b_m < \infty$ and $|\sum_{k=1}^{n} a_{m,k}| \leq b_m$ for each n and each m. Show that, if the series $\sum_{k=1}^{\infty} a_{m,k}$ converges for each m, then $\sum_{m=1}^{\infty} \sum_{k=1}^{\infty} a_{m,k} = \sum_{k=1}^{\infty} \sum_{m=1}^{\infty} a_{k,m}$. (Note: (1) The hypothesis about $\{b_m\}_{m \geq 1}$ is stronger than convergence of $\sum_{m=1}^{\infty} |\sum_{k=1}^{\infty} a_{m,k}|$. The latter is not sufficient for the purpose at hand, in view of Problem 1.2.P1. (2) A proof of the result of the present problem by using metric spaces is found in Feldman [3; p. 2]. What has been stated there has a stronger but more palatable version of the hypothesis about $\{b_m\}$, in that it is assumed that $\sum_{m=1}^{\infty} \sum_{k=1}^{\infty} |a_{m,k}| < \infty$, so that $\sum_{k=1}^{\infty} a_{m,k}$ not only converges for each m but does so absolutely.)

Solution: This is an immediate consequence of Problem 5.1.P3: Choose X in that problem to be $\mathbb{N}$, replace "$x \in X$" by "$m \in \mathbb{N}$", choose μ to be counting measure, the function f to be a and $\phi(x)$ to be b_m. Then ϕ is integrable because $\sum_{m=1}^{\infty} b_m < \infty$, and what has been denoted by $\phi_n(x)$ in the solution there is now $\sum_{k=1}^{n} a_{m,k}$.

5.2 Integration with the Counting Measure

5.2.P1. Complete the proof of Theorem 5.2.3 by showing for any function $a: X \times Y \to \mathbb{R}^{*+}$ that

$$\sum_{(x,y)\in X\times Y} a(x,y) \le \sum_{x\in X}\sum_{y\in Y} a(x,y).$$

Solution: Let $F \subseteq X \times Y$ be finite. Then the subsets $A \subseteq X$ and $B \subseteq Y$ given by

$$A = \{x \in X: (x,y) \in F \text{ for some } y \in Y\},$$
$$B = \{y \in Y: (x,y) \in F \text{ for some } x \in X\}$$

("projections" on X and Y) are both finite and $F \subseteq A \times B$, a finite subset of $X \times Y$. Now,

$$\sum_{(x,y)\in F} a(x,y) \le \sum_{(x,y)\in A\times B} a(x,y) = \sum_{x\in A}\sum_{y\in B} a(x,y) \le \sum_{x\in X}\sum_{y\in Y} a(x,y).$$

Since F is an arbitrary finite subset of $X \times Y$, we obtain

$$\sum_{(x,y)\in X\times Y} a(x,y) \le \sum_{x\in X}\sum_{y\in Y} a(x,y).$$

5.2.P2. Let $f: X \times Y \to \mathbb{R}$ be a real-valued function on the Cartesian product $X \times Y$ and suppose that the unconditional sum $\sum_{(x,y)\in X\times Y}^{\text{un}} f(x,y)$ exists (necessarily finite by definition). Show that (i) $\sum_{y\in Y}^{\text{un}} f(x,y)$ exists for each $x \in X$, (ii) $\sum_{x\in X}^{\text{un}} f(x,y)$ exists for each $y \in Y$, and (iii) $\sum_{x\in X}^{\text{un}} \sum_{y\in Y}^{\text{un}} f(x,y) = \sum_{y\in Y}^{\text{un}} \sum_{x\in X}^{\text{un}} f(x,y)$.
Solution: Since the unconditional sum $\sum_{(x,y)\in X\times Y}^{\text{un}} f(x,y)$ exists, it follows from Theorem 5.2.7 that the sums $\sum_{(x,y)\in X\times Y} f^+(x,y)$ and $\sum_{(x,y)\in X\times Y} f^-(x,y)$ are both finite. Now, by Theorem 5.2.3,

$$\sum_{(x,y)\in X\times Y} f^+(x,y) = \sum_{x\in X}\sum_{y\in Y} f^+(x,y) \quad \text{and} \quad \sum_{(x,y)\in X\times Y} f^-(x,y) = \sum_{x\in X}\sum_{y\in Y} f^-(x,y).$$

Therefore the repeated sums on the right sides are both finite. Consequently, the sums $\sum_{y\in Y} f^+(x,y)$ and $\sum_{y\in Y} f^-(x,y)$ are finite for each x. Together with Remark 5.2.6(d), this implies that $\sum_{y\in Y}^{\text{un}} f(x,y)$ exists for each $x \in X$. This completes the argument for (i). The argument for (ii) is analogous.

By Theorem 5.2.7, we have

$$
\begin{aligned}
\sum_{(x,y)\in X\times Y}^{\text{un}} f(x,y) &= \sum_{(x,y)\in X\times Y} f^+(x,y) - \sum_{(x,y)\in X\times Y} f^-(x,y) \\
&= \sum_{x\in X}\sum_{y\in Y} f^+(x,y) - \sum_{x\in X}\sum_{y\in Y} f^-(x,y) \quad \text{by Theorem 5.2.3} \\
&= \sum_{x\in X}^{\text{un}} \sum_{y\in Y} f^+(x,y) - \sum_{x\in X}^{\text{un}} \sum_{y\in Y} f^-(x,y) \quad \text{by Remark 5.2.6(b)} \\
&= \int \sum_{y\in Y} f^+(x,y)dx - \int \sum_{y\in Y} f^-(x,y)dx \quad \text{by Theorem 5.2.7} \\
&= \int \sum_{y\in Y}^{\text{un}} f^+(x,y)dx - \int \sum_{y\in Y}^{\text{un}} f^-(x,y)dx \quad \text{by Remark 5.2.6(b)} \\
&= \int \left(\sum_{y\in Y}^{\text{un}} f^+(x,y) - \sum_{y\in Y}^{\text{un}} f^-(x,y) \right) dx \\
&= \int \left(\sum_{y\in Y}^{\text{un}} f(x,y) \right) dx \quad \text{by (i) and Theorem 5.2.7} \\
&= \sum_{x\in X}^{\text{un}} \sum_{y\in Y}^{\text{un}} f(x,y) \quad \text{by Theorem 5.2.8.}
\end{aligned}
$$

By a similar argument, $\sum_{(x,y)\in X\times Y}^{\text{un}} f(x,y) = \sum_{y\in Y}^{\text{un}} \sum_{x\in X}^{\text{un}} f(x,y)$.

5.2.P3. (Cf. Problem 5.1.P4) Suppose the real-valued function $f: X \times Y \to \mathbb{R}$ has the property that there exists a map $\phi: X \to \mathbb{R}^+$ such that $\sum_{x\in X} \phi(x) < \infty$ and $|\sum_{y\in F} f(x,y)| \le \phi(x)$ for each finite subset $F \subseteq Y$ and each $x \in X$. Show that (i) $\sum_{x\in X}^{\text{un}} f(x,y)$ exists for each $y \in Y$, $\sum_{y\in Y}^{\text{un}} f(x,y)$ exists for each $x \in X$, (ii) the repeated sum $\sum_{x\in X}^{\text{un}} \sum_{y\in Y}^{\text{un}} f(x,y)$ exists, and (iii) if the other repeated sum $\sum_{y\in Y}^{\text{un}} \sum_{x\in X}^{\text{un}} f(x,y)$ also exists, then the two repeated sums are equal: $\sum_{x\in X}^{\text{un}} \sum_{y\in Y}^{\text{un}} f(x,y) = \sum_{y\in Y}^{\text{un}} \sum_{x\in X}^{\text{un}} f(x,y)$.
Solution: Since $|\sum_{y\in F} f(x,y)| \le \phi(x) < \infty$ for each finite subset $F \subseteq Y$, it follows for each $y \in Y$ upon choosing $F = \{y\}$ that $\sum_{x\in X} |f(x,y)| \le \sum_{x\in X} \phi(x) < \infty$ and hence via Theorem 5.2.1 and Theorem 5.2.8 that $\sum_{x\in X}^{\text{un}} f(x,y)$ exists and also that it equals $\int f(x,y)dx$.

Next, consider an arbitrary $x \in X$. Since $|\sum_{y\in F} f(x,y)| \le \phi(x) < \infty$ for each finite subset $F \subseteq Y$, it is immediate that the same is true when f is replaced either by f^+ or by f^-. Therefore $\sum_{y\in Y} f^+(x,y), \sum_{y\in Y} f^-(x,y) \le \phi(x) < \infty$. It now follows on the basis of Remark 5.2.6(d) that $\sum_{y\in Y}^{\text{un}} f(x,y)$ exists. This shows that $\sum_{y\in Y}^{\text{un}} f(x,y)$ exists for each $x \in X$. This proves (i).

From the second assertion in (i), it follows that, for each $x \in X$ and each $n \in \mathbb{N}$, there exists a finite $F_n \subseteq Y$ such that whenever $G_n \subseteq Y$ is a finite subset containing F_n, the inequality $|\sum_{y \in G_n} f(x,y) - \sum_{y \in Y}^{\text{un}} f(x,y)| < \frac{1}{n}$ holds. Define a sequence of functions ϕ_n on X in terms of an *arbitrary* such sequence $\{G_n\}_{n \geq 1}$ of sets by taking $\phi_n(x) = \sum_{y \in G_n} f(x,y)$. Then $\{\phi_n\}_{n \geq 1}$ is a sequence of real-valued functions measurable with respect to the σ-algebra of all subsets of X, which converges everywhere to $\sum_{y \in Y}^{\text{un}} f(x,y)$. Moreover,

$$|\phi_n(x)| = \left|\sum_{y \in G_n} f^+(x,y) - \sum_{y \in G_n} f^-(x,y)\right| \leq \left|\sum_{y \in G_n} f^+(x,y)\right| + \left|\sum_{y \in G_n} f^-(x,y)\right| \leq 2\phi(x)$$

for every $x \in X$ and every $n \in \mathbb{N}$. Hence the Dominated Convergence Theorem 3.2.6 implies that $\sum_{y \in Y}^{\text{un}} f(x,y) = \lim_{n \to \infty} \phi_n(x)$ is integrable with respect to counting measure on X, so that by virtue of Theorem 5.2.8, it follows that $\sum_{x \in X}^{\text{un}} \sum_{y \in Y}^{\text{un}} f(x,y)$ exists. Thus, (ii) stands proved.

In order to justify (iii), we begin by noting that by virtue of Theorem 5.2.8, it also follows that

$$\sum_{x \in X}^{\text{un}} \sum_{y \in Y}^{\text{un}} f(x,y) = \int \left(\sum_{y \in Y}^{\text{un}} f(x,y)\right) dx.$$

Moreover, the Dominated Convergence Theorem also implies that

$$\int \left(\sum_{y \in Y}^{\text{un}} f(x,y)\right) dx = \int \left(\lim_{n \to \infty} \phi_n(x)\right) dx = \lim_{n \to \infty} \left(\int \phi_n(x) dx\right) = \lim_{n \to \infty} \int \sum_{y \in G_n} f(x,y) dx$$

$$= \lim_{n \to \infty} \sum_{y \in G_n} \int f(x,y) dx = \lim_{n \to \infty} \sum_{y \in G_n} \sum_{x \in X}^{\text{un}} f(x,y).$$

Taking advantage of the fact that this equality holds for an arbitrary sequence $\{G_n\}_{n \geq 1}$ of finite sets G_n such that $F_n \subseteq G_n \subseteq Y$, we can now choose G_n, pursuant to the existence of $\sum_{y \in Y}^{\text{un}} \sum_{x \in X}^{\text{un}} f(x,y)$, in such a manner as to satisfy

$$\left|\sum_{y \in G_n} \sum_{x \in X}^{\text{un}} f(x,y) - \sum_{y \in Y}^{\text{un}} \sum_{x \in X}^{\text{un}} f(x,y)\right| < \frac{1}{n} \quad \text{for every } n \in \mathbb{N}.$$

Then the foregoing equality leads to $\int(\sum_{y \in Y}^{\text{un}} f(x,y)) dx = \sum_{y \in Y}^{\text{un}} \sum_{x \in X}^{\text{un}} f(x,y)$. Since it has already been established that $\sum_{x \in X}^{\text{un}} \sum_{y \in Y}^{\text{un}} f(x,y) = \int(\sum_{y \in Y}^{\text{un}} f(x,y)) dx$, the required equality of repeated sums shines forth.

6.1 Algebras and Monotone Classes of sets

6.1.P1. Suppose the set X consists of three elements a, b, c and B is the family of subsets $\{a\}$ and $\{b, c\}$. List all the subsets in (i) the smallest algebra of sets that contains B and (ii) the monotone class generated by B.
Solution: (i) $\emptyset$, $\{a\},\{b, c\},X$ (ii) $\{a\},\{b, c\}$.

6.1.P2. Let X consist of four elements a, b, c, d and S consist of the subsets of X described in unorthodox notation as $\emptyset$, a, bcd, bc, ad, $abcd$. Determine (i) whether S is an algebra, (ii) whether S is a monotone class, (iii) what the monotone class generated by S is, and (iv) what the σ-algebra generated by S is.
Solution: (i) S is not closed under unions and is therefore not an algebra. (ii) Being a finite family, it is certainly a monotone class. (iii) Consequently, the monotone class generated by it is itself. (iv) When S is enlarged by including abc and its complement d, what we obtain can be verified to be an algebra and therefore a σ-algebra in view of its finiteness. Hence this enlarged family is the σ-algebra generated by S.

6.1.P3. Let X be an uncountable set and B be the family of all subsets that are either finite or have finite complement. Show that B is an algebra of subsets. Is it a σ-algebra?
Solution: Surely B contains $\emptyset$ and is closed under complements. To show that it is closed under unions, consider any $A, B \in B$. If both are finite, then so is $A \cup B$ and therefore $A \cup B \in B$. If one of them is not finite, then its complement must be finite, which implies that the complement of the bigger set $A \cup B$ is finite and hence that $A \cup B \in B$. Thus B is an algebra. However, it is not a σ-algebra, because a countably infinite subset fails to be in B despite being a countable union of subsets belonging to B.

6.1.P4. Let $\mathcal{A}$ be an algebra of subsets of a nonempty set X. If there are finitely many sets in $\mathcal{A}$, show that the number of sets in it is 2^n, where n is the number of nonempty sets B in the algebra (called 'atoms') having the property that the only set $B_1 \in \mathcal{A}$ satisfying $\emptyset \subset B_1 {\subseteq} B$ must be B.
Solution: Our first step will be show that every nonempty set A in the algebra contains an atom. Suppose $A \in \mathcal{A}$ contains no atom. Then there exists an $A_1 \in \mathcal{A}$ satisfying $\emptyset \subset A_1 \subset A$. Now, A_1 also contains no atom and hence there exists an $A_2 \in \mathcal{A}$ satisfying $\emptyset \subset A_2 \subset A_1$. Proceeding in this manner, we obtain infinitely many sets $A_1, A_2, \ldots$ in $\mathcal{A}$, contradicting the finiteness of $\mathcal{A}$. This contradiction establishes what was promised in the first step.

The second step is to note that any two distinct atoms must be disjoint. The reason is that nonemptiness of their intersection would, by definition of an atom, make it equal to each one of the atoms, contradicting their distinctness.

The third step is to show that every nonempty set A in the algebra is the union of all the atoms it contains. In accordance with the first step, A does contain at least one atom and they must be finitely many because $\mathcal{A}$ is finite. Let B be their union. Then $A \supseteq B \in \mathcal{A}$ and the contention is that $A = B$. If not, then $C = A \backslash B {\subseteq} A$ is a nonempty set in $\mathcal{A}$ and is disjoint from B. In accordance with the first step, there exists an atom $A_0 {\subseteq} C {\subseteq} A$. This implies that A_0 is an atom contained in A but disjoint from B, which is not possible because of the way B was set up.

The fourth and final step is to show that there is a bijection between the family of all subsets $\Im$ of the n-element set $\mathcal{T}$ of all atoms and the algebra $\mathcal{A}$. As this family of subsets has 2^n elements, it will follow that $\mathcal{A}$ also has 2^n elements. Consider the map from the family into $\mathcal{A}$ given by $\Im \rightarrow \cup\Im$. Recall that $\Im = \emptyset \Rightarrow \cup\Im = \emptyset$. Therefore we know from the third step that the map is surjective. It remains only to show that it is injective. Let $\Im_1$ and $\Im_2$ be distinct subsets of $\mathcal{T}$, i.e. distinct sets of atoms. One of them must contain an atom A_0 that does not belong to the other. To be specific, suppose $A_0 \in \Im_1$ but $A_0 \notin \Im_2$. Our second step guarantees that A_0 is disjoint from every atom belonging to $\Im_2$ and hence also from $\cup\Im_2$. Since $\cup\Im_1 \supseteq A_0 \neq \emptyset$, it follows that $\cup\Im_1 \neq \cup\Im_2$.

6.1.P5. Let $\mathcal{F}$ be a σ-algebra of subsets of X generated by a subfamily $\mathcal{H}$ and let $A \in \mathcal{F}$. Show that the induced σ-algebra $\mathcal{F}_A$ is the same as the σ-algebra generated by the family $\mathcal{H}_A = \{H \cap A : H \in \mathcal{H}\}$ of intersections of the sets of $\mathcal{H}$ with A.

Solution: Let $\mathcal{G}$ be the σ-algebra of subsets of A generated by $\mathcal{H}_A$. We have to show that $\mathcal{G} = \mathcal{F}_A$.

Since $\mathcal{H} \subseteq \mathcal{F}$, it is clear that $\mathcal{H}_A \subseteq \mathcal{F}_A$ and hence that $\mathcal{G} \subseteq \mathcal{F}_A$. We shall show that $\mathcal{G} \supseteq \mathcal{F}_A$.

For any family $\mathcal{S}$ consisting of subsets of X, the symbol '$\mathcal{S}_A$' will, in conformity with the symbols '$\mathcal{H}_A$' and '$\mathcal{F}_A$', denote the family consisting of subsets of A that are intersections with A of sets belonging to $\mathcal{S}$. In symbols,

$$\mathcal{S}_A = \{S \cap A : S \in \mathcal{S}\}.$$

Note that for any collection $\mathfrak{M}$ of families $\mathcal{S}$ consisting of subsets of X,

$$\bigcap_{S \in \mathfrak{M}} \mathcal{S}_A \supseteq \left(\bigcap_{S \in \mathfrak{M}} \mathcal{S}\right)_A.$$

Take $\mathfrak{M}$ to be the collection of all σ-algebras of subsets of X that contain $\mathcal{H}$. Then $\cap_{S \in \mathfrak{M}} \mathcal{S}$ is given to be the same as $\mathcal{F}$ and therefore the right side of the above inclusion is the induced σ-algebra $\mathcal{F}_A$. Consequently, the inclusion can be stated as

$$\bigcap_{S \in \mathfrak{M}} \mathcal{S}_A \supseteq \mathcal{F}_A.$$

It remains only to be shown that the intersection on the left side here equals $\mathcal{G}$.

We first show that it contains $\mathcal{G}$. Let $\mathcal{S}$ be an arbitrary family belonging to $\mathfrak{M}$, i.e. an arbitrary σ-algebra of subsets of X that contains $\mathcal{H}$. Then $\mathcal{S}_A$ is clearly a σ-algebra of subsets of A and contains $\mathcal{H}_A$. Therefore $\mathcal{S}_A \supseteq \mathcal{G}$. Since this is true for any $\mathcal{S} \in \mathfrak{M}$ whatsoever, it follows that

$$\bigcap_{S \in \mathfrak{M}} \mathcal{S}_A \supseteq \mathcal{G}.$$

With the aim of proving the reverse inclusion, consider an arbitrary σ-algebra $\mathcal{K}$ of subsets of A that contains $\mathcal{H}_A$. Let $\mathcal{S}'$ be the family of those subsets of X whose intersection with A belongs to $\mathcal{K}$, so that $\mathcal{S}'_A = \mathcal{K}$. Also, $\mathcal{S}'$ is a σ-algebra of subsets of X that contains $\mathcal{H}$, which is to say, $\mathcal{S}' \in \mathfrak{M}$. Therefore the intersection of all such $\mathcal{K}$ must contain the intersection on the left side of the foregoing inclusion. However, the intersection of all such $\mathcal{K}$ is nothing but $\mathcal{G}$, and therefore the preceding statement means that the reverse of the foregoing inclusion holds. Thus,

$$\bigcap_{\mathcal{S} \in \mathfrak{M}} \mathcal{S}_A = \mathcal{G},$$

which is all that remained to be shown.

6.2 Defining a Product σ-algebra

6.2.P1. Let $\mathcal{F}$ and $\mathcal{G}$ be σ-algebras of subsets of X and Y respectively. Suppose $X = \bigcup_{n=1}^{\infty} X_n$ and $Y = \bigcup_{k=1}^{\infty} Y_k$, where $\{X_n\}$ and $\{Y_k\}$ are disjoint sequences of sets. Then $\mathcal{F}_n = \{F \cap X_n : F \in \mathcal{F}\}$ and $\mathcal{G}_n = \{G \cap Y_k : G \in \mathcal{G}\}$ are σ-algebras of subsets of X_n and Y_k respectively. (If X_n and Y_k are all measurable, $\mathcal{F}_n$ and $\mathcal{G}_k$ are the induced σ-algebras in the sense of Sect. 4.4.) Show that every $E \in \mathcal{F} \times \mathcal{G}$ satisfies

$$E \cap (X_n \times Y_k) \in \mathcal{F}_n \times \mathcal{G}_k \text{ for every } n \text{ and } k$$

Solution: Let $\mathcal{H} = \{E \subseteq X \times Y : E \cap (X_n \times Y_k) \in \mathcal{F}_n \times \mathcal{G}_k\}$. By Problem 2.1.P9, this is a σ-algebra. First we note that $\mathcal{H}$ contains every rectangle $A \times B$. This is because, by definition of $\mathcal{F}_n$ and $\mathcal{G}_k$,

$$A \cap X_n \in \mathcal{F}_n, \quad B \cap Y_k \in \mathcal{G}_k$$

and hence

$$(A \times B) \cap (X_n \times Y_k) = (A \cap X_n) \times (B \cap Y_k) \in \mathcal{F}_n \times \mathcal{G}_k$$

for every n and k. Since $\mathcal{H}$ is a σ-algebra containing every rectangle, it follows by definition of $\mathcal{F} \times \mathcal{G}$ that $\mathcal{H} \supseteq F \times \mathcal{G}$. By definition of $\mathcal{H}$, this inclusion implies the assertion we seek to prove.

6.2.P2. Prove Proposition 6.2.6.

Solution: It is sufficient to prove that $P \backslash Q$ is an elementary set whenever P and Q are rectangles, because the rest will then follow by Proposition 6.2.5.

Let $P = A \times B$ and $Q = C \times D$, where $A, C \in \mathcal{F}$ and $B, D \in \mathcal{G}$. Then

$$P \backslash Q = ((A \backslash C) \times B) \cup ((A \cap C) \times (B \backslash D)),$$

which is an elementary set.

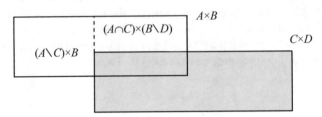

Illustration of the set-theoretic equality in the proof of Proposition 6.2.6

6.2.P3. Let $\mathcal{F}$ and $\mathcal{G}$ be the σ-algebras consisting of all subsets of X and Y respectively, where X and Y are both countable. Show that $\mathcal{F} \times \mathcal{G}$ consists of all subsets of $X \times Y$.

Solution: Let (a, b) be any point in $X \times Y$. Since $\mathcal{F}$ and $\mathcal{G}$ consist of all subsets of X and Y respectively, the sets $\{a\} \subseteq X$ and $\{b\} \subseteq Y$ belong to $\mathcal{F}$ and $\mathcal{G}$ respectively. Therefore $\{a\} \times \{b\} = \{(a, b)\}$ is a rectangle. By Definition 6.2.3, it must belong to $\mathcal{F} \times \mathcal{G}$. But $X \times Y$ is countable and therefore so is any subset of it. Therefore any subset of it is a countable union of sets of the form $\{(a, b)\}$. It follows that it must be in $\mathcal{F} \times \mathcal{G}$.

6.2.P4. Let $\mathcal{F}$ and $\mathcal{G}$ be σ-algebras of subsets of X and Y respectively. Suppose $A \in \mathcal{F}$ and $B \in \mathcal{G}$. Then the σ-algebra $(\mathcal{F} \times \mathcal{G})_{A \times B}$ induced by the product σ-algebra $\mathcal{F} \times \mathcal{G}$ on $A \times B$ is the same as the product $\mathcal{F}_A \times \mathcal{G}_B$ of the induced σ-algebras $\mathcal{F}_A$ and $\mathcal{G}_B$. In particular, if $f: X \times Y \to \mathbb{R}^{*+}$ is $(\mathcal{F} \times \mathcal{G})$-measurable then the restriction of f to $A \times B$ is $(\mathcal{F}_A \times \mathcal{G}_B)$-measurable.

Solution: Let $\mathcal{H}$ be the family of rectangles R in $X \times Y$. By definition, $\mathcal{F} \times \mathcal{G}$ is the σ-algebra generated by $\mathcal{H}$. By the result of Problem 6.1.P5, the σ-algebra $(\mathcal{F} \times \mathcal{G})_{A \times B}$ is the same as the σ-algebra generated by the family $\mathcal{H}_A = \{R \cap (A \times B): R \in \mathcal{H}\}$ of intersections of the sets of $\mathcal{H}$ with $A \times B$. Now, $\mathcal{F}_A \times \mathcal{G}_B$ is, by definition, the σ-algebra generated by the rectangles in $A \times B$. It is sufficient therefore to show that $\mathcal{H}_A$ is precisely the family of rectangles in $A \times B$.

It is obvious that a rectangle in $A \times B$ is a rectangle in $X \times Y$, which means it equals $R \cap (A \times B)$, where $R \in \mathcal{H}$. Now consider any rectangle $P \times Q$ in $X \times Y$. By definition, $R = P \times Q$, where $P \in \mathcal{F}$ and $Q \in \mathcal{G}$. Then $P \cap A \in \mathcal{F}_A$ and $Q \cap B \in \mathcal{G}_B$ and therefore $(P \cap A) \times (Q \cap B)$ is a rectangle in $A \times B$. However, $(P \cap A) \times (Q \cap B) = (P \times Q) \cap (A \times B) = R \cap (A \times B) \in \mathcal{H}_A$.

6.3 Sections of a Subset and of a Function

6.3.P1. Let X, Y and V be as in Example 5.1.3. For each $x \in X$, describe the set V_x and for each $y \in Y$, describe the set V^y. If $f = \chi_V$, describe the function f_x for each $x \in X$.

Solution: By Definition 6.3.1, $V_x = \{y \in Y: (x,y) \in V\}$ and by the definition of V in Example 5.1.3, $V = \{(x,y) \in [0,1] \times [0,1]: x = y\}$. Therefore $V_x = \{y \in Y: x = y\} = \{x\}$. Similarly, $V^y = \{y\}$. By Definition 6.3.6, the function f_x has domain Y and satisfies $f_x(y) = f(x,y)$ for all $y \in Y$. Since $f(x,y) = \chi_V(x,y)$, we have $f_x(y) = \chi_V(x,y) = 1$ if $(x,y) \in V$ and 0 otherwise. Taking into account what V is given to be, we find that $f_x(y) = 1$ if $y = x$ and 0 otherwise. (In other words, $f_x = \chi_{\{x\}}$.)

6.4 Defining a Product Measure

6.4.P1. Prove Theorem 6.4.4.
Solution: Since the function $v(E_x)$ defined on X is nonnegative and measurable by Proposition 6.4.2, it has an integral, though not necessarily finite. We need only show that $\mu \times v$ is a measure, because the σ-finiteness will then follow trivially.

Clearly, $(\mu \times v)(\emptyset) = 0$ and $(\mu \times v)(E) \geq 0$ for every $E \in \mathcal{F} \times \mathcal{G}$. Consider a sequence $\{E_n\}_{n \geq 1}$ of disjoint sets of $\mathcal{F} \times \mathcal{G}$. The sets in the sequence $\{(E_n)_x\}_{n \geq 1}$ of their sections are disjoint and consequently, on the basis of the Monotone Convergence Theorem 2.3.2,

$$(\mu \times v)(\bigcup_{n=1}^{\infty} E_n) = \int_X v((\bigcup_{n=1}^{\infty} E_n)_x)d\mu(x) = \int_X v(\bigcup_{n=1}^{\infty} (E_n)_x)d\mu(x)$$

$$= \int_X \sum_{n=1}^{\infty} v((E_n)_x)d\mu(x) = \sum_{n=1}^{\infty} \int_X v((E_n)_x)d\mu(x) = \sum_{n=1}^{\infty} (\mu \times v)(E_n).$$

Hence $\mu \times v$ is countably additive and therefore a measure.
6.4.P2. For any positive integer n, set $I_j = [\frac{j-1}{n}, \frac{j}{n}]$. For the set V of Example 5.1.3, show that $V = \cap_{n=1}^{\infty} V_n$, where $V_n = \cup_{j=1}^{n} (I_j \times I_j)$.
Solution: Consider any point $(x,x) \in V$. For each n, we have $[0,1] = \cup_{j=1}^{n} I_j$ and therefore there exists a j such that $x \in I_j$. This j satisfies $(x,x) \in I_j \times I_j$, so that $x \in V_n$. Since this holds for every n, it follows that $x \in \cap_{n=1}^{\infty} V_n$. This shows that $V \subseteq \cap_{n=1}^{\infty} V_n$. Now consider any $(x,y) \in [0,1] \times [0,1]$ such that $x \neq y$ and we shall argue that $(x,y) \notin \cap_{n=1}^{\infty} V_n$. We may take $x > y$. Choose any integer $n > 1/(x-y)$. If there were to exist a j such that $(x,y) \in I_j \times I_j$, then we would have $x \in I_j$ and $y \in I_j$, which implies $x - y < 1/n$, a contradiction. This means there can be no such j. In other words, for this particular n, there is no j such that $(x,y) \in I_j \times I_j$, i.e. $(x,y) \notin \cup_{j=1}^{n} (I_j \times I_j) = V_n$. The existence of such an integer n means that $(x,y) \notin \cap_{n=1}^{\infty} V_n$.

6.5 The Tonelli and Fubini Theorems

6.5.P1. Let $X = Y = [1, \infty)$ with Lebesgue measure m. Suppose $f: X \times Y \to \mathbb{R}$ satisfies

$$f(0,0) = 0 \quad \text{and} \quad f(x,y) = \frac{x^2 - y^2}{(x^2 + y^2)^2}.$$

Show that f cannot be integrable even if it is measurable.

Solution: The function that maps $y \in [1, \infty)$ into $\frac{1}{y^2}$ has Riemann integral $1 - \frac{1}{B}$ over any interval $[1, B]$ and therefore has the same Lebesgue integral by Theorem 4.3.4. It follows on applying the Monotone Convergence Theorem 2.3.2 that it is has Lebesgue integral 1 over $[1, \infty)$. Next, the section f_x satisfies $|f_x| \leq \frac{1}{y^2}$, which makes it Lebesgue integrable over $[1, \infty)$. Therefore by Theorem 4.4.2, the Lebesgue integral may be computed as the improper integral, which can be done by noting that $\frac{\partial}{\partial y}\frac{y}{x^2 + y^2} = \frac{x^2 - y^2}{(x^2 + y^2)^2}$, thereupon obtaining the value $-\frac{1}{1+x^2}$. It follows that one repeated integral is given by

$$\int_X dm(x) \int_Y f(x,y)dm(y) = \int_X \left(-\frac{1}{1+x^2}\right) dm(x) = \int_1^\infty \left(-\frac{1}{1+x^2}\right) dx = -\frac{\pi}{4}.$$

For the other repeated integral, an analogous computation leads to a value that turns out to be the negative of what has just been computed. Thus the two repeated integrals do not agree, and therefore Fubini's Theorem 6.5.5 yields the conclusion that f cannot be integrable.

6.5.P2. If X and Y are both countable, show that the product measure of the counting measures on them is the counting measure on $X \times Y$.

Solution: Denote by μ and v the counting measures on X and Y respectively. Their domains $\mathcal{F}$ and $\mathcal{G}$ are the σ-algebras of all subsets of X and Y respectively. We know from Problem 6.2.P3 that the product σ-algebra $\mathcal{F} \times \mathcal{G}$ on $X \times Y$ consists of all subsets. It is sufficient to show for each $(a, b) \in X \times Y$ that $(\mu \times v)\{(a,b)\} = 1$. For any $x \in X$, the section $\{(a,b)\}_x$ is $\{b\}$ if $x = a$ and $\emptyset$ if $x \neq a$. Consequently, $v\{(a,b)\}_x = v\{b\} = 1$ if $x = a$ and 0 if $x \neq a$. In other words, the function on X given by $x \to v\{(a,b)\}_x$ is the characteristic function of $\{a\}$. Hence $\int_X v\{(a,b)\}_x d\mu(x) = \mu\{a\} = 1$. By Definition 6.4.3, this means $(\mu \times v)\{(a,b)\} = 1$.

6.5.P3. Show that repeated integrals with respect to counting measures can be unequal even if the function on the product space is measurable.

Hint: See Problem 1.2.P1.

Solution: Consider the function $a: \mathbb{N} \times \mathbb{N} \to \mathbb{R}$ of Problem 1.2.P1. As seen in that problem, the repeated sums are unequal. However, the repeated sums are precisely the repeated integrals with respect to counting measure. Since the product σ-algebra consists of all subsets by Problem 6.2.P3, the function a is measurable. (Fubini's Theorem 6.5.5 now implies that a cannot be integrable with respect to counting measure on $\mathbb{N} \times \mathbb{N}$; however, the lack of integrability can also be obtained directly from the definition of a by taking into account the infinitude of the points (p, p) where it takes the value 1, or alternatively, the infinitude of the points $(p + 1, p)$.)

6.5.P4. Give an example of two functions on a measure space that are equal almost everywhere, one is measurable but the other is not.

Solution: Let $X = \{a, b, c\}$, $\mathcal{F} = \{\emptyset, \{a, b\}, \{c\}, X\}$ and the measure μ on $\mathcal{F}$ satisfy $\mu(\{a, b\}) = 0$. Consider the functions f and g on X such that $f(a) = f(b) = 1, f(c) = 2$ and $g(a) = 0, g(b) = 1, g(c) = 2$. Then f is measurable but g is not (because $X(g = 0) = \{a\} \notin \mathcal{F}$). But $f(x) = g(x)$ for $x \notin \{a, b\}$ and $\mu(\{a, b\}) = 0$. (Note that it is crucial here that the set $\{a, b\}$ has measure 0 and has a nonmeasurable subset $\{a\}$, a phenomenon that cannot occur with Lebesgue measure by Remark 4.2.3(c).)

6.5.P5. Suppose $\{f_n\}_{n \geq 1}$ is a sequence of measurable functions such that $\lim_{n \to \infty} f_n(x)$ exists for all x in some measurable set of convergence A. Show that the function defined to agree with the limit on A and to be 0 elsewhere is measurable. Show also that two different measurable functions obtained in this manner by using two different sets of convergence, both having complement of measure 0, agree a.e.

Solution: Suppose A is a measurable set such that $\lim_{n \to \infty} f_n(x)$ exists for all $x \in A$. Denote by f the function that is defined to agree with the limit on A and to be 0 on A^c. We have to show that f is measurable. But $f(x) = f\chi_A(x) = \lim_{n \to \infty} f_n \chi_A(x)$ for all x, whether in A or in A^c. Each $f_n \chi_A$ is measurable, and hence the limit f is measurable. If B is another such measurable set and $\mu A^c = \mu B^c = 0$, then the function obtained from B in the same manner agrees with f on $C = A \cap B$ and $\mu C^c = \mu(A^c \cup B^c) = 0$.

6.5.P6. Let $(X, \mathcal{F}, \mu)$ and $(Y, \mathcal{G}, \nu)$ be measure spaces and $f: X \to \mathbb{R}, \phi: X \to \mathbb{R}$ be measurable functions such that $f = \phi$ a.e. Suppose $g: Y \to X$ has the property that $F \in \mathcal{F} \Rightarrow g^{-1}(F) \in \mathcal{G}$, so that the compositions $f \circ g: Y \to \mathbb{R}$ and $\phi \circ g: Y \to \mathbb{R}$ are both measurable. Is it necessary that $f \circ g = \phi \circ g$ a.e.?

Solution: No. It can happen that $(f \circ g)(y) \neq (\phi \circ g)(y)$ for all $y \in Y$. Let f be the characteristic function of a nonempty subset A of X having measure 0 and g be any measurable function with range contained in A. Then $f = \phi$ a.e., where $\phi = 0$ everywhere, but $(f \circ g)(y) = 1 \neq 0 = (\phi \circ g)(y)$ for all $y \in Y$.

6.5.P7. Let $f: [0, 1] \times [0, 1] \to \mathbb{R}$ be the function given by

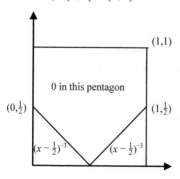

(1,1)

0 in this pentagon

$(0, \frac{1}{2})$

$(1, \frac{1}{2})$

$(x - \frac{1}{2})^{-3}$ $(x - \frac{1}{2})^{-3}$

$f(x, y) = (x - \frac{1}{2})^{-3}$ if $0 < y < |x - \frac{1}{2}|$ and $f(x, y) = 0$ otherwise.

Show that the repeated integral $\int_{[0,1]} dx \int_{[0,1]} f(x,y)dy$ does not even exist but the repeated integral $\int_{[0,1]} dy \int_{[0,1]} f(x,y)dx$ exists.

Solution: Suppose $x \in [0,\frac{1}{2})$. Then $|x - \frac{1}{2}| = \frac{1}{2} - x > 0$ and therefore, $y < |x - \frac{1}{2}| \Leftrightarrow x - \frac{1}{2} < y < \frac{1}{2} - x \Leftrightarrow y < \frac{1}{2} - x$, considering that $x - \frac{1}{2} < 0$ and $y > 0$. Consequently,

$$\int_{[0,1]} f(x,y)dy = \int_{[0,\frac{1}{2}-x]} \left(x - \frac{1}{2}\right)^{-3} dy = -\left(x - \frac{1}{2}\right)^{-2} \quad \text{for } x \in [0,\tfrac{1}{2}).$$

Now suppose $x \in (\frac{1}{2}, 1]$. Then $|x - \frac{1}{2}| = x - \frac{1}{2} > 0$ and therefore, $y < |x - \frac{1}{2}| \Leftrightarrow \frac{1}{2} - x < y < x - \frac{1}{2} \Leftrightarrow y < x - \frac{1}{2}$, considering that $\frac{1}{2} - x < 0$ and $y > 0$. Consequently,

$$\int_{[0,1]} f(x,y)dy = \int_{[0,x-\frac{1}{2}]} \left(x - \frac{1}{2}\right)^{-3} dy = -\left(x - \frac{1}{2}\right)^{-2} \quad \text{for } x \in (\tfrac{1}{2}, 1].$$

From the last two computations of integrals, we obtain

$$\left(\int_{[0,1]} f(x,y)dy\right)^+ = \left(x - \frac{1}{2}\right)^{-2}\chi_{[\frac{1}{2},1]}(x) \text{ and } \left(\int_{[0,1]} f(x,y)dy\right)^-$$
$$= \left(x - \frac{1}{2}\right)^{-2}\chi_{[0,\frac{1}{2}]}(x).$$

It is now elementary that $\int_{[0,1]} dx(\int_{[0,1]} f(x,y)dy)^+$ and $\int_{[0,1]} dx(\int_{[0,1]} f(x,y)dy)^-$ are both ∞.

For all $x \in [0, 1]$, we have $|x - \frac{1}{2}| \leq \frac{1}{2}$, and therefore $\frac{1}{2} \leq y \leq 1$ implies $y \geq |x - \frac{1}{2}|$ and hence $f(x,y) = 0$. It follows that $\int_{[0,1]} f(x,y)dx = 0$ for $\frac{1}{2} \leq y \leq 1$ So, consider $0 < y < \frac{1}{2}$. We write

$$\int_{[0,1]} f(x,y)dx = \int_{[0,\frac{1}{2}]} f(x,y)dx + \int_{[\frac{1}{2},1]} f(x,y)dx.$$

Once again, suppose $x \in [0,\frac{1}{2})$. Then $|x - \frac{1}{2}| = \frac{1}{2} - x > 0$ and therefore, $y < |x - \frac{1}{2}| \Leftrightarrow x - \frac{1}{2} < y < \frac{1}{2} - x \Leftrightarrow y < \frac{1}{2} - x$, considering that $x - \frac{1}{2} \leq 0$ and $y > 0$. However, $y < \frac{1}{2} - x \Leftrightarrow x < \frac{1}{2} - y$ and we know that $\frac{1}{2} > \frac{1}{2} - y > 0$. Consequently,

$$\int_{[0,\frac{1}{2}]} f(x,y)dx = \int_{[0,\frac{1}{2}-y]} \left(x - \frac{1}{2}\right)^{-3} dx = -\frac{1}{2}\left(x - \frac{1}{2}\right)^{-2}\Big|_0^{\frac{1}{2}-y} = -\frac{1}{2}y^2 + 2.$$

Now suppose $x \in (\frac{1}{2}, 1]$. Then $|x - \frac{1}{2}| = x - \frac{1}{2} > 0$ and therefore, $y < |x - \frac{1}{2}| \Leftrightarrow$ $\frac{1}{2} - x < y < x - \frac{1}{2} \Leftrightarrow y < x - \frac{1}{2}$, considering that $\frac{1}{2} - x \leq 0$ and $y > 0$. However, $y < x - \frac{1}{2} \Leftrightarrow x > \frac{1}{2} + y$ and, we know that $\frac{1}{2} < \frac{1}{2} + y < 1$. Consequently,

$$\int_{[\frac{1}{2}, 1]} f(x, y) dx = \int_{[\frac{1}{2} + y, 1]} \left(x - \frac{1}{2}\right)^{-3} dx = \frac{1}{2}\left(x - \frac{1}{2}\right)^{-2}\Big|_1^{\frac{1}{2}+y} = \frac{1}{2}y^2 - 2.$$

From the last two computations of integrals, we obtain $\int_{[0,1]} f(x, y) dx = 0$ also for $0 < y < \frac{1}{2}$. It follows that $\int_{[0,1]} dy \int_{[0,1]} f(x, y) dx = 0$.

6.6 Borel Sets

6.6.P1. (Cf. Problem 6.5.P4) Suppose f and g are functions on $[a, b]$ or $\mathbb{R}$ such that $f = g$ a.e. If f is Lebesgue measurable, show that g is also Lebesgue measurable.
Solution: Let $E \subseteq [a, b] = X$, or $E \subseteq \mathbb{R} = X$, have measure 0 and $f(x) = g(x)$ for $x \notin E$. For any real t, we have

$$X(g > t) = (X(g > t) \cap E)) \cup (X(g > t) \cap E^c)$$
$$= (X(g > t) \cap E) \cup (X(f > t) \cap E^c).$$

Now $(X(f > t) \cap E^c)$ is Lebesgue measurable because f is Lebesgue measurable. The set $(X(g > t) \cap E)$ is also Lebesgue measurable because it is a subset of E and therefore has measure 0, which makes it Lebesgue measurable by Remark 4.2.3(c).
6.6.P2. Prove Lemma 6.6.5.
Solution: Since E is measurable, its Lebesgue measure $m(E)$ is the same as its Lebesgue outer measure $m^*(E)$, which is given to be finite. As noted in Remark 4.1.4, there exists a sequence $\{I_n\}_{n \geq 1}$ of open intervals such that

$$E \subseteq \bigcup_{n=1}^{\infty} I_n \tag{Sol.11}$$

and $\sum_{n=1}^{\infty} \ell(I_n) \leq m^*(E) + \eta$, i.e. $\sum_{n=1}^{\infty} m(I_n) \leq m(E) + \eta$, which implies

$$m\left(\bigcup_{n=1}^{\infty} I_n\right) \leq \sum_{n=1}^{\infty} m(I_n) \leq m(E) + \eta < \infty.$$

Together with (Sol.11), this inequality leads to $m\left(\left(\cup_{n=1}^{\infty} I_n\right) \setminus E\right) \leq \eta$.
6.6.P3. Let $\mathcal{F}$ be a σ-algebra of subsets of a nonempty set X and $F: X \to \mathbb{R}$ be an $\mathcal{F}$-measurable map. Show that any Borel set A satisfies $F^{-1}(A) \in \mathcal{F}$.

Solution: Take $\mathcal{G}$ to be the family of subsets $\{B \subseteq \mathbb{R}: F^{-1}(B) \in \mathcal{F}\}$ of $\mathbb{R}$. We have to show that every Borel set belongs to $\mathcal{G}$. By Problem 2.1.P11, in which we take $Y = \mathbb{R}$, the family $\mathcal{G}$ is a σ-algebra. By Definition 2.1.4 of $\mathcal{F}$-measurability, every interval of the form (t, ∞) belongs to $\mathcal{G}$. Since $\mathcal{G}$ is a σ-algebra, every open interval belongs to $\mathcal{G}$. By Definition 6.6.1 of Borel sets, every Borel set belongs to $\mathcal{G}$.

7.1 Integrability and Step Functions

7.1.P1. A subset $U \subseteq \mathbb{R}$ is said to be *open* if for every $u \in U$, there exists a $\delta > 0$ such that $(u - \delta, u + \delta) \subseteq U$. (Obviously, any open interval is an open subset; an interval that is an open subset is an open interval; any union of open subsets (or open intervals) is an open subset.) Show that any open subset $U \subseteq \mathbb{R}$ is a countable union of disjoint open intervals.

Solution: This is trivial for $U = \emptyset$ because $\{\emptyset\}$ is a countable family of disjoint open intervals having union $\emptyset$. So, assume $U \neq \emptyset$. For each $u \in U$, let K_u be the union of all open intervals that are contained in U and have u as an element. The definition of open subset shows that some such interval must always exist. Then $u \in K_u \subseteq U$. In order to see why K_u must be an interval, suppose $x < z < y$, where $x, y \in K_u$. By definition, there is exist open intervals $I \subseteq K_u$ and $J \subseteq K_u$ such that $u \in I \cap J$ and $x \in I, y \in J$. Now it follows as in Exercise 7.1.2 that $z \in I \cup J \subseteq K_u$. Thus K_u is always an interval. Being a union of open subsets of $\mathbb{R}$, it must again be an open subset and hence an open interval. It is straightforward that U is the union of the family $\{K_u: u \in U\}$ of open intervals.

We shall show that that the intervals in the family are disjoint. In fact, if K_u and K_v are not disjoint, then the open set $K_u \cup K_v$ is an interval by Example 7.1.2 and hence an open interval; but it is contained in U and has both u and v as elements, which renders it a subset of K_u as well as of K_v, which is possible only if it equals both K_u and K_v.

That the intervals are countable in number is an immediate consequence of the observation that every nonempty open interval has to contain a rational number.

7.1.P2. Let $X = [0, 1]$ and consider the intervals $\left[\frac{j-1}{k}, \frac{j}{k}\right]$, $1 \leq j \leq k, k \in \mathbb{N}$, arranged in some order (without repetition). Show that the sequence $\{f_n\}_{n \geq 1}$ of characteristic functions on X of the respective intervals does not converge at any point, although their integrals converge to zero.

Solution: The number of intervals of length $\frac{1}{k}$ or greater is finite (in fact, $\frac{1}{2}k(k+1)$). Therefore, regardless of the order in which the intervals may be arranged, for every $k \in \mathbb{N}$, there exists an $N \in \mathbb{N}$ such that all intervals from the Nth one onwards have length less than $\frac{1}{k}$, so that the integral of f_n is less than $\frac{1}{k}$ when $n \geq N$. This shows that the integrals converge to zero. On the other hand, for any $x \in [0, 1]$ and any $k > 2$, one or two of the intervals $\left[\frac{j-1}{k}, \frac{j}{k}\right]$, $1 \leq j \leq k$, must include x while others must exclude x, which shows that $f_n(x) = 1$ for some arbitrarily large n while $f_n(x) = 0$ for another arbitrarily large n.

7.1.P3. (a) Both the figures below show a stick and a second stick with the overlap of their shadows underneath. They also show a third stick whose shadow overlaps with some part of the existing overlap of the shadows of the first two sticks (i.e. all three shadows together have some overlap). In the figure on the left, if the second stick is removed, the combined shadow of the remaining two sticks is the same as that of all three sticks. In the figure on the right, if the first stick is removed, the combined shadow of the remaining two sticks is the same as that of all three sticks. Question: Can you place three sticks in such a way that all three shadows together have some overlap but if any one among the three sticks is removed, the combined shadow of the remaining two sticks is no longer the same as that of all three sticks?

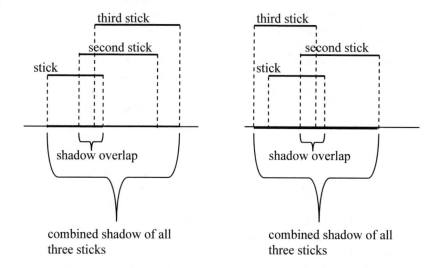

(b) Suppose $\mathcal{F}$ is a finite nonempty family of bounded intervals. Show that there exists a subfamily $\mathcal{G} \subseteq \mathcal{F}$ such that $\cup \mathcal{G} = \cup \mathcal{F}$ and no interval of $\mathcal{G}$ is contained in the union of two other intervals of $\mathcal{G}$.

(c) Can the finiteness hypothesis of part (b) be dropped?

Solution: (a) The interpretation of the question in terms of intervals is as follows. Can there be three bounded intervals such that the intersection of all three is nonempty but none is contained in the union of the other two? We shall argue that there cannot be. In other words, given three bounded intervals with a nonempty intersection, one of the three must be contained in the union of the other two.

To begin with observe that, for a bounded interval of positive length, a number a is its left endpoint if and only if it contains no number less than a but contains infinitely many numbers of the form $a + \frac{1}{n}$, where $n \in \mathbb{N}$. Analogously for the right endpoint.

If one of the three intervals in question has length zero, then the matter is trivial because then the interval must be equal to the intersection and therefore be contained in both the other intervals. We may therefore assume that all three intervals are of positive length. On applying the result of Exercise 7.1.2 twice, we find that the union of all the three intervals is also an interval, which must of course be bounded and have positive length. Let its left endpoint be a and right endpoint be $b > a$. In view of the observation in the preceding paragraph, the union then contains no number less than a but contains infinitely many numbers of the form $a + \frac{1}{n}$, each of which must of course belong to one or the other of the three intervals. It follows that at least one among the three must contain infinitely many such numbers but none less than a, and in view of the above observation once again, must have left endpoint a. Name one such interval as A. When the union contains a, the interval A can be so chosen as to contain a. First consider the case when the other endpoint of A is b. Then the interiors of both the other intervals are contained in A. In the event that neither of them contains b, both are contained in A and the required conclusion follows. In the event that one of them contains b, call that one B, the remaining interval C is contained in $A \cup B$ and the required conclusion is found to hold. Next consider the case when the other endpoint of A is not b. Then, as argued regarding a, the number b must be the right endpoint of some other interval, which we shall call B. As before, when the union contains b, the interval B can be so chosen as to contain b. Now, $A \cup B$ is an interval. Its left endpoint must be a and its right endpoint must be b. Thus the interior of $A \cup B$ is (a, b). Moreover, if a [resp. b] is in the union of all three intervals, then it is in A [resp. B]. This implies that $A \cup B$ coincides with the union $A \cup B \cup C$ of all three intervals. Therefore $C \subseteq A \cup B \cup C = A \cup B$. This shows that C is contained in the union of the other two.

(b) Induction on the number of intervals in $\mathcal{F}$. If the number of intervals in $\mathcal{F}$ is less than three, there is nothing to prove. If there are precisely three intervals in $\mathcal{F}$, then the result about intervals proved in (a) immediately yields the required subfamily $\mathcal{G}$. Now suppose the result is true for every family containing k intervals (induction hypothesis), and consider a family $\mathcal{F}$ containing $k + 1$ intervals. If $\mathcal{F}$ already has the property that no interval of $\mathcal{F}$ is contained in the union of two other intervals of $\mathcal{F}$, then we may choose $\mathcal{G} = \mathcal{F}$. Otherwise, one of the intervals of $\mathcal{F}$ is contained in the union of two others. The subfamily $\mathcal{F}_0$ obtained by deleting that interval from $\mathcal{F}$ has the property that it contains k intervals and $\cup \mathcal{F}_0 = \cup \mathcal{F}$. By the induction hypothesis, $\mathcal{F}_0$ contains a subfamily $\mathcal{G}$ such that no interval of it is contained in the union of two others and $\cup \mathcal{G} = \cup \mathcal{F}_0$. Then $\mathcal{G}$ is a subfamily of $\mathcal{F}$ such that no interval of it is contained in the union of two others and $\cup \mathcal{G} = \cup \mathcal{F}_0 = \cup \mathcal{F}$.

(c) No. Let $\mathcal{F}$ be the family of open intervals $\left(\frac{1}{n}, 3 - \frac{1}{n} \right)$. Then $\cup \mathcal{F} = (0, 3)$. Moreover, this interval cannot be the union of any finite subfamily of $\mathcal{F}$ while any infinite subfamily has the property that each interval of the subfamily is contained in another one, à fortiori in the union of two others.

7.2 The Vitali Covering Theorem

7.2.P1. Let C be a Vitali covering of a subset $A \subseteq \mathbb{R}$ and $\mathcal{J}$ be a family of closed intervals such that the union of their interiors contains A. Show that the family

$$C_0 = \{I \cap J : I \in C, J \in \mathcal{J}\}$$

of closed intervals consisting of the intersections of intervals belonging to C with intervals belonging to $\mathcal{J}$ is a Vitali covering of A.

Solution: Consider any $x \in A$ and any $\varepsilon > 0$. Since C is a Vitali covering of A, there exists an $I \in C$ for which $x \in I$ and $0 < m(I) < \varepsilon$. And since the union of the interiors of the intervals belonging to $\mathcal{J}$ contains A, there also exists a $J \in \mathcal{J}$ such that x belongs to the interior J° of J. Then $x \in I \cap J^\circ \subseteq I \cap J \in C_0$ and $m(I \cap J) \leq m(I) < \varepsilon$. On the other hand, $I \cap J^\circ$ is an intersection of open intervals and is nonempty because it contains x. Therefore it is an interval of positive length, which means it has positive measure. Since $I \cap J^\circ \subseteq I \cap J$, it follows that $m(I \cap J) > 0$.

7.2.P2. If I and J are intervals of finite positive lengths and $I \cap J \neq \emptyset$, show that the distance of any point in I from the center of J is at most $m(I) + \frac{1}{2}m(J)$.

Solution: Denote the center of J by c. Since $I \cap J \neq \emptyset$, there exists some $z \in I \cap J$. As $z \in J$, we have $|z - c| \leq \frac{1}{2}m(J)$. As $z \in I$, any $x \in I$ satisfies $|x - z| \leq m(I)$. It follows for any $x \in I$ that $|x - c| \leq |x - z| + |z - c| \leq m(I) + \frac{1}{2}m(J)$.

7.3 Dini Derivatives

7.3.P1. Prove the following partial converse of the "consequence" in Remark 7.3.6 in the case of $D^+ f(c)$: If $\alpha \in \mathbb{R} \cup \{\infty, -\infty\}$ and there exists a sequence $\{h_n\}_{n \geq 1}$ of positive numbers as in the remark (in which case $\alpha \neq \infty$ of course), then $\alpha \leq D^+ f(c)$.

Solution: We may assume that $\{h_n\}_{n \geq 1}$ is decreasing. Consider any $\delta > 0$. Since there exists an n with $0 < h_n < \delta$, the set Q_δ contains numbers greater than α, so that $\sup Q_\delta > \alpha$. Since this inequality holds for all $\delta > 0$, it follows that $\alpha \leq \inf\{\sup Q_\delta : 0 < \delta\} = D^+ f(c)$.

7.3.P2. Show that $\alpha < D^+ f(c)$ if and only if for some $\beta > \alpha$ there is a sequence $\{h_n\}_{n \geq 1}$ of positive numbers such that

$$h_n \to 0 \quad \text{and} \quad \frac{f(c + h_n) - f(c)}{h_n} > \beta \text{ for every } n \in \mathbb{N}.$$

Solution: Suppose $\alpha < D^+ f(c)$. Then there exists a $\beta > \alpha$ such that $\beta < D^+ f(c)$. Remark 7.3.6 now implies that the required sequence $\{h_n\}_{n \geq 1}$ of positive numbers exists. The converse follows trivially from Problem 7.3.P1.

7.3.P3. Give an example of continuous functions f and g on an interval containing 0 such that $D^+f(0) + D^+g(0) \neq D^+(f+g)(0)$.

Solution: Let f be the function on $\mathbb{R}$ defined by $f(0) = 0$ and $f(x) = |x|\cos\frac{1}{x}$ for $x \neq 0$. When $h > 0$, the difference quotient works out to be $\cos\frac{1}{h}$, which has the value 1 when $h = (2\pi n)^{-1}$ and $n \in \mathbb{N}$. Also $\cos\frac{1}{h} \leq 1$ for all $h \neq 0$. One can always choose n such that $(2\pi n)^{-1} < \delta$. So, $\sup Q_\delta(f) = 1$. It follows that $D^+f(0) = \inf\{\sup Q_\delta(f): 0 < \delta\} = 1$. On the other hand, the difference quotient $\cos\frac{1}{h}$ has the value -1 when $h = (\pi(2n+1))^{-1}$ and $n \in \mathbb{N}$. Also $\cos\frac{1}{h} \geq -1$ for all $h \neq 0$. One can always choose n such that $(\pi(2n+1))^{-1} < \delta$. So, $\inf Q_\delta(f) = -1$. It follows that $D_+f(0) = \sup\{\inf Q_\delta: 0 < \delta\} = -1$. Now let $g = -f$. By Exercise 7.3.7, we have $D^+g(0) = -D_+f(0) = 1$. Therefore $D^+f(0) + D^+g(0) = 2$. However, $f + g = 0$ and therefore $D^+(f+g)(0) = 0$.

7.3.P4. If $g = -f$, show that $D^-g(c) = -D_-f(c)$.

Solution: It is immediate from the definition that $Q_{-\delta}(g) = -Q_{-\delta}(f)$. It follows for any $\delta > 0$ that $\sup Q_{-\delta}(g) = -\inf Q_{-\delta}(f)$. Therefore $\inf\{\sup Q_{-\delta}(g): 0 < \delta\} = -\sup\{\inf Q_{-\delta}(f): 0 < \delta\}$. By Definition 7.3.4, this means $D^-g(c) = -D_-f(c)$.

7.3.P5. Suppose f and g are defined on a common interval I and $c \in I^\circ$. Show that $D_+(f+g)(c) \geq D_+f(c) + D_+g(c)$, provided that the sum on the right side is meaningful. What about $D^+(f+g)(c) \leq D^+f(c) + D^+g(c)$?

Solution: Obviously, we may assume that $D_+f(c) \neq -\infty \neq D_+g(c)$.

Consider arbitrary α and β such that $\alpha < D_+f(c)$ and $\beta < D_+g(c)$. Then there exists a $\Delta > 0$ such that $\inf Q_\Delta(f) > \alpha$ and $\inf Q_\Delta(g) > \beta$. It follows for $0 < h < \Delta$ that

$$\frac{f(c+h) - f(c)}{h} > \alpha \quad \text{and} \quad \frac{g(c+h) - g(c)}{h} > \beta,$$

which implies that every number in the set $Q_\Delta(f+g)(c)$ is greater than $\alpha + \beta$. Therefore $\inf Q_\Delta(f+g)(c) \geq \alpha + \beta$. The existence of such a $\Delta > 0$ implies that $\sup\{\inf Q_\delta(f+g)(c): \delta > 0\} \geq \alpha + \beta$. Thus, $D^+(f+g)(c) \geq \alpha + \beta$. Since this holds for arbitrary α and β such that $\alpha < D_+f(c)$ and $\beta < D_+g(c)$, the desired inequality follows.

The inequality asked about holds provided that the sum on the right side is meaningful.

7.4 The Lebesgue Differentiability Theorem

7.4.P1. For an increasing function f on a bounded open interval I, show by a direct argument, as in the proof of the Lebesgue Differentiability Theorem 7.4.1, that the set $A = \{x \in I: D_+f(x) < D^-f(x)\}$ has Lebesgue outer measure 0.

Solution: The set A is the union of the countably many sets $A_{p,q} = \{x \in I: D_+f(x) < p < q < D^-f(x)\}$, where $p, q \in \mathbb{Q}$. It is sufficient therefore to show that each $A_{p,q}$ has outer measure 0. The increasing nature of the function f guarantees that all Dini derivatives are nonnegative and hence $A_{p,q} \neq \emptyset$ only if $p > q > 0$. So we need consider only those p, q for which this inequality holds.

With this in view, take any $A_{p,q}$ with $p > q > 0$ and consider an arbitrary $\varepsilon > 0$.

Being a subset of the bounded interval I, the set $A_{p,q}$ has finite outer measure and therefore there exists a countable union G of open intervals such that $A_{p,q} \subseteq G$ and $m^*(A_{p,q}) \leq m(G) < m^*(A_{p,q}) + \varepsilon$. By Remark 7.3.6, to each $x \in A_{p,q}$ correspond intervals $[x, x+h] \subseteq G$ with $h > 0$ and arbitrarily small in absolute value such that $\frac{f(x+h)-f(x)}{h} < p$, i.e. such that

$$f(x+h) - f(x) < ph.$$

Since this holds for each $x \in A_{p,q}$, the intervals constitute a Vitali covering of $A_{p,q}$. By Exercise 7.2.5, we may restrict ourselves to only those intervals that are contained in G. By the Vitali Covering Theorem 7.2.7, there exist finitely many disjoint intervals $[x_j, x_j + h_j]$, $1 \leq j \leq J$, in the Vitali covering such that

$$m^*(A_{p,q} \setminus \bigcup_{j=1}^{J} [x_j, x_j + h_j]) < \varepsilon. \tag{Sol.12}$$

The choice of the intervals of the Vitali covering ensures that

$$f(x_j + h_j) - f(x_j) < ph_j \quad \text{for every } j. \tag{Sol.13}$$

Now, observe that since the disjoint intervals are contained in G, we have $\sum_{j=1}^{J} h_j \leq m(G)$. Since $p > q > 0$, taken in tandem with (Sol.13), the preceding observation leads to

$$\sum_{j=1}^{J} (f(x_j + h_j) - f(x_j)) < p \cdot \sum_{j=1}^{J} h_j \leq p \cdot m(G) < p \cdot (m^*(A_{p,q}) + \varepsilon). \tag{Sol.14}$$

In conjunction with (Sol.12), the obvious equality

$$m^*(A_{p,q} \setminus \bigcup_{j=1}^{J} (x_j, x_j + h_j)) = m^*(A_{p,q} \setminus \bigcup_{j=1}^{J} [x_j, x_j + h_j]),$$

gives rise to

$$m^*(A_{p,q} \setminus \bigcup_{j=1}^{J} (x_j, x_j + h_j)) < \varepsilon. \tag{Sol.15}$$

By Remark 7.3.6, to each $y \in A_{p,q} \cap \bigcup_{j=1}^{J} (x_j, x_j + h_j)$ correspond intervals $[y + h', y] \subseteq \bigcup_{j=1}^{J} (x_j, x_j + h_j)$ with arbitrarily small $h' < 0$ such that $\frac{f(y+h')-f(y)}{h'} > q$, i.e. such that

$$f(y+h') - f(y) < qh'.$$

Since this holds for each $y \in A_{p,q} \cap \cup_{j=1}^{J}(x_j, x_j + h_j)$, the intervals constitute a Vitali covering of $A_{p,q} \cap \cup_{j=1}^{J}(x_j, x_j + h_j)$. By the Vitali Covering Theorem 7.2.7, there exist finitely many disjoint intervals $[y_k + h'_k, y_k]$, $1 \le k \le K$, in the Vitali covering such that

$$m^*\left((A_{p,q} \cap \bigcup_{j=1}^{J}(x_j, x_j + h_j)) \setminus \bigcup_{k=1}^{K}[y_k + h'_k, y_k]\right) < \varepsilon. \qquad \text{(Sol.16)}$$

The choice of the intervals of the Vitali covering ensures that

$$f(y_k + h'_k) - f(y_k) < qh'_k \quad \text{for every } k. \qquad \text{(Sol.17)}$$

Since the Vitali covering consists of intervals contained in $\cup_{j=1}^{J}(x_j, x_j + h_j)$, we have

$$\bigcup_{k=1}^{K}[y_k + h'_k, y_k] \subseteq \bigcup_{j=1}^{J}(x_j, x_j + h_j). \qquad \text{(Sol.18)}$$

We claim that

$$-\sum_{k=1}^{K} h'_k > m^*(A_{p,q}) - 2\varepsilon. \qquad \text{(Sol.19)}$$

To prove (Sol.19), let us denote $\cup_{j=1}^{J}(x_j, x_j + h_j)$ by $\mathcal{J}$ and $\cup_{k=1}^{K}[y_k + h'_k, y_k]$ by $\mathcal{K}$. Then

$$m^*(\mathcal{K}) = m(\mathcal{K}) = -\sum_{k=1}^{K} h'_k. \qquad \text{(Sol.20)}$$

Also, $\mathcal{K} \subseteq \mathcal{J}$ in view of (Sol.18), so that $\mathcal{J}^c = \mathcal{J}^c \cap \mathcal{K}^c$. Together with (Sol.15), this implies

$$m^*(A_{p,q} \cap \mathcal{J}^c \cap \mathcal{K}^c) = m^*(A_{p,q} \cap \mathcal{J}^c) < \varepsilon.$$

From (Sol.16), we have

$$m^*((A_{p,q} \cap \mathcal{J}) \cap \mathcal{K}^c) < \varepsilon.$$

Since

$$\left(A_{p,q} \cap \mathcal{J}^c \cap \mathcal{K}^c\right) \cup \left(\left(A_{p,q} \cap \mathcal{J}\right) \cap \mathcal{K}^c\right) = \left(\left(A_{p,q} \cap \mathcal{K}^c\right) \cap \mathcal{J}^c\right) \cup \left(\left(A_{p,q} \cap \mathcal{K}^c\right) \cap \mathcal{J}\right)$$
$$= A_{p,q} \cap \mathcal{K}^c,$$

the preceding two inequalities yield

$$m^*\left(A_{p,q} \cap \mathcal{K}^c\right) < 2\varepsilon.$$

In combination with the fact that $A_{p,q} = \left(A_{p,q} \cap \mathcal{K}\right) \cup \left(A_{p,q} \cap \mathcal{K}^c\right)$, the above inequality leads us to

$$m^*\left(A_{p,q}\right) \leq m^*\left(A_{p,q} \cap \mathcal{K}\right) + m^*\left(A_{p,q} \cap \mathcal{K}^c\right)$$
$$\leq m^*(\mathcal{K}) + m^*\left(A_{p,q} \cap \mathcal{K}^c\right)$$
$$< m^*(\mathcal{K}) + 2\varepsilon$$

and hence $m^*\left(A_{p,q}\right) < -\sum_{k=1}^{K} h_k' + 2\varepsilon$ by (Sol.20). This proves our claim (Sol.19).

It follows from (Sol.18) that each interval $\left[y_k + h_k', y_k\right]$ on the left side therein is contained in the union of finitely many disjoint intervals on the right side. The nature of an interval necessitates that each interval $\left[y_k + h_k', y_k\right]$ is actually contained in a single interval on the right side. This fact and the increasing nature of the function f together justify the inequality [recall that $h_k' < 0$]

$$\sum_{j=1}^{J} \left(f(x_j + h_j) - f(x_j)\right) \geq \sum_{k=1}^{K} \left(f(y_k) - f(y_k + h_k')\right).$$

By (Sol.17) and (Sol.19), this implies

$$\sum_{j=1}^{J} \left(f(x_j + h_j) - f(x_j)\right) > -q \cdot \sum_{k=1}^{K} h_k' > q \cdot \left(m^*(A_{p,q}) - 2\varepsilon\right).$$

Therefore by (Sol.14),

$$p \cdot \left(m^*(A_{p,q}) + \varepsilon\right) > q \cdot \left(m^*(A_{p,q}) - 2\varepsilon\right).$$

Since this has been established for an arbitrary $\varepsilon > 0$, it follows that $q \cdot m^*\left(A_{p,q}\right) \leq p \cdot m^*\left(A_{p,q}\right)$, which is only possible if $m^*\left(A_{p,q}\right) = 0$, because $p < q$.

This completes the demonstration that $m^*(A) = 0$, where $A = \{x \in I : D_+ f(x) < D^- f(x)\}$.

7.4.P2. Assuming that the result $m^*\{x \in I : D^+ f(x) > D_- f(x)\} = 0$ established in the course of proving the Lebesgue Differentiability Theorem 7.4.1 is valid for decreasing functions, deduce the result of Problem 7.4.P1.

250 Solutions

Solution: Let $g = -f$. Then g is decreasing and therefore $m^*\{x \in I: D^+ g(x) > D_- g(x)\} = 0$. By Exercise 7.3.7, $D^+ g(x) = -D_+ f(x)$ and by Problem 7.3.P4, $D_- g(x) = -D^- f(x)$. Therefore $\{x \in I: D^+ g(x) > D_- gx)\} = \{x \in I: D_+ f(x) < D^- f(x)\}$.

7.4.P3. Suppose f and g are defined on a common interval I and $c \in I^\circ$ is such that g has a finite right derivative $g'_+(c)$ at c. Show that $D^+ (f + g)(c) = D^+ f(c) + g'_+(c)$.

Solution: By hypothesis, $D^+ g(c)g'_+(c)$. Therefore from Problem 7.3.P5, we know that $D^+ (f + g)(c) \le D^+ f(c) + g'_+(c)$. We shall prove the reverse inequality.

Consider any $\alpha < D^+ f(c)$. By Remark 7.3.6, there exists a sequence $\{h_n\}_{n \ge 1}$ of positive numbers such that

$$h_n \to 0 \quad \text{and} \quad \frac{f(c + h_n) - f(c)}{h_n} > \alpha \text{ for every } n \in \mathbb{N}.$$

Given any $\beta < g'_+(c) = D^+ g(c)$, the sequence $\{h_n\}_{n \ge 1}$ must also satisfy

$$\frac{g(c + h_n) - g(c)}{h_n} > \beta \text{ for sufficiently large } n.$$

The sequence obtained by deleting the first few terms of $\{h_n\}_{n \ge 1}$ then has limit 0 and satisfies

$$\frac{(f + g)(c + h_n) - (f + g)(c)}{h_n} > \alpha + \beta \text{ for all } n.$$

By Problem 7.3.P1, we obtain $D^+ (f + g)(c) \ge \alpha + \beta$. Since this is valid for any $\alpha < D_+ f(c)$ and $\beta < D_+ g(c) = g'_+(c)$, the reverse inequality we seek is an immediate consequence.

7.6 Bounded Variation

7.6.P1. Show that the product of two functions of bounded variation is of bounded variation.

Solution: First of all, a function $f: [a, b] \to \mathbb{R}$ of bounded variation is bounded. This is because, for any $x \in [a, b]$, the two points a and x constitute a partition of $[a, x]$ and therefore, by definition of total variation, $V_a^x(f) \ge |f(x) - f(a)|$, which implies $f|f(x)| \le |f(a)| + V_a^x(f) \le |f(a)| + V_a^b(f)$.

Now let $f: [a, b] \to \mathbb{R}$ and $g: [a, b] \to \mathbb{R}$ both be of bounded variation and be bounded in absolute value by A and B respectively. Consider an arbitrary partition $P: af = x_0 < x_1 < \cdots < x_n = bof [a, b]$. Then for any integer j satisfying $1 \le j \le n$, we have

$$\left|(fg)(x_j) - (fg)(x_{j-1})\right| \le A\left|g(x_j) - g(x_{j-1})\right| + B\left|f(x_j) - f(x_{j-1})\right|$$

and therefore $S(fg, P) \le A \cdot S(g, P) + B \cdot S(f, P) \le A \cdot V_a^b(g) + B \cdot V_a^b(f)$. Since this holds for every partition P, it follows that $V_a^b(fg) \le A V_a^b(g) + B V_a^b(f) < \infty$.

7.6.P2. Suppose $a < b$ and the bounded function $f: [a, b] \to \mathbb{R}$ is decreasing on $(a, b]$, so that the right limit $f(a+)$ exists. Show that

$$V_a^b(f) = f(a+) - f(b) + |f(a+) - f(a)|.$$

Solution: If $f(a) \ge f(a+)$, then the function is decreasing on the closed interval $[a, b]$ and the equality in question is seen to be true, both sides being equal to $f(a) - f(b)$. So, we assume that $f(a) < f(a+)$. Then the equality in question becomes

$$V_a^b(f) = f(a+) - f(b) + f(a+) - f(a).$$

Consider an arbitrary $x \in (a, b]$. In view of Proposition 7.6.6(d), we have $V_a^b(f) = V_a^x(f) + V_x^b(f)$. Since f is decreasing on $(a, b]$, we also have $V_x^b(f) = f(x) - f(b)$. Therefore $V_a^b(f) = V_a^x(f) + f(x) - f(b) = (V_a f)(x) + f(x) - f(b)$. As this holds for an arbitrary $x \in (a, b]$, we have $V_a^b(f) = (V_a f)(a+) + f(a+) - f(b)$. It remains to show that $(V_a f)(a+) = f(a+) - f(a)$.

Take any positive $\varepsilon < f(a+) - f(a)$. There exists some $\delta > 0$ such that $0 < x - a < \delta \Rightarrow 0 \le f(a+) - f(x) < \frac{\varepsilon}{2}$. For any $\xi \in (a, b]$ such that $\xi - a < \delta$, we shall show that $V_a^\xi(f) = (V_a f)(\xi)$ satisfies

$$(f(a+) - f(a)) - \varepsilon < V_a^\xi(f) < (f(a+) - f(a)) + \varepsilon.$$

Consider an arbitrary partition $P: a < x_1 < \cdots < x_n = \xi$ of $[a, \xi]$. Since $a < x_1 \le \xi$, therefore $0 < x_1 - a \le \xi - a < \delta$ and hence

$$0 \le f(a+) - f(x_1) < \frac{\varepsilon}{2} \tag{Sol.21}$$

and

$$0 \le f(a+) - f(\xi) < \frac{\varepsilon}{2}. \tag{Sol.22}$$

As $\varepsilon < f(a+) - f(a)$, the second inequality in (Sol.21) yields

$$f(x_1) > f(a+) - \frac{\varepsilon}{2} \tag{Sol.23}$$

$$> f(a+) - (f(a+) - f(a)) = f(a). \tag{Sol.24}$$

Moreover,

$$f(\xi) \le f(x_1) \le f(a+)$$

because $a < x_1 \le \xi$. Together with (Sol.22), this implies

$$0 \le f(x_1) - f(\xi) \le f(a+) - f(\xi) < \frac{\varepsilon}{2}. \qquad\qquad \text{(Sol.25)}$$

Since f is decreasing on $[x_1, \xi]$ if $x_1 < \xi$, we have

$$\begin{aligned} S(f, P) &= |f(x_1) - f(a)| + (f(x_1) - f(\xi)) \\ &= (f(x_1) - f(a)) + (f(x_1) - f(\xi)) \text{by (Sol.24).} \end{aligned} \qquad \text{(Sol.26)}$$

Therefore by (Sol.25) and the first inequality in (Sol.21),

$$S(f, P) < (f(x_1) - f(a)) + \frac{\varepsilon}{2} \le (f(a+) - f(a)) + \frac{\varepsilon}{2}.$$

This holds for an arbitrary partition P of $[a, \xi]$ and consequently,

$$(V_a f)(\xi) = V_a^\xi(f) \le (f(a+) - f(a)) + \frac{\varepsilon}{2} < (f(a+) - f(a)) + \varepsilon.$$

On the other hand, it also follows from (Sol.26), (Sol.23) and the first inequality in (Sol.22) that

$$\begin{aligned} S(f, P) &> \left(\left(f(a+) - \frac{\varepsilon}{2}\right) - f(a)\right) + \left(\left(f(a+) - \frac{\varepsilon}{2}\right) - f(a+)\right) \\ &= (f(a+) - f(a)) - \varepsilon. \end{aligned}$$

This leads to

$$(V_a f)(\xi) = V_a^\xi(f) > (f(a+) - f(a)) - \varepsilon.$$

7.6.P3. If $f : [a, b] \to \mathbb{R}$ has a bounded derivative on the open interval (a, b), show that it is of bounded variation on the closed interval $[a, b]$.
Solution: Let M be an upper bound for the absolute value of the derivative. Take any $c \in (a, b)$. Then by the Mean Value Theorem, $|f(x)| \le |f(c)| + M(b - a)$ for any $x \in (a, b)$. Denote $|f(c)| + M(b - a)$ by K and $\max\{|f(a)|, |f(b)|\}$ by L.

Now, let $P: a = x_0 < x_1 < \cdots < x_n = b$ be any partition of $[a, b]$ with $n \ge 3$. By what has been proved in the preceding paragraph,

$$\left|f(x_j) - f(x_{j-1})\right| \le \left|f(x_j)\right| + \left|f(x_{j-1})\right| \le L + K \quad \text{for } j = 1 \text{ and } j = n.$$

Another consequence of the Mean Value Theorem is that $a<y<z<b$ implies $|f(z)-f(y)|\leq M(z-y)$, so that

$$\sum_{j=1}^{n-1}|f(x_j)-f(x_{j-1})|\leq M(b-a).$$

Adding the two inequalities displayed above yields $S(f,P)\leq L+K+M(b-a)$. This has been derived on the assumption that P contains at least four points (i.e. $n\geq 3$). Since any partition has a refinement consisting of at least four points, the inequality must actually be valid for all partitions. Upon taking the supremum over all partitions, we arrive at $V_a^b(f)\leq L+K+M(b-a)$.

7.6.P4. If $f:[a,b]\rightarrow\mathbb{R}$ is of bounded variation, show that it has one sided limits at each point and that, for any $c\in[a,b)$, the equality

$$V_a\langle f\rangle(c+)-V_a\langle f\rangle(c)=|f(c+)-f(c)|$$

holds. (In particular, f is right continuous at c if and only if $V_a\langle f\rangle$ is.)

Solution: Every monotone function has one sided limits at each point and therefore the same is true of a sum of two monotone functions.

To simplify the notation, let us denote $V_a\langle f\rangle$ simply by V.

First we shall prove that $V(c+)-V(c)\leq|f(c+)-f(c)|$.

Let $V(c+)-V(c)>0$ and consider an arbitrary $\alpha<V(c+)-V(c)$. The inequality in question will follow if we can show that $|f(c+)-f(c)|\geq\alpha$. To achieve this, it is sufficient to prove that, given any $\beta<\alpha$ and any $\delta>0$, there exists a $t\in(c,c+\delta)$ such that $|f(t)-f(c)|>\beta$. With this goal in view, consider an arbitrary $\beta<\alpha$ and an arbitrary $\delta>0$. Since $V(c+)$ exists, there must exist an $s\in(c,b)$ satisfying

$$c<x\leq s\quad\Rightarrow\quad V(s)-V(x)<\alpha-\beta.$$

And since $\alpha<V(c+)-V(c)$, the number s can be so chosen so as to satisfy also

$$V(s)-V(c)>\alpha.$$

This inequality implies that there is a partition P of $[c,s]$ such that $S(f,P)>\alpha$. By refining P if necessary, we may assume that P has at least three points in it and that the point t immediately to the right of c lies in $(c,c+\delta)$. Let $S'(f,P)$ denote the sum of all terms in $S(f,P)$ except $|f(t)-f(c)|$. Then $S'(f,P)=S(f,P)-|f(t)-f(c)|>\alpha-|f(t)-f(c)|$. But on the other hand, the inequality $S'(f,P)\leq V(s)-V(t)<\alpha-\beta$ must obtain. Consequently, we have $\alpha-\beta>\alpha-|f(t)-f(c)|$, which yields $|f(t)-f(c)|>\beta$. The existence of the required kind of $t\in(c,c+\delta)$ has thus been established, thereby completing the argument why $V(c+)-V(c)\leq|f(c+)-f(c)|$.

We shall also prove the reverse inequality, which is easy to do.

Recall from the proof of Theorem 7.6.8 that the functions $g = V + f$ and $h = V - f$ are both increasing, Therefore $g(c+) - g(c) \geq 0$ and $h(c+) - h(c) \geq 0$. However,

$$g(c+) - g(c) = (V(c+) - V(c)) + (f(c+) - f(c))$$

and

$$h(c+) - h(c) = (V(c+) - V(c)) - (f(c+) - f(c)).$$

The reverse inequality we seek now rolls in.

7.7 Absolute Continuity

7.7.P1. (a) Show that a continuous function on $[0, 1]$ which is absolutely continuous on $[\eta, 1]$ for every positive η less than 1 need not be absolutely continuous on $[0, 1]$.

(b) Show that a continuous function on $[0, 1]$ which is absolutely continuous on $[\eta, 1]$ for every positive η less than 1 and is of bounded variation must be absolutely continuous on $[0, 1]$.

Solution: (a) The function f given by $f(x) = x \cos^2(\pi/2x)$ for $x \neq 0$ and $f(0) = 0$ is continuous on $[0, 1]$. Since it has a bounded derivative on $[\eta, 1]$ for every positive η less than 1, it is absolutely continuous on every such interval. But as seen in Example 7.6.4, it is not even of bounded variation on $[0, 1]$ and therefore by Theorem 7.7.2 certainly not absolutely continuous.

(b) Call the function f and consider any $\varepsilon > 0$. By Problem 7.6.P4, $V_0\langle f \rangle$ is continuous at 0, and therefore there exists some positive $\eta < 1$ such that $0 \leq V_0\langle f \rangle(\eta) < \frac{\varepsilon}{2}$. It follows for any finitely many disjoint intervals $(a_1, b_1), (a_2, b_2), \ldots, (a_k, b_k) \subseteq [0, \eta]$ that $\sum_{j=1}^k |f(b_j) - f(a_j)| \leq V_0\langle f \rangle(\eta) < \frac{\varepsilon}{2}$. The given function f is absolutely continuous on $[\eta, 1]$ and therefore there exists a $\delta > 0$ satisfying the condition in the definition of absolute continuity but with $\frac{\varepsilon}{2}$ in place of ε and $[\eta, 1]$ in place of $[a, b]$.

Consider finitely many disjoint open subintervals of $[0, 1]$ with total length less than δ. If one of them contains η, we may break up the corresponding closed subinterval into two parts, one containing η and the points to the left of it and the other containing η and the points to the right of it. By doing so, the kind of sum we need to consider may increase or remain the same but not decrease. Also, the corresponding open subintervals continue to be disjoint and each is contained either in $[0, \eta]$ or in $[\eta, 1]$. Denote the former, if any, by $(a_1, b_1), (a_2, b_2), \ldots, (a_k, b_k)$ and the latter, if any, by $(\alpha_1, \beta_1), (\alpha_2, \beta_2), \ldots, (\alpha_p, \beta_p)$. If intervals of one of these kinds do not actually occur, then the sums relating to them are empty and therefore understood to be 0. With this proviso, we have $\sum_{j=1}^{k} |f(b_j) - f(a_j)| < \frac{\varepsilon}{2}$, and since we started with subintervals with total length δ, we also have $\sum_{r=1}^{p} (\beta_r - \alpha_r) < \delta$ The choice of δ now leads to $\sum_{r=1}^{p} |f(\beta_r) - f(\alpha_r)| < \frac{\varepsilon}{2}$. It follows that

$$\sum_{j=1}^{k} |f(b_j) - f(a_j)| + \sum_{r=1}^{p} |f(\beta_r) - f(\alpha_r)| < \varepsilon.$$

Consequently, f is seen to be absolutely continuous on $[0, 1]$.

7.7.P2. (a) Show that the function $g: [0, 1] \to \mathbb{R}$ given by $g(x) = x^{\frac{1}{2}}$ is absolutely continuous.

(b) Show that the function $f: [0, 1] \to \mathbb{R}$ given by $f(x) = x^2 \cos^4(\pi/2x)$ for $x \neq 0$ and $f(0) = 0$ is absolutely continuous.

(c) Show that a composition of absolutely continuous functions need not be absolutely continuous.

Solution: (a) The function is increasing on $[0, 1]$ and is absolutely continuous on $[\eta, 1]$ for every positive η less than 1, by reason of having a bounded derivative on such an interval. We deduce absolute continuity on $[0, 1]$ by invoking Problem 7.7.P1(b).

(b) The function in question has a bounded derivative on $[0, 1]$.

(c) Consider the functions f and g above. The composition $g \circ f$ is given by $(g \circ f)(x) = x \cos^2(\pi/2x)$ for $x \neq 0$ and $f(0) = 0$. This function is not of bounded variation, as seen in Example 7.6.4, and hence by Theorem 7.7.2, it cannot be absolutely continuous.

7.7.P3. (a) Suppose $f: [a, b] \to [\alpha, \beta]$ and $g: [a, b] \to \mathbb{R}$ are absolutely continuous. If f is increasing, show that $g \circ f: [a, b] \to \mathbb{R}$ is absolutely continuous. Does the same conclusion hold if the function f is decreasing instead?

(b) Give an example of a function that has an unbounded derivative, is not monotone, but is absolutely continuous.

Solution: (a) Let $\varepsilon > 0$ be arbitrary. Choose $\delta' > 0$ in accordance with the definition of absolute continuity for g on $[\alpha, \beta]$. Then choose $\delta > 0$ in accordance with the definition of absolute continuity for f on $[a, b]$ but with δ' in place of ε. Consider finitely many disjoint open subintervals $(a_1, b_1), (a_2, b_2), \ldots, (a_k, b_k)$ of $[a, b]$ with total length less than δ. That is, $\sum_{j=1}^{k} (b_j - a_j) < \delta$. By choice of δ, it follows that $\sum_{j=1}^{k} |f(b_j) - f(a_j)| < \delta'$. Since f is increasing, $f(a_j) \leq f(b_j)$ for every j, and besides, the intervals $(f(a_j), f(b_j))$ (some of which may be empty because $f(a_j)$ can be equal to $f(b_j)$) must be disjoint. Considering that $\sum_{j=1}^{k} |f(b_j) - f(a_j)| < \delta'$, the choice of δ' forces $\sum_{j=1}^{k} |(g \circ f)(b_j) - (g \circ f)(a_j)| < \varepsilon$.

Consequently, $g \circ f \colon [a, b] \to \mathbb{R}$ is seen to be absolutely continuous.

The same conclusion holds if the function f is decreasing instead; the intervals $(f(a_j), f(b_j))$ have to be replaced by the intervals $(f(b_j), f(a_j))$.

(b) The function on $[0, 1]$ that maps $x \in (0, 1]$ into $x^2 \cos(\pi/2x)$ and maps 0 into 0 is absolutely continuous, because it has a bounded derivative and the function described by $x^{\frac{1}{2}}$ is also absolutely continuous by Problem 7.7.P2(a). Moreover, the latter is increasing. It follows by part (a) that the composition, which maps $x \in (0, 1]$ into $x \cos(\pi/2x^{\frac{1}{2}})$ and maps 0 into 0, is absolutely continuous. However, its derivative maps $x \in (0, 1]$ into

$$\cos(\pi/2x^{\frac{1}{2}}) + (\pi/4)x^{-\frac{1}{2}}\sin(\pi/2x^{\frac{1}{2}}).$$

Therefore the value of the derivative when $x = (4k+1)^{-2}$ is $(\pi/4)(4k+1)$, and when $x = (4k+3)^{-2}$, it is $-(\pi/4)(4k+3)$. Thus the derivative takes large positive values as well as large negative values as x approaches 0, with the consequence that the function has an unbounded derivative and is not monotone.

7.7.P4. If $f \colon [a, b] \to \mathbb{R}$ is absolutely continuous on each of the subintervals $[a, c]$ and $[c, b]$, where $a < c < b$, show that it is absolutely continuous on $[a, b]$.

Solution: Take any $\varepsilon > 0$. Let δ_1 be as in the definition of absolute continuity for f on the subinterval $[a, c]$ but with $\frac{\varepsilon}{2}$ in place of ε, and δ_2 the same on the subinterval $[c, b]$. Set $\delta = \min\{\delta_1, \delta_2\} > 0$.

Consider finitely many disjoint open subintervals of $[a, b]$ with total length less than δ. If none of them contains c, then each one is contained either in $[a, c]$ or in $[c, b]$. Denote the former, if any, by $(a_1, b_1), (a_2, b_2), \ldots, (a_k, b_k)$ and the latter, if any, by $(\alpha_1, \beta_1), (\alpha_2, \beta_2), \ldots, (\alpha_p, \beta_p)$. If intervals of one of these kinds do not actually occur, then the sums relating to them are empty and therefore understood to be 0. With this proviso, we have $\sum_{j=1}^{k}(b_j - a_j) < \delta_1$ and also $\sum_{j=1}^{p}(\beta_r - \alpha_r) < \delta_2$, because we started with subintervals with total length less than δ. The choice of δ_1 and δ_2 now leads to $\sum_{j=1}^{k}|f(b_j) - f(a_j)| < \frac{\varepsilon}{2}$ and $\sum_{r=1}^{p}|f(\beta_r) - f(\alpha_r)| < \frac{\varepsilon}{2}$. It follows that

$$\sum_{j=1}^{k}\left|f(b_j) - f(a_j)\right| + \sum_{r=1}^{p}\left|f(\beta_r) - f(\alpha_r)\right| < \varepsilon.$$

Now consider finitely many disjoint open subintervals of $[a, b]$ with total length less than δ, one of them containing c. We may replace the subinterval by two parts of it, one containing all its points that are to the left of c and the other containing all its points that are to the right of c. By doing so, the kind of sum we need to consider may increase or remain the same but not decrease (see Exercise 7.7.2). Also, the open subintervals continue to be disjoint and each is contained either in $[a, c]$ or in $[c, b]$. We can now arrive at the desired inequality by proceeding as in the preceding paragraph.

Thus f is seen to be absolutely continuous on $[a, b]$.

7.7.P5. Given a finite number of points in (a, b), show that there exists an absolutely continuous function on $[a, b]$ that fails to be differentiable at the given points.

Solution: Define a function to be 1 and 0 alternately at a, the given points in increasing order and at b; extend it by linear interpolation to the interiors of the intervals formed by the points at which it has just been defined.. Then it is absolutely continuous on each of the corresponding closed intervals, considering that it has a constant derivative in each interior. It is not differentiable at the given points but is absolutely continuous on $[a, b]$ by Problem 7.7.P4.

7.8 The Integral of the Derivative

7.8.P1. (a) Let f be a real-valued function on $[a, b]$ such that its values on (a, b) are linear interpolations between its values at the endpoints. If one among $f(a)$ and $f(b)$ is positive and the other nonnegative, show that $f(x)$ is positive for all $x \in (a, b)$.

(b) Let f and g be real-valued functions on $[a, b]$ such that $f(a) \geq g(a)$, $f(b) \geq g(b)$, one of the inequalities being strict, and let the values of both functions on (a, b) be linear interpolations between their values at the endpoints. Show that $f(x) > g(x)$ for all $x \in (a, b)$.

Solution: (a) By the linear interpolation hypothesis,

$$f(x) - f(a) = (f(b) - f(a))(b - a)^{-1}(x - a) \quad \text{for all } x \in [a, b].$$

Let $x \in (a, b)$, so that $x - a > 0$ and $0 < (b - a)^{-1}(x - a) < 1$. Then if $f(b) - f(a) > 0$, we have $f(x) > f(a) \geq 0$. If $f(b) - f(a) = 0$, we have $f(x) = f(a) = f(b) > 0$. If $f(b) - f(a) < 0$, then in view of the fact that $0 < (b - a)^{-1}(x - a) < 1$, we have $(f(b) - f(a))(b - a)^{-1}(x - a) > f(b) - f(a)$, and hence $f(x) - f(a) > f(b) - f(a)$, so that $f(x) > f(b) \geq 0$.

(b) It is easy to check that $f - g$ also has the property that its values on (a, b) are linear interpolations between its values at the endpoints. The desired conclusion is straightforward from the result of part (a).

7.8.P2. (a) Let f be a real-valued function on $[a, b]$ such that its values on (a, b) are linear interpolations between its values at the endpoints and let $a < c < b$. Show that the function has the same interpolation property over the intervals (a, c) and (c, b).

(b) Let f and g be real-valued functions on $[a, b]$ such that $f(a) \geq g(a), f(b) \geq g(b)$, and the values of g on (a, b) are linear interpolations between its values at the endpoints. Also, let $a < c < b$ and the values of f on each of (a, c) and (c, b) be linear interpolations between its values at the endpoints of the respective intervals. If $f(c) > g(c)$, show that $f(x) > g(x)$ for all $x \in (a, b)$.

Solution: (a) By the linear interpolation hypothesis,

$$f(x) - f(a) = (f(b) - f(a))(b-a)^{-1}(x-a) \text{ for all } x \in [a,b]. \quad \text{(Sol.27)}$$

Taking $x = c$, we get

$$f(c) - f(a) = (f(b) - f(a))(b-a)^{-1}(c-a), \quad \text{(Sol.28)}$$

and hence

$$(f(b) - f(a))(b-a)^{-1} = (f(c) - f(a))(c-a)^{-1}.$$

Using this in (Sol.27), we obtain an equality valid for $x \in [a, b]$ and whose validity on $[a, c]$ asserts that the values of f on (a, c) are linear interpolations between its values at the endpoints of (a, c).

Upon subtracting (Sol.28) from (Sol.27), we obtain

$$f(x) - f(c) = (f(b) - f(a))(b-a)^{-1}(x-c) \quad \text{for all } x \in [a,b].$$

Taking $x = b$ in this, we get

$$f(b) - f(c) = (f(b) - f(a))(b-a)^{-1}(b-c),$$

and hence

$$(f(b) - f(a))(b-a)^{-1} = (f(b) - f(c))(b-c)^{-1}.$$

Using this in (Sol.27), we obtain an equality valid for $x \in [a, b]$ and whose validity on $[c, b]$ asserts that the values of f on (c, b) are linear interpolations between its values at the endpoints of (c, b).

(b) By part (a), the values of g on (a, c) are linear interpolations between its values at the endpoints. Since $f(a) \geq g(a)$ and $f(c) > g(c)$, it follows from Problem 7.8.P1(b) that $f(x) > g(x)$ for all $x \in (a, c)$. Similarly, $f(x) > g(x)$ for all $x \in (c, b)$. It is given that $f(c) > g(c)$. Thus $f(x) > g(x)$ for all $x \in (a, b)$.

7.8.P3. Suppose $0 < \alpha < 1 < \beta$ and $\{\tau_j\}_{j \geq 1}$ is a sequence of numbers such that, for each j, either $0 < \tau_j \leq \alpha$ or $\tau_j \geq \beta$. Let $\sigma_n = \Pi_{j=1}^n \tau_j$. Show that σ_n cannot have a finite positive limit.

Solution: Suppose to the contrary that $\lim_{n \to \infty} \sigma_n = \sigma$, a finite positive number.

The hypothesis on α and β implies $0 < \max\{\alpha, \beta^{-1}\} < 1$. Call it γ. Let γ_0 be any number such that $\gamma < \gamma_0 < 1$ and set $\lambda = \gamma_0/\gamma > 1$. Then $\gamma_0 \sigma < \sigma < \lambda \sigma$. By our supposition, there must exist an $n \in \mathbb{N}$ such that

$$\gamma_0\sigma < \sigma_n < \lambda\sigma \quad \text{as well as} \quad \gamma_0\sigma < \sigma_{n+1} < \lambda\sigma.$$

If $\tau_{n+1} \leq \alpha$, then the second part of the first double inequality implies

$$\sigma_{n+1} < \lambda\sigma\alpha \leq \lambda\sigma\alpha = \sigma\gamma_0,$$

contradicting the first part of the second double inequality, whereas if $\tau_{n+1} \geq \beta$, the first part of the first double inequality implies

$$\sigma_{n+1} > \gamma_0\sigma\beta \geq \gamma_0\sigma\gamma^{-1} = \lambda\sigma,$$

contradicting the second part of the second double inequality. The contradiction thus obtained shows that our supposition is untenable.

7.8.P4. Given that $f: [a,b] \to \mathbb{R}$ is absolutely continuous, show that the function $V_a f: [a,b] \to \mathbb{R}$ defined by $(V_a f)(\xi) = V_a^\xi(f)$ for all $\xi \in [a,b]$ is also absolutely continuous. Can one draw the inference that an absolutely continuous function is the sum of an absolutely continuous increasing function and an absolutely continuous decreasing function?

Solution: By Corollary 7.8.4, there exists an integrable $g: [a,b] \to \mathbb{R}$ such that

$$f(x) - f(a) = \int_{[a,x]} g \quad \text{for all } x \in [a,b]$$

and the function $\phi: [a,b] \to \mathbb{R}$ defined by

$$\phi(x) - \phi(a) = \int_{[a,x]} |g| \quad \text{for all } x \in [a,b]$$

is absolutely continuous. Since $|f(x_2) - f(x_1)| \leq |\phi(x_2) - \phi(x_1)| = \phi(x_2) - \phi(x_1)$ whenever $[x_1, x_2] \subseteq [a,b]$, it follows that we also have $|(V_a f)(x_1) - (V_a f)(x_2)| \leq |\phi(x_1) - \phi(x_2)|$ whenever $[x_1, x_2] \subseteq [a,b]$. Therefore the absolute continuity of ϕ yields the absolute continuity of $V_a f$.

The inference can indeed be drawn, because a sum of absolutely continuous functions is absolutely continuous.

8.1 A Surjection from the Cantor Set to [0, 1]

8.1.P1. Suppose $x = \sum_{k=1}^{\infty} \alpha_k b^{-k}$ and $y = \sum_{k=1}^{\infty} \beta_k b^{-k}$ are b-expansions and there exists a positive integer j such that $\alpha_j > \beta_j$ and $\alpha_k = \beta_k$ for $k < j$ (vacuously true if $j = 1$). If there does not exist any positive integer J such that $\alpha_k = 0$ for every $k > J$, show that $x > y$.

Solution: By (1) of Proposition 8.1.1, the inequality $x \geq y$ must hold. If x were to be equal to y, then by (2) of Proposition 8.1.1, j would have the property that $\alpha_k = 0$ for every $k > j$. This would contradict the present hypothesis. So, $x > y$.

8.1.P2. If $y, z \in [a, b]$ and $|y - z| = b - a$, show that one among y and z is a and the other is b. (Ignore the trivial case when $a = b$.)

Solution: Since $|y - z| > 0$, we know $y \neq z$. Suppose $y < z$. Then the hypothesis is that $b - a = z - y$. In conjunction with the inequality $a \leq y$, this yields $b \leq z$. But $z \leq b$ and therefore $z = b$. In combination with the equality $b - a = z - y$, this yields $y = a$.

8.1.P3. Let $\{\beta_k\}_{k \geq 1}$ be a sequence in which each term is either 0 or 1.

(a) Then $\sum_{k=1}^{\infty} 2 \cdot \beta_k 3^{-k} \in [0, 1]$ and the series is a 3-expansion (called *ternary*) of the sum.

(b) If the first n terms of the sequence $\{\beta_k\}_{k \geq 1}$ of digits in any one ternary expansion $x = \sum_{k=1}^{\infty} 2 \cdot \beta_k 3^{-k}$ of the number x agree with the first n terms of $\phi^{-1}(u)$, where $u \in C$, then x is in the same component of C_n as u is.

(c) If $\phi^{-1}(u) = \{\alpha_k\}_{k \geq 1}$, then $\sum_{k=1}^{\infty} 2 \cdot \alpha_k 3^{-k}$ is a ternary expansion of u.

(d) A number in $[0, 1]$ belongs to C if and only if it has a ternary expansion in which every digit is either 0 or 2.

(e) If a number in $[0, 1]$ has a ternary expansion in which every digit is either 0 or 2, then it has no other ternary expansion of the same kind.

Solution: (a) $0 \leq \sum_{k=1}^{\infty} 2 \cdot \beta_k 3^{-k} \leq 2 \sum_{k=1}^{\infty} 3^{-k} = 2\left(\frac{1}{3}\right)\left(1/\left(1 - \frac{1}{3}\right)\right) = 1$. Since $2\beta_k$ is an integer satisfying $0 \leq 2\beta_k \leq (3 - 1)$, the series is a ternary expansion of its sum.

(b) Regardless of the matter of ternary expansions, we note to begin with that $x \in [a, b]$ is in the left third $[a, a + \frac{1}{3}(b - a)]$ of $[a, b]$ if and only if $x - a \leq \frac{1}{3}(b - a)$; similarly, $x \in [a, b]$ is in the right third $[a + \frac{2}{3}(b - a), b]$ if and only if $x - a \geq \frac{2}{3}(b - a)$.

Let $\phi^{-1}(u) = \{\alpha_k\}_{k \geq 1}$. If $\beta_1 = \alpha_1$, then by separating the first term in the given ternary expansion of x, we obtain the double inequality

$$2 \cdot \alpha_1 3^{-1} = 2 \cdot \beta_1 3^{-1} \leq x \leq 2 \cdot \beta_1 3^{-1} + \sum_{k=2}^{\infty} 2 \cdot 3^{-k} = 2 \cdot \alpha_1 3^{-1} + \frac{1}{3}.$$

If u is in the component $[0, \frac{1}{3}]$ of C_1, then $\alpha_1 = 0$ and the second half of the above double inequality shows that $x \leq \frac{1}{3}$, so that x is found to be in the same component as u. If on the other hand u is in the component $[\frac{2}{3}, 1]$ of C_1, then $\alpha_1 = 1$ and the first half of the above double inequality shows that $x \geq \frac{2}{3}$, so that x is once again found to be in the same component as u.

The assertion at issue has thus been proved to hold for $n = 1$.

Now assume as induction hypothesis that the assertion at issue holds for some n, and let the first $n + 1$ terms of the sequence $\{\beta_k\}_{k\geq 1}$ of digits in the ternary expansion $x = \sum_{k=1}^{\infty} 2 \cdot \beta_k 3^{-k}$ agree with the first $n + 1$ terms of $\phi^{-1}(u)$. Then certainly the first n terms of the sequence $\{\beta_k\}_{k\geq 1}$ of digits in the ternary expansion $x = \sum_{k=1}^{\infty} 2 \cdot \beta_k 3^{-k}$ agree with the first n terms of $\phi^{-1}(u)$. Note that the same happens to be true regarding

$$y = \sum_{k=1}^{n} 2 \cdot \alpha_k 3^{-k} \quad \text{and} \quad z = \sum_{k=1}^{n} 2 \cdot \alpha_k 3^{-k} + \sum_{k=n+1}^{\infty} 2 \cdot 3^{-k}.$$

The induction hypothesis therefore implies that not only x but also y and z are in the same component of C_n as u is. Now, $z - y = \sum_{k=n+1}^{\infty} 2 \cdot 3^{-k} = 3^{-n}$, which is the length of the component. It follows from Problem 8.1.P2 that y is the left endpoint of the component. So, the component is $[a, b]$, where $a = y = \sum_{k=1}^{n} 2 \cdot \alpha_k 3^{-k}$ and $b - a = 3^{-n}$. Also, $x - a = \sum_{k=n+1}^{\infty} 2 \cdot \beta_k 3^{-k}$, where $\beta_{n+1} = \alpha_{n+1}$. Separating the first term in the series for $x - a$, we obtain the double inequality

$$2 \cdot \alpha_{n+1} 3^{-n-1} \leq x - a \leq 2 \cdot \alpha_{n+1} 3^{-n-1} + 2 \cdot (3^{-n-2})\left(1/(1-\tfrac{1}{3})\right)$$
$$= 2 \cdot \alpha_{n+1} 3^{-n-1} + 3^{-n-1}.$$

The component of C_{n+1} that u is in is either the left third of the component $[a, b]$ of C_n, in which case $\alpha_{n+1} = 0$, or the right third of the component $[a, b]$ of C_n, in which case $\alpha_{n+1} = 1$. If u is in the left third of the component $[a, b]$ of C_n, then $\alpha_{n+1} = 0$ and the second half of the above double inequality shows that $x - a \leq 3^{-n-1} = (\tfrac{1}{3})3^{-n} = \tfrac{1}{3}(b - a)$, which implies that x is also in the left third and hence is in the same component of C_{n+1} as u. If u is in the right third of the component $[a, b]$ of C_n, then $\alpha_{n+1} = 1$ and the first half of the above double inequality shows that $x - a \geq 2 \cdot 3^{-n-1} = 2 \cdot (\tfrac{1}{3})3^{-n} = \tfrac{2}{3}(b - a)$, which implies that x is also in the right third and hence is again in the same component of C_{n+1} as u. This shows that in both the possible cases, x is in the same component of C_{n+1} as u.

(c) In view of (b), a number having $\sum_{k=1}^{\infty} 2 \cdot \alpha_k 3^{-k}$ as one ternary expansion has the property that, for every n, it is in the same component of C_n as u. Therefore its distance from u is at most 3^{-n} for every n.

(d) It was shown in (c) that every number in C has a ternary expansion in which every digit is either 0 or 2. Conversely, a ternary expansion in which every digit is either 0 or 2 is of the form $\sum_{k=1}^{\infty} 2 \cdot \alpha_k 3^{-k}$, where every α_k is either 0 or 1. Set $u = \phi(\{\alpha_k\}_{k\geq 1}) \in C$. Then $\phi^{-1}(u) = \{\alpha_k\}_{k\geq 1}$, and hence by (c), $\sum_{k=1}^{\infty} 2 \cdot \alpha_k 3^{-k}$ is a ternary expansion of u. Therefore the number with ternary expansion $\sum_{k=1}^{\infty} 2 \cdot \alpha_k 3^{-k}$ is the same as u, which belongs to C.

(e) Consider two distinct ternary expansions $\sum_{k=1}^{\infty} \alpha_k b^{-k}$ and $\sum_{k=1}^{\infty} \alpha_k' b^{-k}$ of a number and suppose every α_k as well as every α_k' is either 0 or 2. Since the expansions are distinct, there must be some k such that $\alpha_k \neq \alpha'_k$. Let j be the smallest among such k. Then $\alpha_k \neq \alpha'_k$ for $k < j$ (vacuously true if $j = 1$) and $\alpha_j \neq \alpha'_j$. We may assume $\alpha_j > \alpha'_j$. Now, (2) of Proposition 8.1.1 yields $\alpha_j - \alpha'_j = 1$, which is not possible because each of α_j and α'_j is either 0 or 2.

8.2 A Bijection from the Cantor Set to [0, 1] and the Cantor Function

8.2.P1. (a) Prove Proposition 8.2.1.
(b) Prove Proposition 8.2.2.

Solution: (a) Let $\sum_{k=1}^{\infty} \alpha_k b^{-k}$ and $\sum_{k=1}^{\infty} \alpha'_k b^{-k}$ be two distinct b-expansions of the same $x \in [0, 1]$. Since they are distinct, there must exist some k such that $\alpha_k \neq \alpha'_k$. It follows that there is a smallest k with this property. Call it j. Then either $\alpha_k > \alpha'_k$ or $\alpha_k < \alpha'_k$. To be specific, suppose $\alpha_k > \alpha'_k$. It then follows by statement (2) of Proposition 8.1.1 that every $k > j$ satisfies $\alpha_k = 0, \alpha'_k = b - 1$. Thus the sequences $\{\alpha_k\}_{k \geq 1}, \{\alpha'_k\}_{k \geq 1}$ are both recurring.

(b) Suppose $x = \sum_{k=1}^{\infty} \alpha_k b^{-k}$ and $\alpha_k = \alpha$ for $k > j$. Then $x = \sum_{k=1}^{j} \alpha_k b^{-k} + \sum_{k=j+1}^{\infty} \alpha_k b^{-k} = \sum_{k=1}^{j} \alpha_k b^{-k} + \alpha \sum_{k=j+1}^{\infty} b^{-k} = \sum_{k=1}^{j} \alpha_k b^{-k} + \alpha b^{-j}(b-1)^{-1}$, which is a rational number.

Ordinarily, the "post inverse" (or "left inverse") g of a function G is understood to have the range of G as its domain and to satisfy the condition that $g(G(x)) = x$ for every x in the domain of G. However, for the present purpose, it will be necessary to broaden the concept somewhat. If $G: [a, b] \to \mathbb{R}$ is strictly increasing, then any function g defined on $[G(a), G(b)]$ and satisfying the aforementioned condition will be called a post inverse. Its values at points of $[G(a), G(b)]$ that are outside the range of G are not uniquely determined unless some additional condition is imposed.

8.2.P2. (Rubel's Lemma [6]) Let $G: [a, b] \to \mathbb{R}$ be a strictly increasing function. Then there is a continuous increasing function $g: [G(a), G(b)] \to \mathbb{R}$ which is a "post inverse" of G, i.e. satisfies $g(G(x)) = x$ for each $x \in [a, b]$, and which is constant on any open subinterval of $[G(a), G(b)]$ contained in the complement of the range of G.

Solution: Let $g: [G(a), G(b)] \to \mathbb{R}$ be the function such that

$$g(y) = \sup\{u \in [a, b]: G(u) \leq y\}.$$

Then it is obvious that the range of g is contained in $[a, b]$.

Let us show first that $g(G(x)) = x$ for each $x \in [a, b]$. With this in mind, consider any $x \in [a, b]$. Since G is strictly increasing, we have $G(u) \leq G(x) \Leftrightarrow u \leq x$. Therefore $\{u: G(u) \leq G(x)\} = \{u: u \leq x\}$; hence $\sup\{u: G(u) \leq G(x)\} = x$. By definition of g, this is the same as saying that $g(G(x)) = x$.

Next, we show that g is increasing. Suppose $y, \eta \in [G(a), G(b)]$ and $y \leq \eta$. Then $G(u) \leq y \Rightarrow G(u) \leq \eta$, and therefore $\{u: G(u) \subseteq y\} \subseteq \{u: G(u) \leq \eta\}$. This implies $\sup\{u: G(u) \leq y\} \leq \sup\{u: G(u) \leq \eta\}$, which means $g(y) \leq g(\eta)$ by definition of g. Thus g is an increasing function.

We proceed to prove continuity. Since $g(G(x)) = x$ for each $x \in [a, b]$, the range of g is the interval $[a, b]$. As we have shown that the function is increasing, its continuity now follows by Remark 8.2.4.

Let I be a subinterval of $[G(a), G(b)]$ that is contained in the complement of the range of G. We consider only nonempty I, because it is trivial that g is constant on $\emptyset$. Suppose $y_1, y_2 \in I$ and $y_1 \leq y_2$. Since g has been shown to be increasing, we know that $g(y_1) \leq g(y_2)$. Consider any $u \in [a, b]$ such that $G(u) \leq y_2$. Since $[y_1, y_2] \subseteq I$ and is therefore contained in the complement of the range of G, it follows that $G(u) < y_1$. This shows that $\{u: G(u) \leq y_2\} \subseteq \{u: G(u) \leq y_1\}$, which implies $g(y_2) \leq g(y_1)$ by definition of g. Consequently, $g(y_1) = g(y_2)$.

8.2.P3. Let G be a strictly increasing real-valued function on the domain $[a, b]$ and g_0 be an increasing function with domain containing the range of G and satisfying $g_0(G(x)) = x$ for each $x \in [a, b]$. (In particular, g_0 could be an increasing post inverse of G.) Show the following.

(a) For each $y \in [G(a), G(b)]$ that belongs to the domain of g_0,

$$g_0(y) = \sup\{u \in [a, b]: G(u) \leq y\}.$$

(b) The increasing post inverse is unique and g_0 agrees with the unique increasing post inverse at points of $[G(a), G(b)]$ that belong to its domain.

Hint: Argue for any $y \in [G(a), G(b)]$ that belongs to the domain of g_0 that $g_0(y)$ lies between the numbers $\sup\{u \in [a, b]: G(u) \leq y\}$ and $\inf\{u \in [a, b]: G(u) \geq y\}$ and that these two numbers are equal.

Solution:

(a) Let $y \in [G(a), G(b)]$ belong to the domain of g_0 and

$$A = \{u \in [a, b]: G(u) \leq y\}.$$

We have to prove that

$$g_0(y) = \sup A.$$

Let $B = \{u \in [a, b] : G(u) \geq y\}$. Then $A \cup B = [a, b]$. Both A and B are nonempty because $a \in A$ and $b \in B$. So, $\sup A$ and $\inf B$ both exist and belong to $[a, b]$. Since it is given that g_0 is increasing, that its domain contains the range of G and that y is in the domain of g_0, we can assert that

$$u \in A \Rightarrow u = g_0(G(u)) \le g_0(y)$$
$$u \in B \Rightarrow u = g_0(G(u)) \ge g_0(y).$$

Hence $\sup A \le g_0(y) \le \inf B$. But if $\sup A < \inf B$, then any $x \in (\sup A, \inf B)$ must belong to $[a, b]$ and also satisfy $x \notin A$ as well as $x \notin B$, which is not possible because $A \cup B = [a, b]$. Therefore $\sup A = \inf B$, which implies that $g_0(y) = \sup A$, as was to be proved.

(b) This is an easy consequence of (a).

8.2.P4. Let G be a strictly increasing real-valued function on the domain $[a, b]$. Show that the function $g: [G(a), G(b)] \to \mathbb{R}$ defined by

$$g(y) = \inf\{u \in [a, b]: G(u) \ge y\}$$

is an increasing post inverse of G.

Solution: It is obvious that the range of g is contained in $[a, b]$.

Let us show first that $g(G(x)) = x$ for each $x \in [a, b]$. With this in mind, consider any $x \in [a, b]$. Since G is strictly increasing, we have $G(u) \le G(x) \Leftrightarrow u \le x$. Therefore $\{u: G(u) \ge G(x)\} = \{u: u \ge x\}$; hence $\inf\{u: G(u) \ge G(x)\} = x$. By definition of g, this is the same as saying that $g(G(x)) = x$.

It remains to show that g is increasing. Suppose $y_1, y_2 \in [G(a), G(b)]$ and $y_1 \le y_2$. Then $G(u) \ge y_2 \Rightarrow G(u) \ge y_1$, and therefore $\{u: G(u) \ge y_2\} \subseteq \{u: G(u) \ge y_1\}$. This implies $\inf\{u: G(u) \ge y_2\} \ge \inf\{u: G(u) \ge y_1\}$, which means $g(y_1) \le g(y_2)$ by definition of g. Thus g is an increasing function.

8.2.P5. Let G_1 and G_2 be strictly increasing real-valued functions on the domain $[a, b]$ and let g_1 and g_2 be their respective (unique) increasing post inverses. If $G_1 = G_2$ almost everywhere, show that $g_1 = g_2$ everywhere on the intersection of their domains, i.e. on $[G_1(a), G_1(b)] \cap [G_2(a), G_2(b)]$.

Solution: Let $y \in [G_1(a), G_1(b)] \cap [G_2(a), G_2(b)]$. Then $\{u \in [a, b]: G_1(u) \le y\}$ and $\{u \in [a, b]: G_2(u) \le y\}$ are both nonempty. Consider $x_1 = \sup\{u \in [a, b]: G_1(u) \le y\}, x_2 = \sup\{u \in [a, b]: G_2(u) \le y\}$. By Problem 8.2.P3, we need only prove that $x_1 = x_2$. This will follow if we derive a contradiction from the assumption that $x_1 < x_2$. So, assume $x_1 < x_2$. By definition of x_2, the interval $(x_1, x_2]$ contains some u_0 such that $G_2(u_0) \le y$. Since G_2 is increasing, it follows that every $u \in (x_1, u_0)$ satisfies $G_2(u) \le y$. Since $G_1 = G_2$ almost everywhere, some $u \in (x_1, u_0)$ must satisfy $G_1(u) = G_2(u) \le y$. Since $x_1 < u$, this contradicts the definition of x_1.

8.2.P6. In the course of proving Theorem 8.2.3, it was established that the mapping κ_0 is strictly increasing on $C \backslash C_0$, where C_0 is the subset of the Cantor set C consisting of those u for which the associated sequence $\phi^{-1}(u) = \{\alpha_k\}_{k \ge 1}$ is recurring. Use this fact in conjunction with Problem 6.6.P3 to show that there exists a measurable set that is not a Borel set.

Hint: Consider the inverse image by κ_0 of the intersection of a nonmeasurable subset of $[0, 1]$ with the range of κ_0.

Solution: Denote the restriction of κ_0 to $C \backslash C_0$ by κ_1. We know from the proof of Theorem 8.2.3 that κ_1 is strictly increasing and has range $[0, 1] \backslash I_0$, where I_0 is the subset of [0, 1] consisting of numbers having a recurring binary expansion. Here C_0 and I_0 are both countably infinite. Since κ_1 is strictly increasing, so is its inverse κ_1^{-1}, which has domain $[0, 1]\backslash I_0$.

Let N_1 be a subset of $\mathbb{R}$ that is not Lebesgue measurable (Corollary 4.3.8). Then at least one among the countably many intersections $N_1 \cap [n, n+1]$, where $n \in \mathbb{Z}$, must fail to be Lebesgue measurable. It follows that $N = \{x \in \mathbb{R}: x+n \in N_1 \cap [n, n+1]\}$ is a subset of [0, 1] that is not Lebesgue measurable. Since I_0 is countable, the same is true of $N \cap I_0$, which implies that it is measurable. Since $N = (N \cap I_0) \cup (N\backslash I_0)$, we infer that $N\backslash I_0$ is not Lebesgue measurable. [(Since $N \subseteq$ [0,1], the set-theoretic difference $N\backslash I_0$ is the same as the intersection $N \cap ([0, 1]\backslash I_0)$.]

Since κ_1^{-1} is a strictly increasing and therefore measurable function on the Lebesgue measurable subset $[0, 1]\backslash I_0$ of $\mathbb{R}$, the inverse image by κ_1^{-1} of any Borel set is measurable by Problem 6.6.P3. Now the inverse image by κ_1^{-1} of $\kappa_1(N\backslash I_0)$ is precisely $N\backslash I_0$, which is not measurable. Therefore $\kappa_1(N\backslash I_0)$ is not a Borel set. However, it is a Lebesgue measurable set because it is a subset of C and the latter has measure 0.

References

1. Craven, B.D.: Lebesgue Measure & Integral. Pitman Publishing Inc, Marshfield, MA (1982)
2. Dunford, N., Schwartz, J.T.: Linear Operators, Part II. Interscience Publishers, New York (1963)
3. Feldman, J.: Interchanging the order of summation. http://www.math.ubc.ca/∼feldman/m321/twosum.pdf. Last accessed 19 June 2016
4. Habil, E.D.: Double sequences and double series. IUG J. Nat. Stud. **14**(1), 1–32 (2006). http://www2.iugaza.edu.ps/ar/periodical/articles/volume%2014-%20Issue%201%20-studies%20-16.pdf. Last accessed 19 June 2016
5. Munkres, J.R.: Analysis on Manifolds. Westview Press, Boulder, Colorado (1991)
6. Rubel, L.A.: Differentiability of monotonic functions. Colloq. Math. **10**, 277–279 (1963)
7. Rudin, W.: Principles of Mathematical Analysis, 3rd edn. McGraw-Hill, New York (1976)
8. Shirali, S.: Analog of L^p with a subadditive measure. Ricerche Mat. **57**, 43–54 (2008)
9. Shirali, S., Vasudeva, H.L.: Introduction to Mathematical Analysis. Narosa, New Delhi (2014)
10. Shirali, S., Vasudeva, H.L.: Multivariable Analysis. Springer, London (2010)
11. Wang, Z., Klir, G.J.: Fuzzy Measure Theory. Plenum Press, New York (1992)

© Springer Nature Switzerland AG 2018
S. Shirali, *A Concise Introduction to Measure Theory*,
https://doi.org/10.1007/978-3-030-03241-8

Index

© Springer Nature Switzerland AG 2018
S. Shirali, *A Concise Introduction to Measure Theory*,
https://doi.org/10.1007/978-3-030-03241-8

Printed in the United States
By Bookmasters